Pulp and Paper Chemistry and Technology   Volume 4
Paper Products Physics and Technology

Edited by Monica Ek, Göran Gellerstedt, Gunnar Henriksson

Pulp and Paper Chemistry and Technology
Volume 4

This project was supported by a generous grant by the Ljungberg Foundation (Stiftelsen Erik Johan Ljungbergs Utbildningsfond) and originally published by the KTH Royal Institute of Technology as the "Ljungberg Textbook".

Paper Products Physics and Technology

Edited by Monica Ek, Göran Gellerstedt,
Gunnar Henriksson

 **DE GRUYTER**

Editors
Dr. Monica Ek
Professor (em.) Dr. Göran Gellerstedt
Professor Dr. Gunnar Henriksson
Wood Chemistry and Pulp Technology
Fibre and Polymer Technology
School of Chemical Science and Engineering
KTH – Royal Institute of Technology
100 44 Stockholm
Sweden

ISBN 978-3-11-021345-4

Bibliographic information published by the Deutsche Nationalbibliothek

The Deutsche Nationalbibliothek lists this publication in the Deutsche Nationalbibliografie; detailed bibliographic data are available in the Internet at http://dnb.d-nb.de.

© Copyright 2009 by Walter de Gruyter GmbH & Co. KG, 10785 Berlin.
All rights reserved, including those of translation into foreign languages. No part of this book may be reproduced or transmitted in any form or by any means, electronic or mechanic, including photocopy, recording, or any information storage retrieval system, without permission in writing from the publisher. Printed in Germany.
Typesetting: WGV Verlagsdienstleistungen GmbH, Weinheim, Germany.
Printing and binding: Hubert & Co. GmbH & Co. KG, Göttingen, Germany.
Cover design: Martin Zech, Bremen, Germany.

Foreword

The production of pulp and paper is of major importance in Sweden and the forestry industry has a profound influence on the economy of the country. The technical development of the industry and its ability to compete globally is closely connected with the combination of high-class education, research and development that has taken place at universities, institutes and industry over many years. In many cases, Swedish companies have been regarded as the initiator of new technology which has started here and successively found a general world-wide acceptance. This leadership in knowledge and technology must continue and be developed around the globe in order for the pulp and paper industry to compete with high value-added forestry products adopted to a modern sustainable society.

The production of forestry products is based on a complex chain of knowledge in which the biological material wood with all its natural variability is converted into a variety of fibre-based products, each one with its detailed and specific quality requirements. In order to make such products, knowledge about the starting material, as well as the processes and products including the market demands must constitute an integrated base. The possibilities of satisfying the demand of knowledge requirements from the industry are intimately associated with the ability of the universities to attract students and to provide them with a modern and progressive education of high quality.

In 2000, a generous grant was awarded the Department of Fibre and Polymer Technology at KTH Royal Institute of Technology from the Ljungberg Foundation (Stiftelsen Erik Johan Ljungbergs Utbildningsfond), located at StoraEnso in Falun. A major share of the grant was devoted to the development of a series of modern books covering the whole knowledge-chain from tree to paper and converted products. This challenge has been accomplished as a national four-year project involving a total of 30 authors from universities, Innventia and industry and resulting in a four volume set covering wood chemistry and biotechnology, pulping and paper chemistry and paper physics. The target reader is a graduate level university student or researcher in chemistry / renewable resources / biotechnology with no prior knowledge in the fields of pulp and paper. For the benefit of pulp and paper engineers and other people with an interest in this fascinating industry, we hope that the availability of this material as printed books will provide an understanding of all the fundamentals involved in pulp and paper-making.

For continuous and encouraging support during the course of this project, we are much indebted to Yngve Stade, Sr Ex Vice President StoraEnso, and to Börje Steen and Jan Moritz, Stiftelsen Erik Johan Ljungbergs Utbildningsfond.

Stockholm, August 2009   Göran Gellerstedt, Monica Ek, Gunnar Henriksson

List of Contributing Authors

Bo Andreasson
Expancel AB
Box 13000
850 13 Sundsvall, Sweden
bo.andreasson@akzonobel.com

Anthony Bristow
Bristow Consulting AB
146 38 Tullinge, Sweden
bristow@beta.telenordia.se

Christer Fellers
Innventia AB
Drottning Kristinas väg 61
Stockholm, Sweden
christer.fellers@innventia.com

Mikael Nygårds
Innventia AB
Box 5604
114 86 Stockholm, Sweden
mikael.nygards@innventia.com

Sören Östlund
KTH Royal Institute of Technology
School of Engineering Sciences
Department of Solid Mechanics
100 44 Stockholm, Sweden
soren@kth.se

Christer Söremark
Christer Söremark AB
Hagav. 10
944 72 Piteå, Sweden
christer@soremark.se

Göran Ström
Innventia AB
Drottning Kristinas väg 61
Stockholm, Sweden
goran.strom@innventia.com

Johan Tryding
Tetra Pak Packing Solutions AB
Ruben Rausingsgata
221 86 Lund, Sweden
johan.tryding@tetrapak.com

Lars Wågberg
KTH Royal Institute of Technology
Chemical Science and Engineering
Fibre and Polymer Technology
100 44 Stockholm, Sweden
wagberg@pmt.kth.se

Torbjörn Wahlström
The Packaging Greenhouse
Axel Johnsons v. 6
652 21 Karlstad, Sweden
torbjorn@thepackaginggreenhouse.com

Contents

1. The Structure of Paper and its Modelling.................................... 1
   Christer Fellers

2. Paper Physics 25
   Christer Fellers

3. Development of Paper Properties during Drying............................. 69
   Torbjörn Wahlström

4. The Interaction of Paper with Water Vapour 109
   Christer Fellers

5. Optical Properties of Pulp and Paper 145
   Anthony Bristow

6. On the Mechanisms behind the Action of Dry Strength and Dry Strength Agents 169
   Lars Wågberg

7. On the Mechanisms Behind the Action of Wet Strength and Wet Strength Agents 185
   Bo Andreasson and Lars Wågberg

8. The Surface of Paper 209
   Anthony Bristow

9. Paper and Printing 233
   Göran Ström

10. Packaging 257
    Christer Söremark and Johan Tryding

11. Laminate Theory for Papermakers... 287
    Christer Fellers

12. Mathematical Modelling and Analysis of Converting and Enduse 315
    Mikael Nygårds and Sören Östlund

Index.. 335

1 The Structure of Paper and its Modelling

Christer Fellers
STFI-Packforsk AB

1.1 Functional Paper Properties 1

1.2 The Complexity of a Model 2

1.3 Modelling at Different Structural Levels 2

1.4 The Physical Nature of Paper 3

1.5 Factors Affecting the Mechanical Properties of Paper 4

1.6 The Nature of Paper 5

1.7 The Paper Structure 6

1.8 Paper Divided into Categories 8

1.9 The Structure of some Paper Products 9

1.10 Properties of Different Materials Compared to Paper 11

1.10.1 General Considerations 11
1.10.2 Textiles 12
1.10.3 Fibre Textiles 12
1.10.4 Polyethylene Film 12

1.11 Standardization 13

1.1 Functional Paper Properties

During manufacturing, converting and end-use of paper there are many demands for its functional mechanical properties. A few illustrative examples of this are given below.

- *Example 1)* Bending stiffness paper is probably the most important mechanical property of paper and carton board. *Figure 1.1* illustrates the obvious importance of bending stiffness for a milk carton.

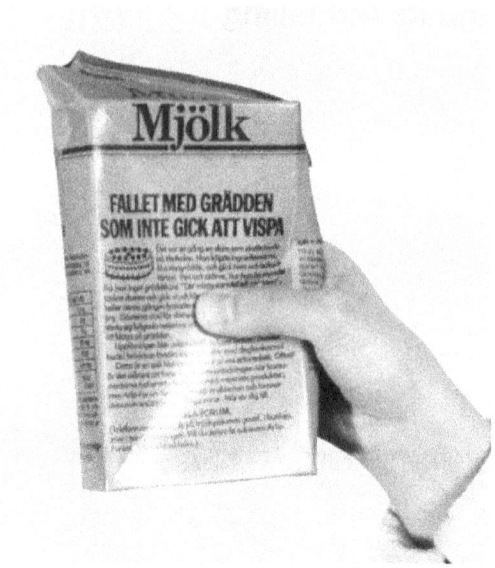

Figure 1.1. The bending stiffness is important for milk cartons as well as most paper grades.

- *Example 2)* During storage of corrugated boxes and cartons, the structure may collapse due to creep in compression forces after a certain time of loading, *Figure 1.2*. The term lifetime is sometimes used to describe the time to break.

It was shown already half a century ago that the lifetime decreases linearly in a semi-logarithmic plot as shown in *Figure 1.3*.

- *Example 3)* During manufacturing, rewinding and printing, thin paper grades may experience web breaks. The breaks may be caused by high loads, defects or inferior fracture properties. *Fig. 1.4* shows a web break in a paper machine.

- *Example 4)* Sacks are expected not to break during filling, transportation and end-use. In the worst case the sack breaks which is shown in the following *Figure 1.5*.

- *Example 5)* During certain printing operations, for example as offset printing, the material is exposed to high stresses in the thickness direction which may lead to delamination, *Figure 1.6*.

Many functional properties may be difficult to evaluate for the purpose of trade and pulp evaluation. However efforts are made to standardise important methods for property evaluation in ISO and national standards.

Figure 1.2. Compressive collapse of corrugated boxes after a long time of loading.

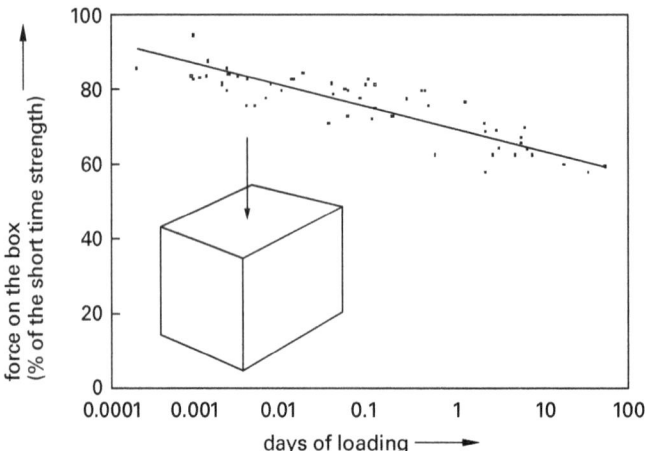

Figure 1.3. The lifetime, the time to failure of corrugated boxes as a function of the time of loading.

Figure 1.4. A web break in a paper machine. (Nordiskafilt).

Figure 1.5. Break of a sack filled with cement.

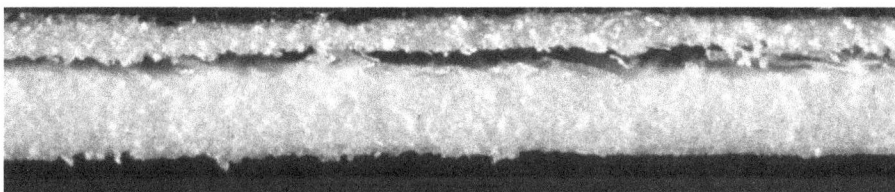

Figure 1.6. Delamination of a carton board in offset printing.

1.2 The Complexity of a Model

In research and development of paper grades it is often useful to model the paper behaviour. Depending on the actual loading situation, the time scale involved and the environmental conditions, different models must be used. For example, if the time scale is short, the model need not include time as a variable. If the deformation is small, the material may be regarded as elastic. The more complicated the loading situation is the more complex the model needs to be.

1.3 Modelling at Different Structural Levels

During product development the paper may be modelled at different structural levels, depending on the purpose of the modelling. The descriptions of the structure may be expressed at different structural levels, the structure hierarchy, *Figure 1.7*.

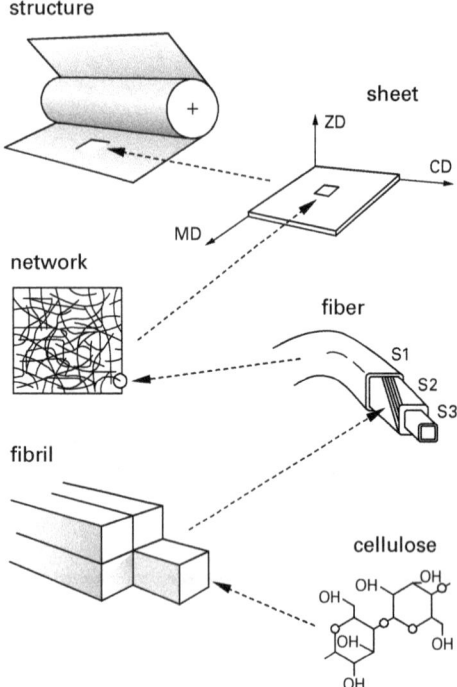

Figure 1.7. The descriptions of the structure may be expressed at different structural levels, the structure hierarchy.

Below are listed some concepts to consider related to modelling:

- At the level of modelling, the properties of the material are considered to be constant in small volumes (homogeneous), although the entire body can be heterogeneous, for example made of several materials or have density variations.
- The structure of the material at smaller scales is neglected
- At the level of modelling, quantities of interest, for example the deformation of the material, are viewed as varying in a continuous matter.
- Continuum models typically predict the "average" response of the material at the level under consideration

Figure 1.8 illustrates that paper often is modelled as a continuum, which means that the structure characteristics are disregarded. In other instances, the relation between the paper properties and the constituents are of interest.

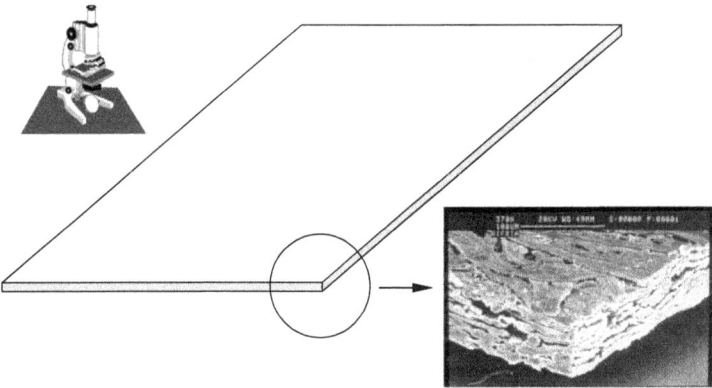

Figure 1.8. Paper is often modelled as a continuum, which means that the structure characteristics are disregarded. In other instances, the relation between the paper properties and the constituents are of interest.

For paper materials, each modelling level has particular advantages and disadvantages. Typical uses:

- *Macro level*: predict behaviour of paper structure, for example boxes
- *Meso level*: Formation studies
- *Micro level*: Predict the influence of fibre and bond properties on sheet (macroscopic) properties
- *Sub-micro level*: Understand the influence of hydrogen bonding on the sheet (macroscopic) properties

Even if the material is modelled as a continuum, different degrees of complexity may be necessary to introduce in order to describe the material behaviour. The more levels of structure hierarchy that are introduced, the more complex will the model be if all features shall be captured. In reality one has to make shortcuts in the modelling. The following degrees of complexity may be introduced, in order from simple to difficult.

- *Elastic*: The material is like a spring -all deformations are recovered when the loading is removed
- *Plastic*: Permanent deformations occur which are non-recoverable
- *Elastic-Plastic*: The material is first elastic, and then plastic
- *Viscoelastic*: The material experiences elastic and viscous (time dependent) deformations
- *Viscoplastic*: Combines viscoelastic and plastic behaviour

For paper (or any material), you pick the most simple constitutive model that can accurately simulate the material behaviour under consideration.

Examples:

- *Elastic*: Short term bending of corrugated board
- *Elastic-Plastic*: Embossing
- *Viscoelastic/Viscoplastic*: Long term stacking of containers

1.4 The Physical Nature of Paper

The physical properties of paper are evaluated by many different methods. This chapter treats the most common of these methods. Testing of paper is carried out for different reasons, for example for production control, as a measure of the properties of the paper in trade between seller and buyer, and for purely functional reasons, where the paper shall function in a printing press, in boxes etc.

Paper belongs to the field of polymeric materials. Its mechanical properties are thus much more similar to different plastic materials than to for example metals. Characteristic of polymeric construction materials is that the reaction of the material to mechanical stresses is time-dependent. And it is not possible, as it is for many metals, to distinguish a purely elastic region at small stresses followed by a well-defined yield point. Already under very small deformations of paper, viscoelastic effects occur, i.e. the properties depend on how quickly the test is carried out. A striking example of this is that the deflection of a unilaterally clamped horizontal paper strip under the influence of gravity increases with time. Under large deformations, paper is also permanently plasticized.

Technological tests therefore require a defined time schedule. It is essential to choose test methods where the time scale is best adapted to the intended practical use of the paper.

As an indication of the magnitude of the rate dependence, it can be mentioned that the strength increases by 5–10 percent when the rate of elongation is increased 10 times. For this reason, all standard methods specify the rate of elongation.

The viscoelastic properties of paper are quantified, for example, by testing at a constant rate of elongation, by a constant stress increase per unit time or by exposing the sample to sinusoidal tensile force cycles at small deformations.

If a paper is deformed and is kept deformed at a certain elongation, the force decreases with time. This phenomenon is called stress relaxation. An example when relaxation phenomena are important in the use of paper is that a sheet of paper, which is left in a typewriter, adopts a tubular appearance.

If a paper is held under a constant force, the elongation increases with time. This phenomenon is called creep. When corrugated cardboard boxes or cartons are stacked in a storeroom, they creep under the compression load. The creep can lead to a collapse of the boxes after a certain time. The concept of lifespan is used sometimes to characterize this phenomenon.

Paper is a hygroscopic material and its properties are dependent on the relative humidity (RH) and on the temperature in the surrounding air. Most paper tests are now carried out at a standardized humidity of 50 per cent RH and at a temperature of 23 °C. In the choice of test method, it must be decided whether this environment is representative for the technical use of the paper. For example, paper at high temperatures and low relative humidity can show brittle failure, while a paper creeps more rapidly in a moist environment.

1.5 Factors Affecting the Mechanical Properties of Paper

The mechanical properties of a paper depend on a number of factors, the most important of which are:

- *Botanical factors:* The structure, length distribution, fibre wall thickness and proportion of lumen in the fibres.
- *Chemical factors:* The degree of delignification, the degree of polymerisation of cellulose DP, the content and type of hemicellulose.
- *Papermaking factors*: Beating, forming, pressing, drying and calendering.
- *Chemical environment:* The presence of electrolytes, polyelectrolytes and surfactant substances.

A theoretical analysis of the mechanical properties of paper is made more difficult by the fact that the heterogeneous fibre material cannot be characterized uniformly. Proposed theories use strongly idealized models.

Most theories for mechanical properties are based on the assumption that these properties can be changed if the bonded area and strength between the fibres are changed. Many methods have been tested in order to measure the bonded area, for example by nitrogen adsorption, measurement of the bonded area between crossing fibres in a microscope, measurement of the light scattering coefficient and determination of the density of the whole sheet.

Optical and mechanical properties of paper are always related, but in different ways depending on the type of pulp and on papermaking unit-operations.

For instance, the light scattering coefficient, which is a measure of the light scattering surface area in the sheet, can for *chemical pulps* under very special conditions be used to suggest the relative degree of bonding.

On the other hand, for *mechanical pulp,* the light scattering coefficient increases with increasing energy input in the refining as a result of a large number of light scattering particles are formed.

Density of a paper is increased by beating or pressing and will increase the chance for bonding since the fibres to come closer to each other.

It is also shown in practice that the density is a good way of characterizing the way in which the structure of the paper influences mechanical properties.

It is possible to find empirically certain relationships between the mechanical properties and density of the sheet as below, Equation (1.1).

$$property = k \cdot \rho^a \qquad (1.1)$$

The exponent a, varies with fibre source and all papermaking conditions.

Although paper fibres have a considerable strength in both the dry and the wet states, paper has a very low strength when it is wet. Thus, it is evident that the strength of paper cannot be related in a simple way to the strength properties of the fibres.

1.6 The Nature of Paper

There is no strict definition of paper. The main component of paper has always been fibrous material. The most common are plant fibres from wood, grasses and cotton. Animal fibres are also used, e.g. wool in roof board and leather in leather board. Synthetic fibres have a certain importance and can be both inorganic such as glass fibres and organic such as synthetic polymers from petroleum products.

Paper can be described as a layered structure consisting of fibres, which are more or less flattened. A large amount of fibre fragments, so-called fine material, is usually present. The fibres are pre-treated through mechanical treatment, beating, and are bonded to each other without the need for other binding substances.

A large flexibility in the fibres and the presence of fine material increases the contact area and improves the bonding ability between the fibres.

Paper contains about one million fibres per gram. Besides the fibrous material, many types of paper contain considerable amounts of filler, usually a natural mineral, e.g. kaolin, finely ground marble, chalk or talcum. The fillers, which are cheaper than the fibres, improve the optical properties and printability of the product. A large number of chemicals are also used as aids in the paper making process.

Besides being our most important information-bearer, paper is an important and unique construction material.

Paper is a cheap product due to the relatively cheap raw material, the fact that the manufacturing process is in principle simple and the fact that no binders need to be added to bind the fibres together. The manufacturing process is also highly automated.

Paper can be re-utilized, it is combustible, it contains no heavy metals or agents toxic to the environment, and it is environment-friendly.

1.7 The Paper Structure

Paper is formed through filtration of a fibre suspension on a wire, after which the wet product is pressed and dried. Due to the forming process and the fact that fibres are much longer than their thickness, the fibres become oriented more or less in the plane of the paper and paper becomes a layered structure. Usually three directions in paper are of interest (*Figure 1.9*), the machine direction MD, the cross-machine direction CD and the thickness direction ZD.

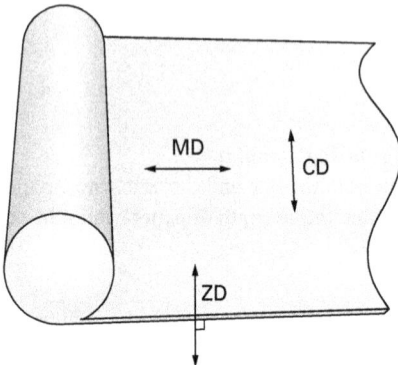

Figure 1.9. Definition of the three directions in paper.

Figure 1.10a, shows schematically the thickness structure of paper formed at a low consistency. A more three-dimensional structure of paper can be achieved if the forming is performed at higher consistency, *Figure 1.10b*.

a)   LC-structure

b)   HC-structure

Figure 1.10. The structure of paper in the thickness direction. a) LC, low consistency, b) HC, high consistency

Since the fibres are much stronger in the longitudinal direction than in the transverse direction paper is much stronger in the plane of the paper than in the thickness direction.

By controlling the sheet forming, different types of more or less marked fibre orientations can be produced. The anisotropy, i.e. the relationship between a property in the machine direction and in the cross-direction, is influenced by the fibre orientation and by the deformation of the paper during the manufacturing process.

The forming consistency, the hydrodynamic conditions during forming and the fibre dimensions and properties determines how well the fibres are distributed in the plane of the paper. The term formation is used for this. *Figure 1.11* shows a paper with bad formation and a good formation. The formation affect the appearance, the printing properties and the strength of the paper. Controlling the formation is an issue every day for the papermaker.

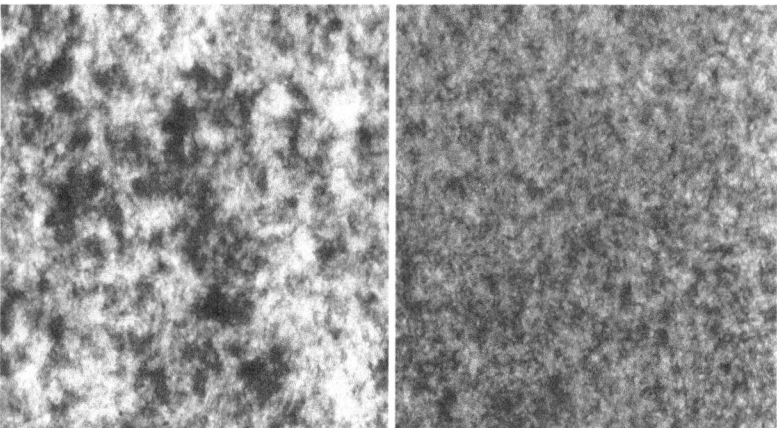

Figure 1.11. A paper with bad and good formation.

The fibre orientation can be controlled in the forming process and influences all physical properties of the paper. In reality, paper never has a completely uniform distribution of the fibres in the plane of the paper, but the fibres flock together in a spectrum of sizes. The lower the forming concentration, the more even is the grammage distribution obtained. The task of the papermaker is to produce as even a sheet as possible. The more evenly formed the sheet is, the better are the properties of the sheet in general, both from a mechanical viewpoint and with regard to surface smoothness and printability.

Different properties have different sensitivities to the forming concentration. Most properties deteriorates with increasing forming concentration. The most sensitive property is the tensile energy absorption index, tensile index, bursting index and strain at break show the same trend but are not quite as sensitive. Tensile stiffness index and compression index are relatively insensitive while the strength in the thickness direction increases with increasing concentration.

Thick paper is called carton board or board. The boundary between paper and carton board is indistinct, except in the customs duties imposed in certain countries. Carton board is the raw material for the production of packages of different kinds and boxes.

Fibre boards, so-called wallboard or hardboard, can be regarded as thick, stiff paper and they are manufactured according to paper technological methods.

Paper need not be a plane product. Egg-packages are manufactured according to normal paper technology, but are shaped in a special mould, which gives the product its final appearance.

Paper is nevertheless normally smooth, feels stiff to the hand and is slightly extensible. It is easy to tear paper.

It is also easy to fold, and the crease becomes permanent. *Figure 1.12* shows a fold failure in a paper. As a consequence of the layered structure of paper, the folding ends up in a delamination between the fibres in the thickness direction and has resulted in a buckling on the compression side of the paper.

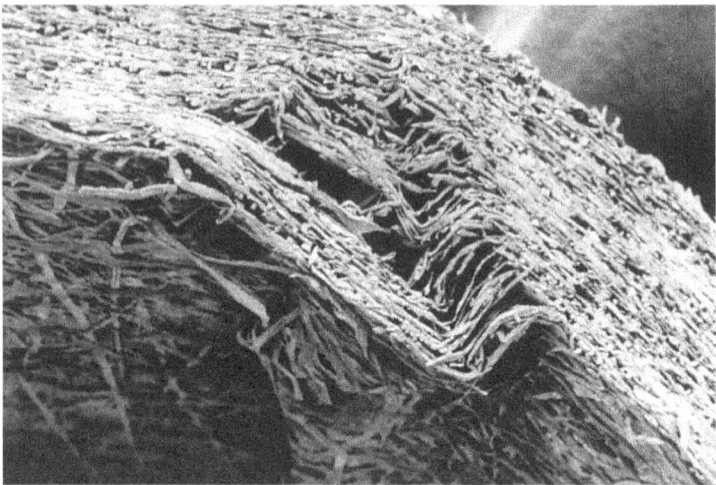

Figure 1.12. Folding failure in paper.

Paper can be folded and reopened many times like a hinge. In the folding of a package, folding marks are made in the carton board, so-called creases, without reducing the tensile strength noticeably, *Figure 1.13* and *Figure 1.14*.

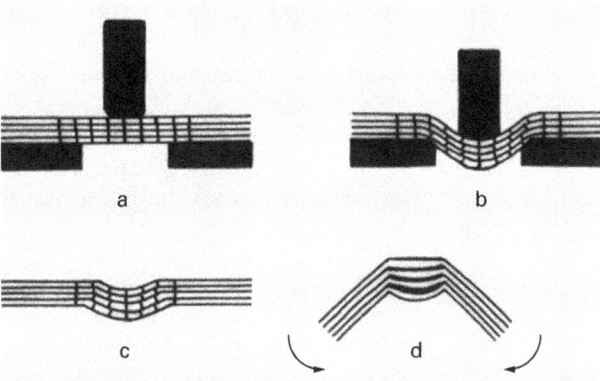

Figure 1.13. Schematic illustration of creasing operation and the bending of the creased carton board.

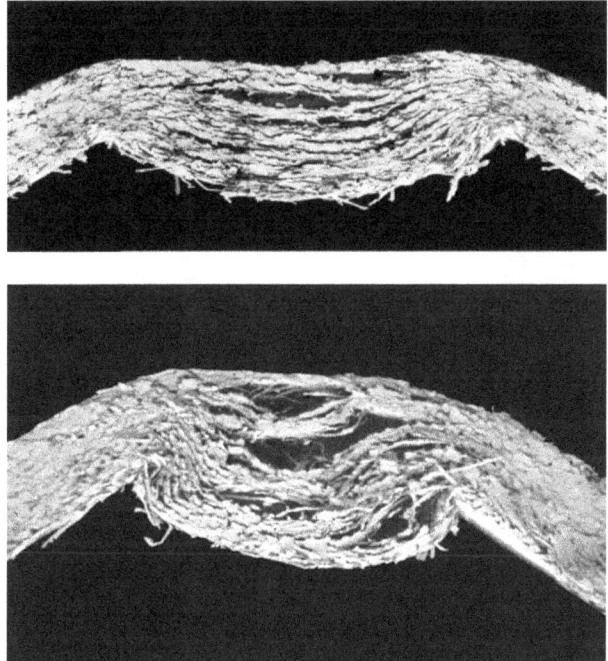

Figure 1.14. The bending of the creased carton board.

If the paper is wrinkled, permanent folds are formed with a rustling sound. It is not possible to bend paper to a double-curved surface without a fold or dent arising.

During drying, the fibres become stiff and are locked into position. The high strength is essentially due to the fact that surface tension forces during the drying draw the fibres together so that so-called hydrogen bonds can be formed between the fibres. These chemical forces are weaker than the forces, which hold the cellulose molecules together in the fibres, but are stronger than most other physical attraction forces.

Most of the deformation under stress comes from the fibres themselves, with relatively small contributions due to a reorganization of the fibres.

Delamination may be a problem during offset printing, plastic coating, corner gluing of boxes and several other converting operations. *Figure 1.15* shows one example, the delamination between the surface layer and the layer below after offset printing. Naturally the product is subject to complaint from the printer.

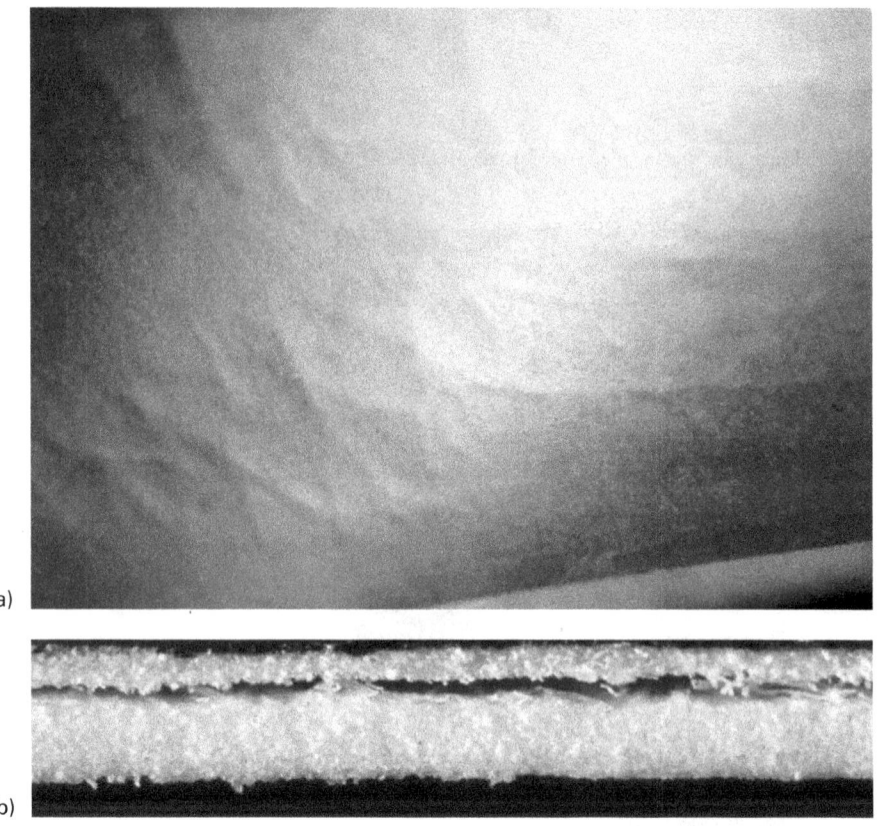

Figure 1.15. Delamination of a paper after offset printing. A) The surface, B) The thickness direction.

Paper loses its strength on wetting since the hydrogen bonds between the fibres are broken and the fibres become soft. Wet strength can be developed with special chemicals.

Although the papermaking process is simple, the mechanical properties of paper can be varied strongly and adapted to end-use requirements.

Pulps of different kinds of fibre and different production procedures give papers with different properties.

If rigid papers are desired, they can be built-up of several layers. According to the I-beam principle, the surface layers should be non-elastic and consist of highly beaten pulps whilst the central layer should be made from pulps with the lowest possible density.

By controlling the drying shrinkage, the paper properties may be changed dramatically.

1.8 Paper Divided into Categories

There is no simple way of dividing papers into different categories. A division of papers can be made according to use as in *Table 1.1*.

Table 1.1. Division of different types of papers according to areas of use.

Thin wrapping papers Bag paper Sack paper Envelope paper	Specialty papers Photographic papers Electric insulating papers cable papers,
Thick wrapping papers Carton board Liquid carton board	presspan for motor winding, transformers Filter papers for air, water and oil Cigarette paper
Printing papers Newsprint (daily newspapers) Magazine paper (magazines, catalogues, advertisements) Uncoated (SC) Coated (LWC)	Gypsum board carton board Wallpaper Book-binding paper Core cardboard Technical drawing papers Emery paper
Fine papers Uncoated (e.g. carton board paper and copy paper) Coated (e.g. advertising paper)	Security paper for banknotes and cheques Greaseproof paper Self-copying paper
Hygienic papers Soft crêpe paper Toilet paper Kitchen paper	

1.9 The Structure of some Paper Products

Figure 1.16–1.24 shows the structure of some paper products, chosen to illustrate the large variety of properties of paper. By changing features of the pulp- and papermaking processes, the structure of paper can be varied across wide limits. The structure is adapted to give suitable properties to a given paper. In some of the figures we see a corner of a paper, showing both the surface and the thickness direction, in other cases the surface structure.

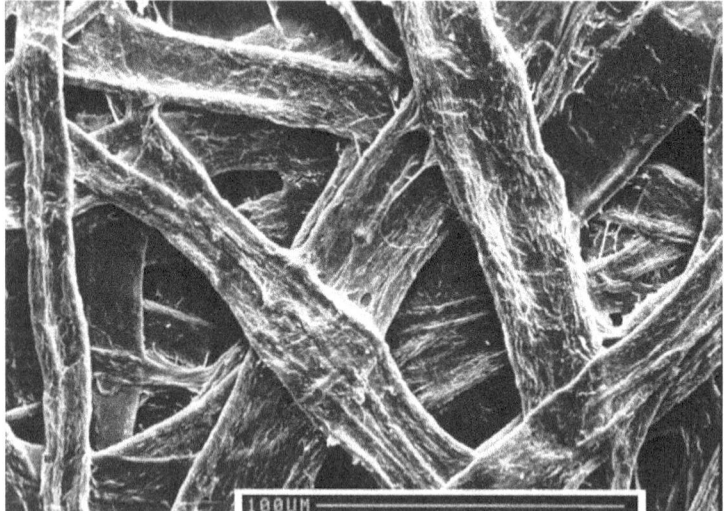

Figure 1.16. Linerboard, the surface-layer in corrugated cardboard, is manufactured mainly from bleached pine Kraft pulp in a yield interval of 47–52 %. The pulp is only slightly beaten. The pulp and structure is chosen mainly to give high compression strength, creep resistance, toughness in converting and delamination resistance.

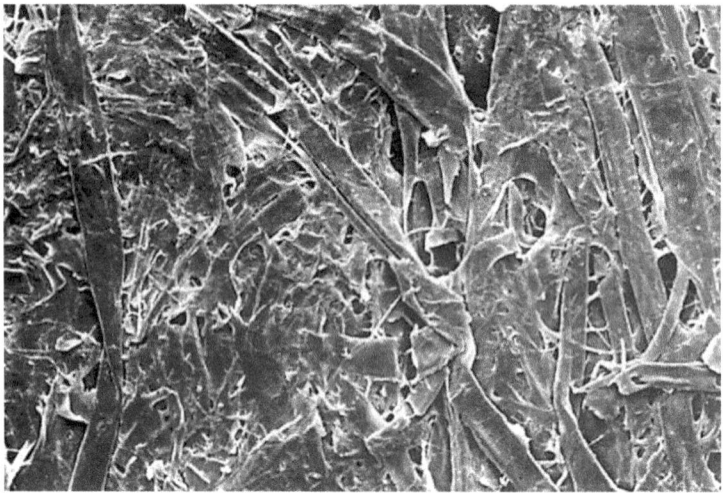

Figure 1.17. Newsprint, may be manufactured from different fibre sources. The most recent pulp is thermomechanical pulp of softwood Still newsprint is manufactured from groundwood and from recycled newspapers. To improve strength properties often the paper is reinforced with an addition, in the order of 5–10 % of chemical softwood pulp. The pulp and structure is chosen mainly to give good runnability during manufacturing and end-use, good opacity and printability.

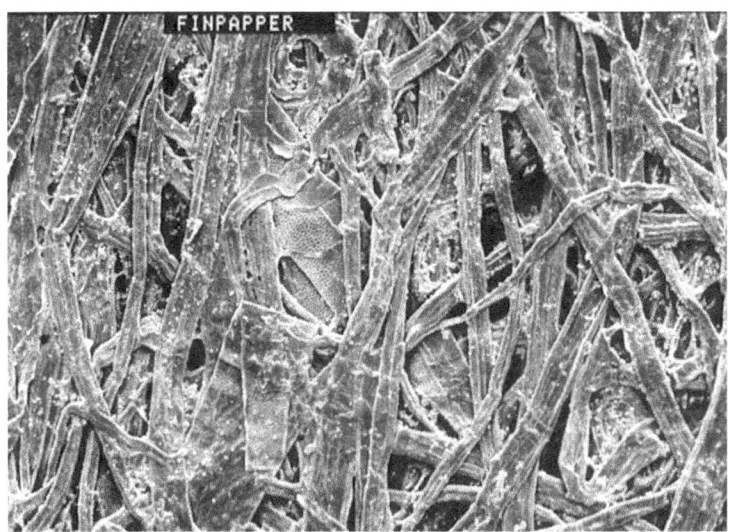

Figure 1.18. Fine paper, consists of chemical, undamaged fibres, which are often a mixture of hardwood and softwood. Note the filler particles. The pulp and structure is chosen mainly to give good opacity and surface properties.

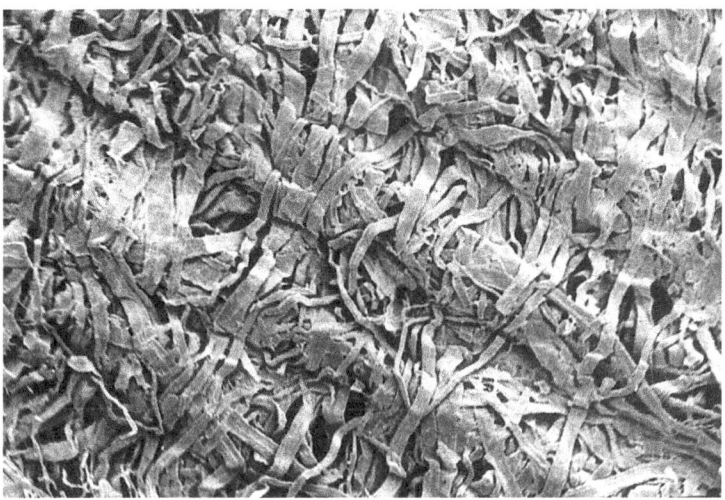

Figure 1.19. Tissue. Creped paper with low grammages, 14–25 g/m^2. The type of fibre can vary. Note the creped structure in the picture. (The magnification is slightly lower than in the other pictures).

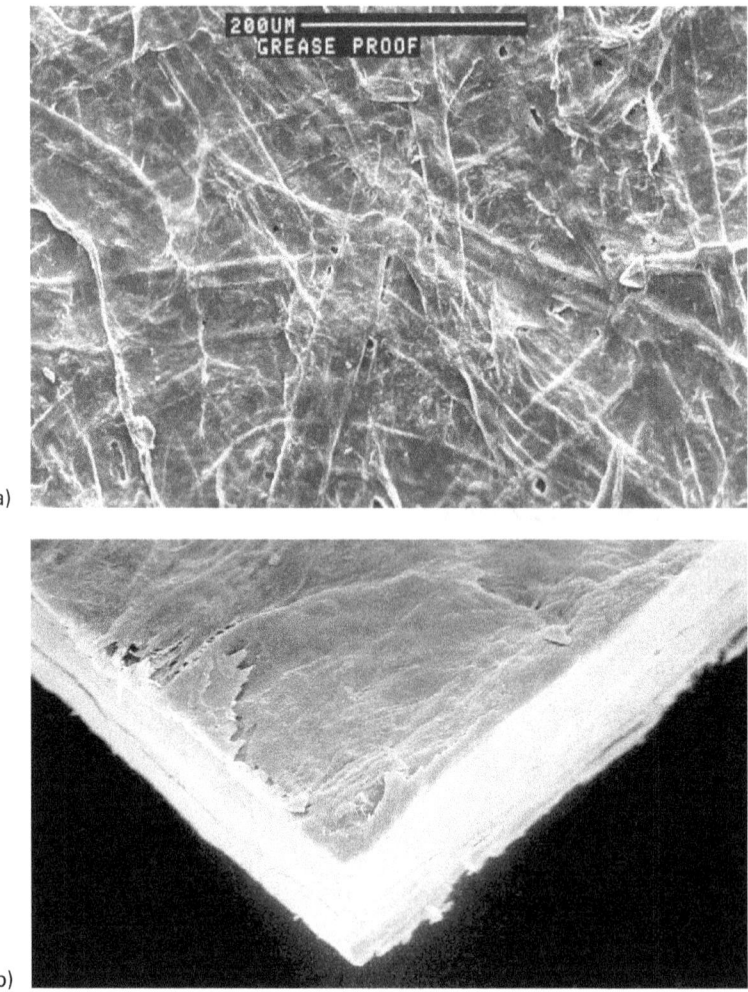

a)

b)

Figure 1.20. Greaseproof paper. The fibres in the greaseproof paper are so well beaten that the fibre structure in the paper has been "erased". The paper becomes transparent, has a low opacity. Greaseproof paper is used e.g. for drawing, baking and sandwich paper.

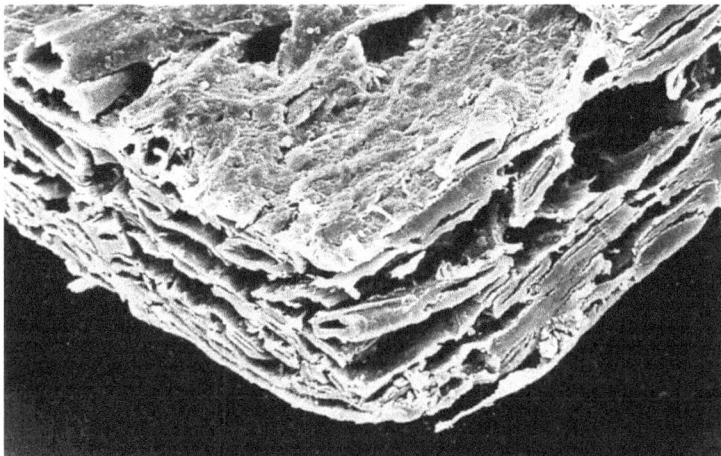

Figure 1.21. Copy paper, often consists of chemical, undamaged fibres, often a mixture of softwood and hardwood. Note the open paper structure. The fibres and structure are chosen mainly to give flatness and, good surface properties.

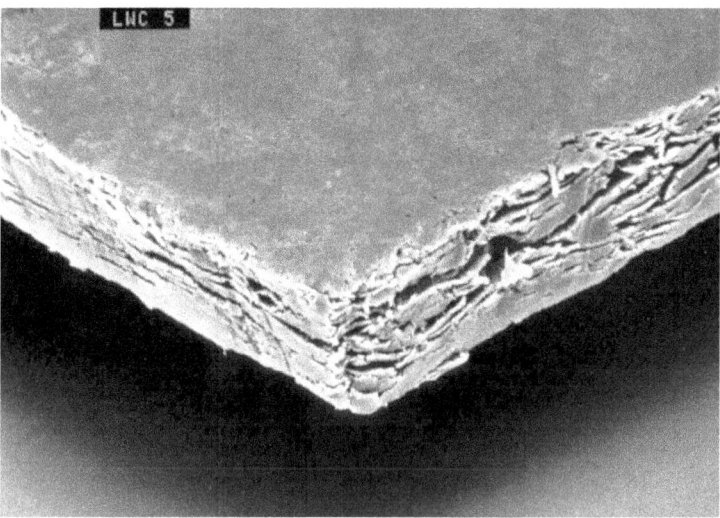

Figure 1.22. LWC, Light Weight Coated paper, consists of mechanical, fibres, reinforced with chemical fibres and with a coating layer. The fibres and structure are chosen mainly to give good runnability in printing presses and to provide a good base paper for the coating.

20

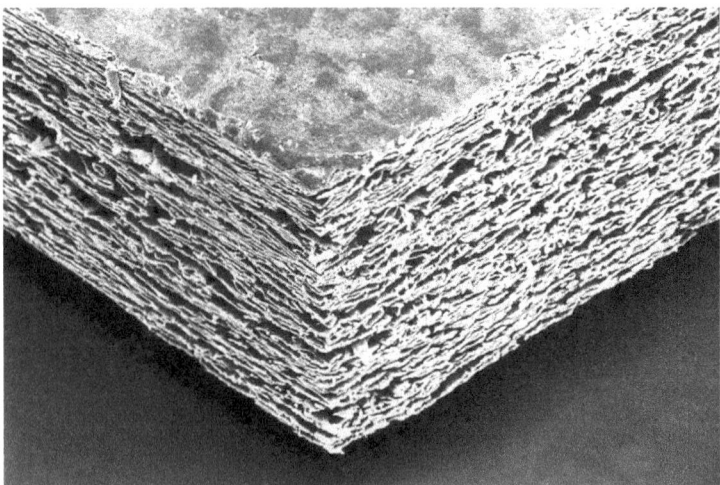

Figure 1.23. Coated carton board. Carton board is manufactured by a multi-layer technology, e.g. with a surface layer of bleached kraft pulp and mechanical pulp in the middle, or with several layers of the same type, e.g. bleached Kraft pulp. The coating layer makes the carton board surface more even and more suitable for high-class print. The fibres and structure is chosen mainly to give bending stiffness, good converting properties and surface properties.

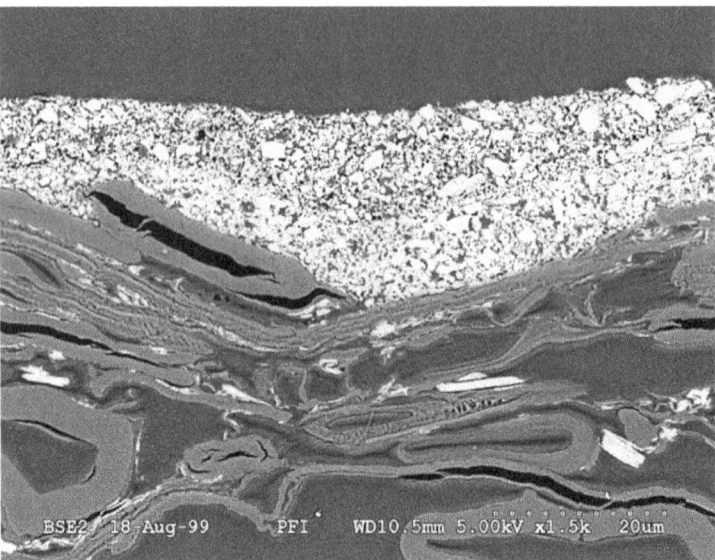

Figure 1.24. The coating layer of a coated carton board seen in the thickness direction. Note the cross-sections of the fibres.

1.10 Properties of Different Materials Compared to Paper

1.10.1 General Considerations

When materials are chosen for instance in airplanes, the important material property often is tensile stiffness per weight, called specific elastic modulus.

For paper this property is called tensile stiffness index. If we then calculate the bending stiffness for different materials based on equal weight (Bending stiffness index), the following results are obtained, *Table 1.2*. It is apparent that paper belongs to the stiffest materials in this respect.

Table 1.2. Material properties.

Material	Elastic modulus E	Density r	Tensile stiffness index $E^w = \dfrac{E}{\rho}$	Bending stiffness index $S^w = \dfrac{S^b}{w^3} = \dfrac{E^w}{12 \cdot \rho^2}$
	MN/m²	kg/m³	M Nm/kg	Nm⁷/kg³
Steel	210 000	7800	25	0,03
Titanium	120 000	4500	25	0,10
Aluminium	73 000	2800	25	0,30
Magnesium	42 000	1700	25	0,70
Glass	73 000	2400	25	0,40
Concrete	15 000	2500	6	0,08
Carbon fibre composites	200 000	2000	100	2,00
Wood in grain direction	14 000	500	25	8,30
Paper, linerboard in MD	15 600	700	22	3,70

Paper differs from stamped felt, usually of cotton, through the manufacturing process. The strength of a stamped felt, e.g. a felt hat, is produced by mechanical treatment of the formed product through stamping and needling in which the fibres get tangled into each other and are bonded by frictional forces. Particle boards are made from chips of wood, which can be randomly distributed or ordered. They are held together with glue.

A piece of textile, a polyethene film and a paper can be separated even with the eyes closed.

All these products are essentially two-dimensional and formed by linear macromolecules. These are also more or less organized into crystalline regions. If the textile material is made of cotton, flax or ramie, it consists of the same material, cellulose, as bleached paper. Why do these materials have different properties?

The answer lies in the construction of the materials, which leads to different types of interaction between the components.

We shall here compare paper with three other types of material made from fibres: textiles, fibre cloth and polyethene film.

1.10.2 Textiles

A textile material is built-up of fibres which are first organized through carding and spinning into a yarn which is thereafter used in a loom or knitting machine. The structures obtained allow a certain freedom of movement to both the yarn and the fibres. The structure therefore reacts to stresses in the first place through rearrangement of the fibres relative to each other, and rearrangement of the fibres in the yarn, and only in a second stage through elastic deformation of the fibres themselves. The range of movement of the fibres is determined mainly by how strongly the yarn is twisted. The mechanical properties can further be varied by letting the yarns run through the textile in different ways. A woven textile can be torn relatively easily in the thread direction, but not in between. Textile materials range from stretchable fibres of the Spandex type to very stiff carbon fibres.

A woven textile has very long threads of twisted fibres, which lie perpendicular to each other. Its strength comes from the fact that these threads run above and below each other so that they are bent. The textile has essentially different mechanical properties in different directions. It is slightly stretchable and strong in the thread directions, but considerably more stretchable and weak in directions in between. Characteristic for a fabric is that it drapes itself, i.e. adapts to the underlying structure. The product is held together by normal forces between the fibres, in the twisted thread and between the crooked threads in the fabric.

A knitted textile is produced from a single thread which runs through more or less complicated loops. The product is almost isotropic, i.e. equally strong and stretchable in all directions. Also in knitted goods, normal forces are the cause of the strength properties of the structure.

The textile feels flexible and soft. If it is wrinkled together and released, it regains its shape. It can be draped over double-curved surfaces without noticeable stretching. If the textile is knitted, it can be easily stretched in all directions, but after a long stretching it stiffens. If it is wrinkled together, no sharp creases are formed. The knitted product is very difficult to tear. If the textile is wetted, it behaves in approximately the same way as when it is dry. Textiles can therefore be washed.

1.10.3 Fibre Textiles

Fibre (non-woven) textiles are manufactured from natural or synthetic fibres which randomly form an essentially two-dimensional product. The fibres can be dispersed in air or water. The normal forces between these randomly ordered fibres are negligible. The product can be given strength by gluing the fibres together in the contact surfaces. This can be achieved either by adding a suitable adhesive or, if the fibres are thermoplastic, by heating so that the fibres stick together. The mechanical properties are determined by the fibre orientation and by the content and type of adhesive. An extreme case occurs if the product contains so much binder that it can be regarded as a fibre-reinforced composite.

1.10.4 Polyethylene Film

A polyethylene film is usually very smooth, it feels flabby and can easily be deformed by stretching so that it conforms to a double-curved surface. If the film is released soon after the deformation, it contracts and returns essentially to its original shape. If the film is folded, no permanent creases are formed. If an attempt is made to tear the film, it is difficult to start the tearing but, once a tear has started, it is very easy to continue tearing. Water has no influence on the mechanical properties.

There is nothing corresponding to the fibres in a plastic film. The structure lacks discrete structural elements. Stresses lead to a distortion of the molecular structure or a sliding of the molecules relative to each other, possibly leading to failure in the long tangled polymer molecules. Ordered, crystalline areas act as stiffening reinforcements. Polymer technology explains how the properties of the products can be modified, e.g. through the choice of polymer, molar weight, degree of crystallization and orientation.

1.11 Standardization

In order to facilitate trade, several standard methods exist for the determination of different paper properties. These standards are issued by organizations, which have members representing manufacturers, buyers, wholesale dealers and researchers. The most important standards in Sweden are listed in *Table 1.3*.

Table 1.3. Important standards for the Swedish industry.

ISO	International Organization of Standardization
ASTM	American Society for Testing and Materials
TAPPI	Technical Association of the Pulp and Paper Industry
CEN	Comité Européen de Normalisation
DIN	Deutsches Institut für Normung
SIS	Standardiseringskommissionen i Sverige
SCAN-Test	Scandinavian Pulp and Paper Testing Methods

Buyers and sellers reach agreements about which different standards shall be applied.

Of interest are ISO Standard 9000 and the corresponding national standard, which constitute a well-established system for quality control during manufacture. The method requires a couple of years' personnel education and the introduction of new routines. Certification is followed by supervision from the certifying organization. More and more buyers turn only to paper mills which meet this standard.

The standardization organ CEN has the task of developing EN-standards, i.e. European standards. In the first place, it seeks to ensure that existing ISO-standards become EN-standards, and thereby automatically national standards within the EU's member countries. The primary interest within CEN is to remove technical trade obstacles in order to reinforce European harmonization. This means that CEN is interested only in standard methods with a commercial interest, i.e. methods which are of interest in the quality assessment of products (e.g. brightness and spots of a pulp but not analyses of white liquor and mesa).

The main secretariat for CEN is in Brussels, whilst the secretariats for the different technical committees (in all more than 150) are in different cities in Europe. A technical committee for pulp and paper has been formed, TC 172, with its secretariat at DIN in Berlin.

CEN has begun to issue EN-standards relating to the pulp and paper industry. When an EN-standard is issued, it automatically becomes a Swedish standard and is issued by SIS, since SIS is Sweden's national standardization organ.

2 Paper Physics

Christer Fellers
STFI-Packforsk AB

2.1 Grammage 26

2.2 Thickness 26

2.3 Density 29

2.4 Stress and Specific Stress 30

2.5 Strain and Poisson's Ratio 30

2.6 Tensile Strength Properties 32

2.7 Tensile Strength of Fibres Measured by Zero Span Technique 34
2.7.1 Zero-Span and Short-Span Tests 34

2.8 Page's Theory for Tensile Strength of Paper 36

2.9 Strain at Break and Tensile Energy Absorption 37

2.10 Elastic Properties 37

2.11 Tensile Stiffness – Theoretical Approach 38

2.12 Ultrasonic Techniques for Stiffness Determination 40

2.13 Plastic Stress-Strain Properties in Tension 42

2.14 Creep Properties in Tension 43

2.15 Compression Properties 45

2.16 Viscoelastic Properties in Compression 49

2.17 Tearing Resistance 51

2.18 Theory for Tearing Resistance 52

2.19 Bursting Strength 53

2.20 Bending Stiffness 55

2.21 Bending Resistance 56

2.22 Fold Number and Folding Endurance 57

2.23 Thickness Direction Strength and Delamination Resistance 58

2.24 Methods that Simulate Converting Operations 60

2.25 Methods for Production Control 61

2.26 Fracture Mechanics 63

2.27 Air Permeance 66

2.1 Grammage

Grammage, w is the mass per unit area. Basis weight is another term sometimes used.
The grammage includes the amount of water in the paper under the existing climatic conditions, which shall be well specified.
"Dry" or "bone-dry" grammage, w_d is sometimes indicated. This means the amount of dry substance per square metre after a standardized drying, which takes place in an oven at 105 °C.
The relationship between the moisture ratio mr (mass of moisture per mass of paper), conditioned grammage w and dry grammage w_d is given by Equation 2.1.

$$w = w_d(1+mr) \qquad (2.1)$$

2.2 Thickness

Thickness is a fundamental property of paper for several reasons. It has a great impact on the bending stiffness of the paper and an even thickness profile important so that a paper reel has an even cross profile, which in turn influences the runnability of the paper in the printing press and in other converting situations. A paper thickness with narrow tolerances is required in book manufacture where thickness errors from hundreds of pages would otherwise be cumulative.
According to SCAN-P 7, the thickness of the paper can be measured in two ways, as the thickness of a single sheet or as the thickness in a stack, *Figure 2.1*. In both cases, the thickness is measured as the distance between two parallel plates having an area of 200 mm² and under a pressure of 100 kPa.
In measurements of the thickness, especially for single sheets, the average thickness of the paper will overrated since the surface unevenness of the paper are included in the measurement. In the measurement of thickness in a stack, the idea is that errors of this kind should be reduced. However, experience shows that surface unevenness effects cannot be completely avoided in

this way. The measurement in a stack is, however, relevant for example in the estimation of how thick a book will be.

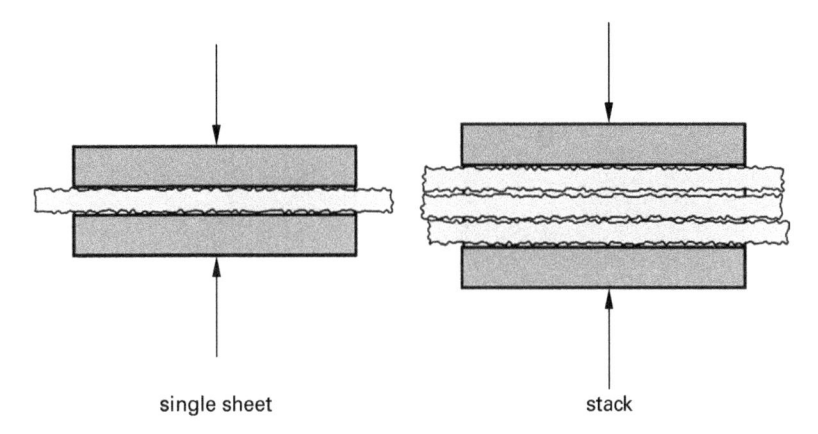

Figure 2.1. Thickness measurement according to SCAN-P 7.

SCAN P-88:01 describes an improved thickness determination for use in bending stiffness calculations and studies of thickness changes in calendering and other applications.

According to this method, the thickness of a paper is scanned between two opposite measurement points so that a thickness profile of the paper is obtained, *Figure 2.2*.

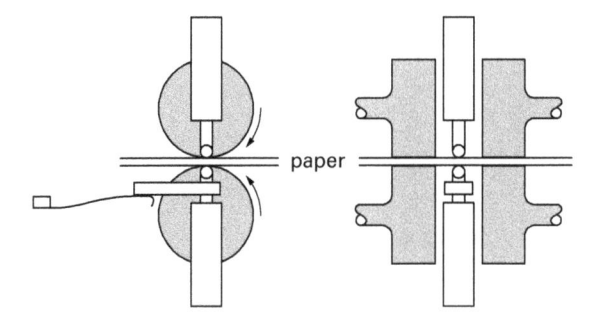

Figure 2.2. Structural thickness measurement according to SCAN P-88:01. Paper and board. Structural thickness and structural density.

Using this method, a thickness scan is performed. *Figure 2.3* shows how the thickness varies along a line across a paper sheet. The mean value is called the structural thickness. The SCAN-P 7 thickness corresponds to a value close to the maximum value of the peaks.

The thickness of some papers can vary substantially and a measurement of thickness as the distance between two flat plates sometimes becomes less relevant.

By adapting the force and diameter of the measurement points, a true mean value of the thickness can be obtained. In measurements on papers manufactured in the same way but with different grammage, a thickness value is thus obtained which is proportional to the grammage.

The method prescribed in SCAN-P 7 overestimates the thickness. In *Figure 2.4* the thickness as a function of grammage is performed on laboratory-made papers with equal surface characteristics. If the thickness of the papers is extrapolated to zero grammage, a measure of the surface unevenness of the paper is obtained.

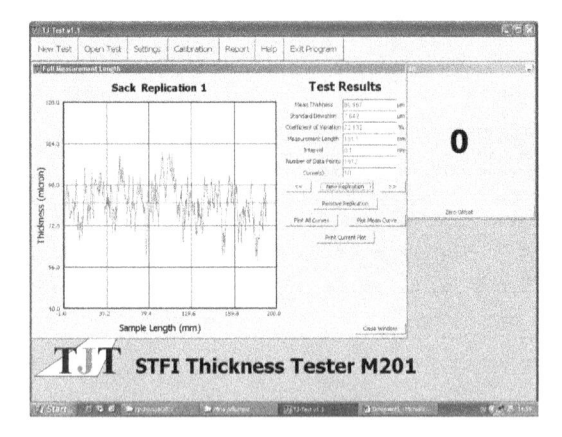

Figure 2.3. Thickness variations in a paper measured with SCAN P:88:01. The mean value is called the structural thickness.

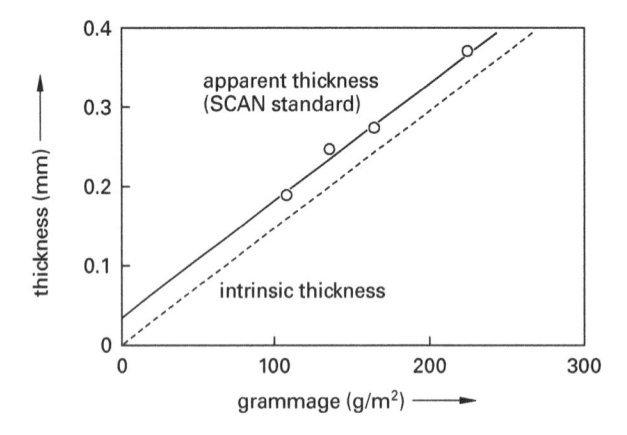

Figure 2.4. Apparent thickness of unbleached kraft paper determined according to the SCAN standard. The intrinsic thickness is given by a parallel shift of the SCAN thickness values to yield zero intercept when extrapolated to zero grammage.

Other methods have been proposed to measure structural thickness. For example, a piece of paper has been submerged in mercury and the volume has been measured and a thickness gauge with rubber-coated surfaces, which adapt to the unevenness of the paper surface has been used. *Table 2.1* shows a comparison of thickness values according to different methods.

The above examples show that a true thickness measurement is a prerequisite for a meaningful density calculation to be used as a description of the structure of paper.

Table 2.1 Percentage deviation in thickness for the SCAN-P 7 methods compared with SCAN P-88:0. The grammage of the sheets is 60 g/m².

Pulp	Four sheets in a stack $\dfrac{SCAN\ P7}{SCAN\ P88}$	Single sheets $\dfrac{SCAN\ P7}{SCAN\ P88}$
Unbleached pine kraft. 800 PFI-mill rev.	+13 %	+21 %
Unbleached birch kraft. 1000 PFI-mill rev.	+18 %	+42 %

2.3 Density

Many constructional materials such as metals and plastics have a homogeneous structure, which means that the density of the material is constant. There are many materials however where the proportion of air can be varied, for example foamed plastics and wood where the density varies between the species of wood. For paper, the density of the material can be varied between ca 300 and 1000 kg/m³. Pure cellulose has a density of ca 1500 kg/m³. The differences in paper properties in this interval are considerable. The possibility of paper technology to control the density means that there is a unique potential to steer the properties of the product to the desired quality. Density is consequently a very important structural parameter of the sheet.

The density of paper is influenced by the type of fibre and all paper technological operations such as fibre separation, beating, pressing, drying and calendering. The density ρ is calculated as the grammage w divided by the thickness t (Equation 2.2).

$$\rho = \frac{w}{t} \ (\text{kg/m}^3) \tag{2.2}$$

The accuracy of the density determination is consequently dependent on the thickness measurement which is difficult to measure unambiguously for a rough and compressible material such as paper, see Section 2.2.

SCAN-P 7 specifies that density of paper shall be evaluated from the thickness measured in a stack, but that for carton board, it shall be evaluated from the thickness of a single sheet. As mentioned earlier, this measurement can give a large error depending on the surface roughness of the paper. Structural density is usually for a meaningful density determination.

Within paper technology, the inverted density is often used, which is called the bulk (Equation 2.3).

$$\text{Bulk} = \frac{1}{\rho} \tag{2.3}$$

The bulk is usually given in cm³/g, which gives a value with a magnitude of 1–2.

2.4 Stress and Specific Stress

In solid mechanics, the concept stress, i.e. force per unit area is a fundamental, accepted way to describe the intensity of loading. For paper, which is a porous material and a material where the pore volume and density can be changed, the concept off stress for paper material ranking and development can lead us in the wrong direction as exemplified in the following example.

The example illustrates a case where the stress is unity in the left figure, *Figure 2.5*. If the paper is calendered and the thickness reduced to half its value the stress becomes twice as high. It is quite obvious that the loading intensity of the fibres has not changed. If the stress is the failure stress, the material to the right is judged as twice as good as the left material, which of course is wrong.

A more realistic way is to use a specific stress or indexed stress, which simply means a division of the force by the width and grammage, *Figure 2.6*. In this way the stress is equal regardless of the pore volume.

Different symbols will be used to distinguish between the different stress concepts.

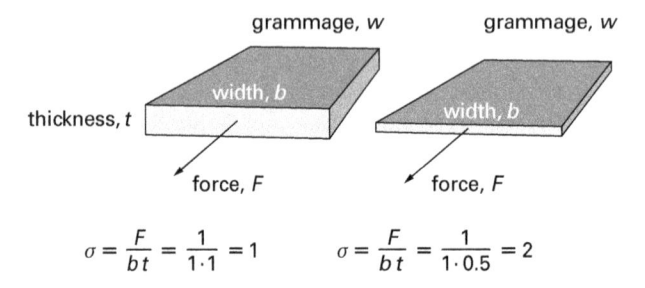

$$\sigma = \frac{F}{bt} = \frac{1}{1 \cdot 1} = 1 \qquad \sigma = \frac{F}{bt} = \frac{1}{1 \cdot 0.5} = 2$$

Figure 2.5. The change in stress with calendering.

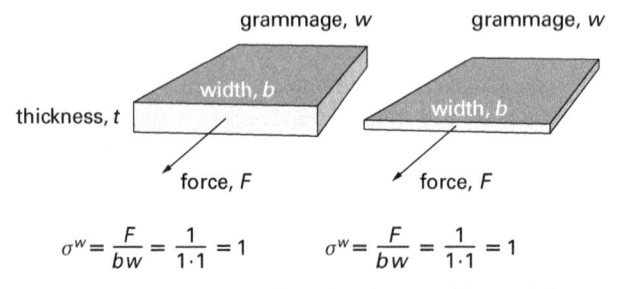

$$\sigma^w = \frac{F}{bw} = \frac{1}{1 \cdot 1} = 1 \qquad \sigma^w = \frac{F}{bw} = \frac{1}{1 \cdot 1} = 1$$

Figure 2.6. The change in specific, indexed stress, with calendering.

2.5 Strain and Poisson´s Ratio

For paper the useful definition of strain, ε, is the elongation divided by the original length. A piece of material is strained in the X-direction and gets the strain ε_x. The strain in the transverse direction becomes ε_y (*Figure 2.7*). For most material the material contracts (exceptions such as cork exist).

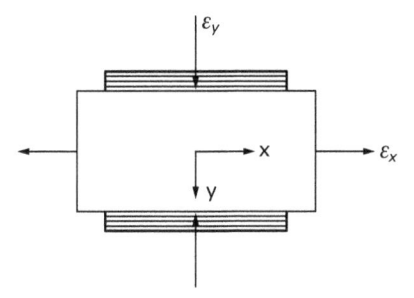

Figure 2.7. Strains in paper.

Poisson´s ratio v_{xy} is uses to tell the relation between these strains, Equation 2.4

$$\varepsilon_x = - v_{xy} \cdot \varepsilon_x \tag{2.4}$$

Tensile testing of paper is carried out by clamping a paper strip, usually with a width $b = 15$ mm and a length $l = 100$ mm at each end, and straining the material until it breaks, as indicated in *Figure 2.8*.

A large number of devices are available for tensile testing with both vertical and horizontal clamping of the paper.

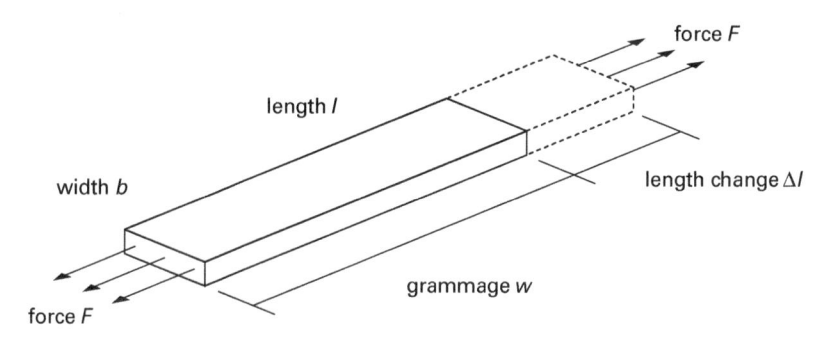

Figure 2.8. Tensile test of paper strip with definitions.

If the stress σ^w is plotted against strain ε (with the definitions given in Equation 2.5) the curve has the appearance shown in *Figure 2.9*. The curve is linear at first and then deviates to reach a maximum which defines the failure point.

$$\sigma^w = \frac{F}{b \cdot w} \qquad \varepsilon = \frac{\Delta l}{l} \tag{2.5}$$

A number of useful properties can be derived from the curve.

- Tensile stiffness index, $E^w = \dfrac{\Delta \sigma^w}{\Delta \varepsilon}$ is the slope of the linear part of the curve.
- Tensile index, σ_T^w is the maximum stress.
- Strain at break, ε_T is the maximum strain.
- Tensile energy absorption index W_T^w is the area under the curve.

These properties will be further discussed below.

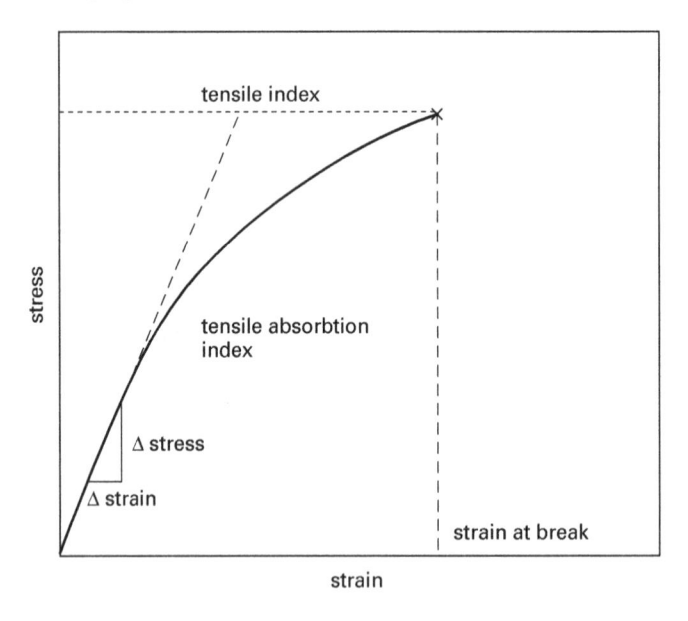

Figure 2.9. A typical stress-strain curve for paper.

2.6 Tensile Strength Properties

The strength of paper due to tensile loading may be expressed in various ways.

- Tensile index is the maximum tensile force per unit width and unit grammage.

$$\sigma_T^w = \frac{F_T}{bw} \quad (\text{Nm/kg}) \tag{2.6}$$

- Tensile strength is the maximum force related to the width of the test piece.

$$\sigma_T^b = \frac{F_T}{b} \quad (\text{N/m}) \tag{2.7}$$

- The tensile failure stress is the maximum tensile force per unit width and unit thickness.

$$\sigma_T = \frac{F_T}{bt} \quad (\text{Pa or N/m}^2) \tag{2.8}$$

In older literature, the concept of breaking length is used. This designates the maximum length which a hanging sheet of paper can have without breaking. The expression can be derived as follows:

$$\sigma_T^l = \frac{F_T}{bt\rho g} \quad (\text{m}) \tag{2.9}$$

The relation between breaking length and tensile index is derived

$$\rho = \frac{w}{t} \text{ and we get}$$

$$\sigma_T^l = \frac{\sigma_T^w}{g} \quad (\text{Nm/kg}) \tag{2.10}$$

Note that

$$\sigma_T^w = \frac{\sigma_T}{\rho} \tag{2.11}$$

σ_T	Maximum stress,	Pa (N/m²)
σ_T^b	Tensile strength	N/m
σ_T^w	Tensile index	Nm/kg
σ_T^l	Breaking length	m
F_T	Maximum force	N
t	Thickness	m
b	Width	m
w	Grammage	kg/m²
ρ	Density	kg/m³
g	Acceleration due to gravity	m/s²

For paper, the tensile index is of the order of 10–100 kNm/kg, *Figure 2.10*. If, for example, a paper roll is rolled out of an air balloon, it can thus hang down with a length of 1–10 km before the paper breaks because of its own weight.

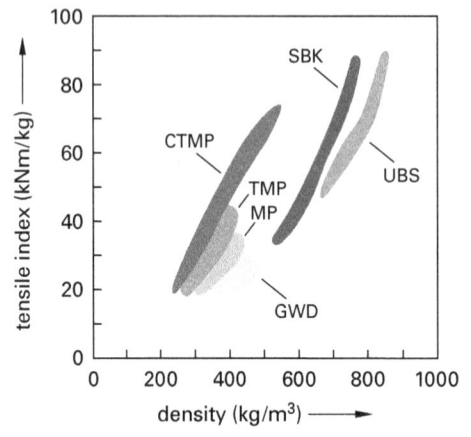

Figure 2.10. Tensile index for different fibre raw materials plotted against density. (Restrained drying). Different density levels are achieved by varying the beating. (Isotropic laboratory papers).

Figure 2.11 shows typical tensile failures in a paper. The failure zone is characterised by elongated and broken fibres. Depending on whether the degree of bonding is low or high in the paper, the character of the failure surface will change from a structure with few broken fibres into a structure where nearly all the fibres are broken.

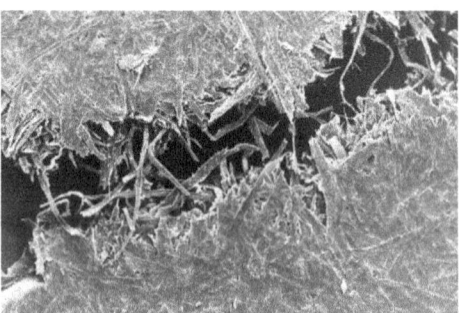

a) low degree of bonding   b) high degree of bonding

Figure 2.11. The tensile failure in paper is characterized by elongated and broken fibres.

2.7 Tensile Strength of Fibres Measured by Zero Span Technique

2.7.1 Zero-Span and Short-Span Tests

Sheet tensile properties are developed by the use of adequately strong fibres and the enhancement of the fibre-fibre bonding properties. Good (or poor) fibre-fibre bonding is achieved primarily by the pulping, bleaching, refining and drying systems. If one could eliminate the influence of bonding in sheet tests, then presumably the fibre mechanical properties could be deduced without resorting to tests on individual fibres. In theory, shortening the span between the two tensile grips to the point where the fibre bonds are insignificant – the ideal "zero" span – means that all forces between grips are transmitted by those fibres, lying at various orientation, that are held within both grips simultaneously. For an in-plane statistically isotropic sheet, the mean fibre strength, expressed as breaking force, is related to the zero-span tensile strength.

The zero-span test has become widely used, and several standards organizations have published test procedures for it; for example, TAPPI issued T 231 cm-96 for the dry zero-span test and T 273 pm-95 for the wet zero-span test. Equipment is readily available as a fixture for the universal testing machine or as a stand-alone system. The zero-span tensile test was introduced to distinguish the difference between the strength of the network and the strength of the fibres themselves. For softwood pulps, zero-span sheet strengths are typically 1.5–3 times as great as the strength values achieved in a conventional tensile test. For hardwood pulps, the ratios tend to be greater.

Figure 2.12 shows a typical curve for the tensile index versus the free span between the clamps in a tensile test. Special clamps are used to grip the paper without slipping. Still, a true "zero span" is difficult to obtain and the values are sometimes extrapolated to a "true" zero span value.

Figure 2.13 shows a schematic drawing of the clamps and the fibre network during a tensile loading.

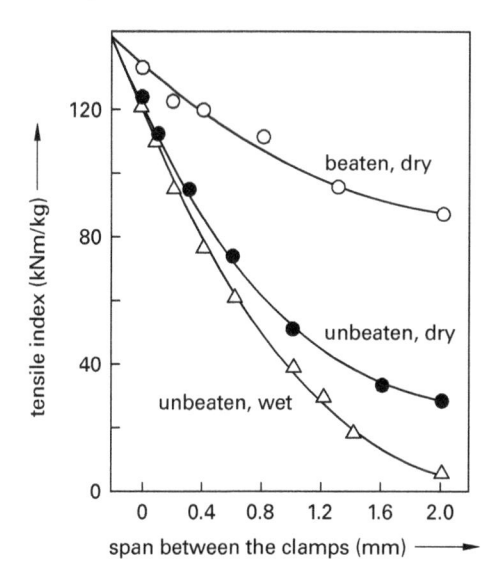

Figure 2.12. A typical curve for the tensile index versus the free span between the clamps.

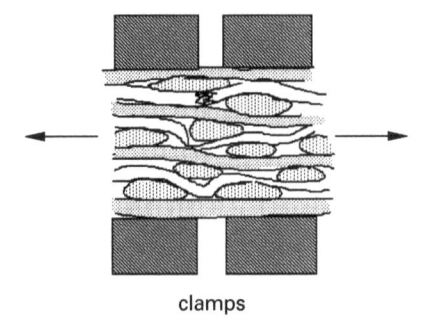

Figure 2.13. A schematic drawing of the clamps and the fibre network during a tensile loading.

2.8 Page´s Theory for Tensile Strength of Paper

Several researchers have attempted to calculate the tensile strength of paper from the fibre strength and bonding ability. The most accepted theory was formulated by Page:

$$\frac{1}{\sigma_T^w} = \frac{9}{8}\cdot\frac{1}{\sigma_{ZS}^w} + \frac{12}{\tau_s l \alpha} \tag{2.12}$$

where
σ_T^w = Tensile index

σ_{ZS}^w = Zero span tensile index

τ_S = Shear stress at break of a fibre-to-fibre bond
α = Bonded area between the fibres per kg of fibres
l = Average fibre length

Due to the extremely complicated interaction between fibres of different dimensions and the influence of different methods for sheet forming, pressing and drying the theory only describes the strength in idealised cases. Equation (2.12) is reported to give good agreement between properties of straight fibres and tensile index of papers dried under restraint.

The value of the equation lies in that it shows the relationship between some important bonding and fibre properties. However, there are limitations for the use of the equation. Many important factors are not taken into account, examples are listed below.

- Formation
- Fibre shape and fibre length distribution
- The true area and failure stress of the bonds

2.9 Strain at Break and Tensile Energy Absorption

Strain at break is of the order of 1–5 % for sheets dried under restraint. For freely dried sheets, strain at break of up to 20 per cent can be obtained.
Tensile Energy Absorption (TEA) index

$$W_T^w = \frac{W_T^b}{w} \text{ (J/kg)} \tag{2.13}$$

where W_T^b is the TEA, the area under a stress-strain (force /width v.s strain) curve (J/m²).
The tensile energy absorption index is of great importance particularly for sack paper and depends to a great degree on the shrinkage of the paper during drying and it can vary within wide limits, in the range of 1–5 kJ/kg.

2.10 Elastic Properties

The in-plane stiffness of paper due to tensile loading may be expressed in a few various ways.
For a homogeneous material, the linear part of the stress-strain curve is described by Hooke's law:

$$\sigma = E \cdot \varepsilon \text{ (N/m}^2\text{)} \tag{2.14}$$

The elastic modulus E is thus a measure of the in-plane stiffness properties of the material.
Because paper has a porous structure, it is in many cases misleading to use the concept of modulus of elasticity for paper as discussed in relation to *Figure 2.5* and *Figure 2.6*.
If both sides of the equation representing Hooke's law are multiplied by the thickness t, we obtained the following equation in which the tensile stiffness is included.

$$\sigma^b = E^b \cdot \varepsilon \text{ (N/m)} \tag{2.15}$$

If both sides of the equation are divided by the grammage, Hooke's law has the following form.

$$\sigma^w = E^w \cdot \varepsilon \text{ (Nm/kg)} \tag{2.16}$$

In some literature, tensile stiffness is often found given as the elasticity modulus expressed in metres. This is analogous to the expression for breaking length. Hooke's law in these units is given by the expression:

$$\sigma^l = E^l \cdot \varepsilon \text{ (m)} \tag{2.17}$$

$$E^l = \frac{E^w}{g} \text{ (m)} \tag{2.18}$$

Note that

$$E^w = \frac{E}{\rho} \qquad (2.19)$$

$\sigma = E \cdot \varepsilon$	Hooke's law for stress	Pa (N/m²)
$\sigma^b = E^b \cdot \varepsilon$	Hooke's law for force per unit width	N/m
$\sigma^w = E^w \cdot \varepsilon$	Hooke's law for force per unit width and grammage	Nm/kg
$\sigma^l = E^l \cdot \varepsilon$	Hooke's law for metres	m
E	Elastic modulus	Pa (N/m²)
E^b	Tensile stiffness	N/m
E^w	Tensile stiffness index	Nm/kg
g	Acceleration due to gravity	m/s²
ρ	Density	kg/m³

The tensile stiffness index for paper is in the range of 1–10 MNm/kg, *Figure 2.14*.
TMP = Thermomechanical pulp
CTMP = Chemimechanical pulp
UBS = Unbleached kraft pulp
USS = Unbleached softwood sulphate
OCC = Recycled corrugated board

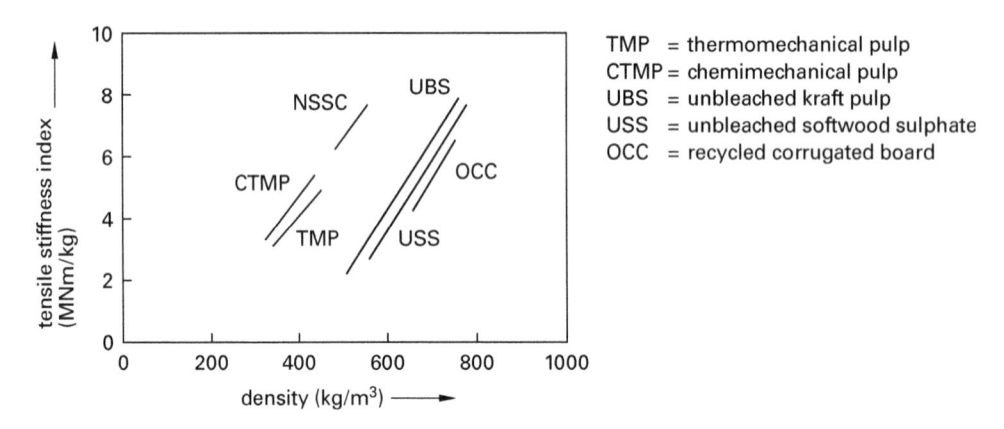

Figure 2.14. Tensile stiffness index of a paper dried under restraint plotted against density for different fibre raw materials. Different density levels have been attained by varying the beating.

2.11 Tensile Stiffness - Theoretical Approach

Tensile stiffness index may be related to the degree of bonding and fibre properties by Seth and Page. *Figure 2.15* shows the tensile stiffness index plotted versus the light scattering coefficient (indicating the state of bonding) for a pulp at different degrees of beating. For each degree of beating, the wet pressing pressure has been increased.

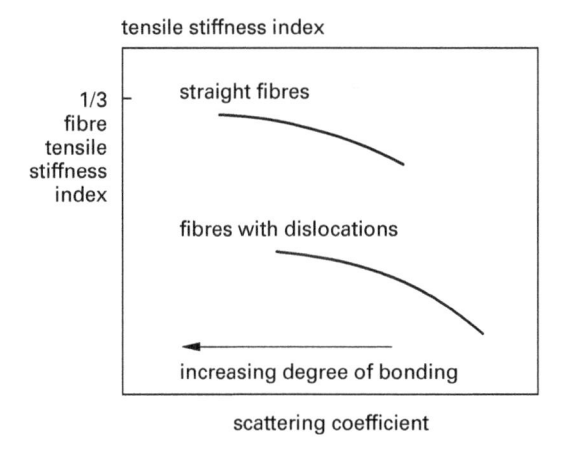

Figure 2.15. Tensile stiffness index versus light scattering coefficient for a bleached kraft pulp of Southern Pine. The beating times are from 0 min for the bottom curve and 50 min for the upper.

The results show that the tensile stiffness index, for a given degree of beating, increases with increasing press pressure up to a plateau value. The plateau value increases with increasing degree of beating. The effects can be explained as follows: *Figure 2.16* shows schematic pictures of the state of bonding of a fibre at two degrees of bonding. At low press pressures, a given fibre is bonded to only a few fibres. At fibre ends, the formed network is not able to transfer forces. At high press pressures, more bonding points are obtained and a better force transfer is obtained.

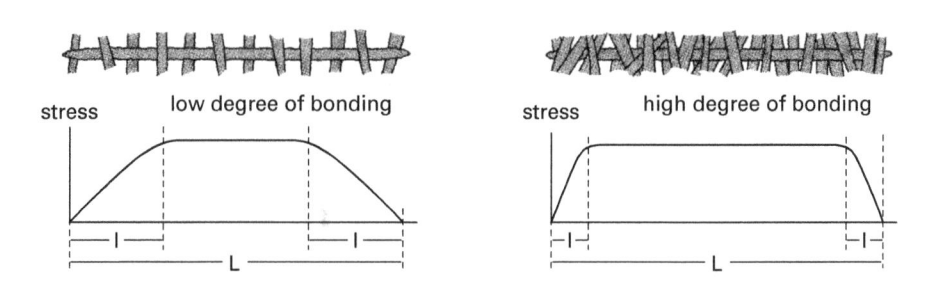

Figure 2.16. Schematic picture of the state of bonding of a fibre at two degrees of bonding.

The influence of the beating can be explained with the help of *Figure 2.16*. Lightly beaten pulps always contain fibres with a certain degree of dislocations, crookedness and micro-compressions. During beating, the fibres swell and these faults have a tendency to be straightened out, and this leads to a higher modulus. It can be shown theoretically that the elasticity modulus in an isotropic sheet can at most be 1/3 of the modulus of the fibre.

For straight well-beaten fibres, the theoretically possible value of 1/3 of the tensile stiffness index of the fibres is attained in an isotropic sheet. When the fibres have dislocations of different kinds, the elasticity modulus is reduced.

2.12 Ultrasonic Techniques for Stiffness Determination

Ultrasound may be used to evaluate the in-plane stiffness of paper. The method is non-destructive and fast. The principle is given below.

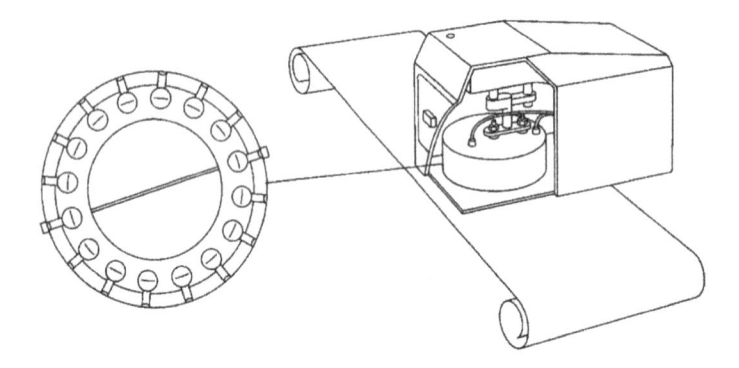

Figure 2. 17. The in-plane stiffness may be evaluated by ultrasonic technique. Transmitters and receivers are placed 100 mm apart in a circle. Ultrasonic pulses are sent from the transmitter and the time needed to reach the receiver is recorded.

The ultrasonic stiffness index measured by ultrasonic technique E_{US}^{w} may be calculated from the sound velocity v as follows.

The value will not be identical to the tensile stiffness index evaluated from tensile tests due to the viscoelastic, time dependence of paper, and the boundary condition,

Depending on the fibre properties and orientation and moisture content, the ratio between the two ways of evaluating stiffness will vary. In all cases the ultrasonic value will be higher.

A typical example is given in *Figure 2.18* for a kraft paper. Properties in MD and CD is given for the tensile stiffness index and for the ultrasonic stiffness index versus moisture ratio.

A very useful application for ultrasonic stiffness is the determination of non-symmetry in paper. *Figure 2.19* shows a paper with slight off-axis symmetry. The method may be used for cross profile studies and product control of different mechanical properties.

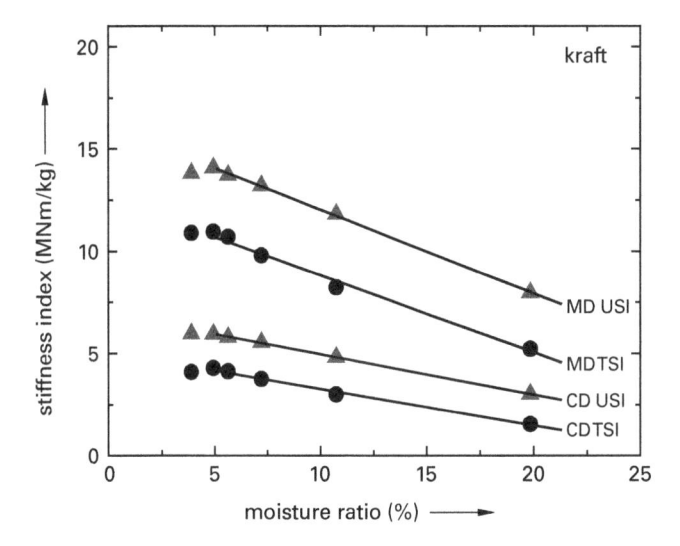

Figure 2.18. Ultrasonic stiffness index for a kraft paper. Properties in MD and CD is giver for the tensile stiffness index and for the ultrasonic stiffness index versus moisture ratio.

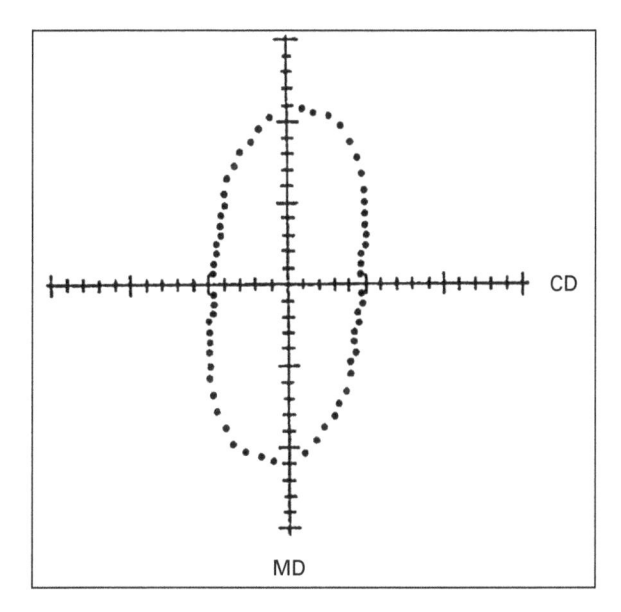

Figure 2.19. A paper and with a slight off-axis symmetry.

2.13 Plastic Stress-Strain Properties in Tension

Figure 2.20 shows stress strain curves for a kraft paper under tension at different strain rates. Both tensile stiffness and tensile strength are influenced by the strain rate.

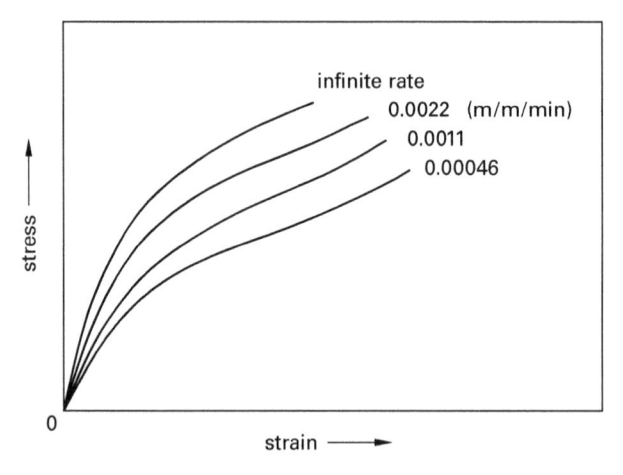

Figure 2.20. Stress-strain diagram for a kraft paper at different strain rates.

If a paper is loaded and unloaded under a tensile load at a constant strain rate, the stress-strain diagram gets a typical appearance as illustrated in *Figure 2.21*.

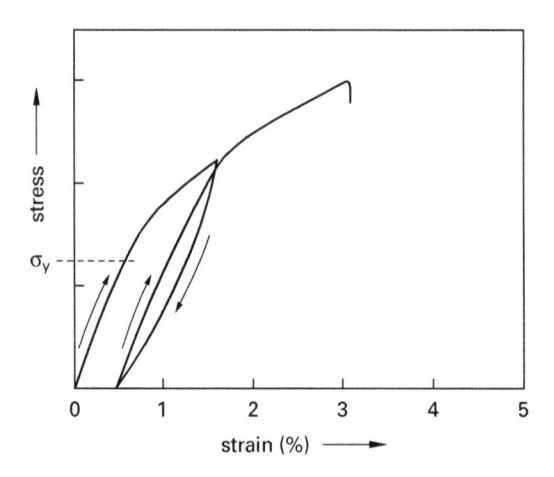

Figure 2.21. The appearance of the stress-strain diagram when a paper is loaded and unloaded.

The curve begins with a straight part. When the direction of strain is reversed at a given force, the force decreases.

On unloading, the strain does not return to zero. The paper exhibits a permanent strain. When the paper is again strained, the force increases up to the force to which the paper was subjected earlier. Thereafter, the stress-strain curve proceeds along the original path.

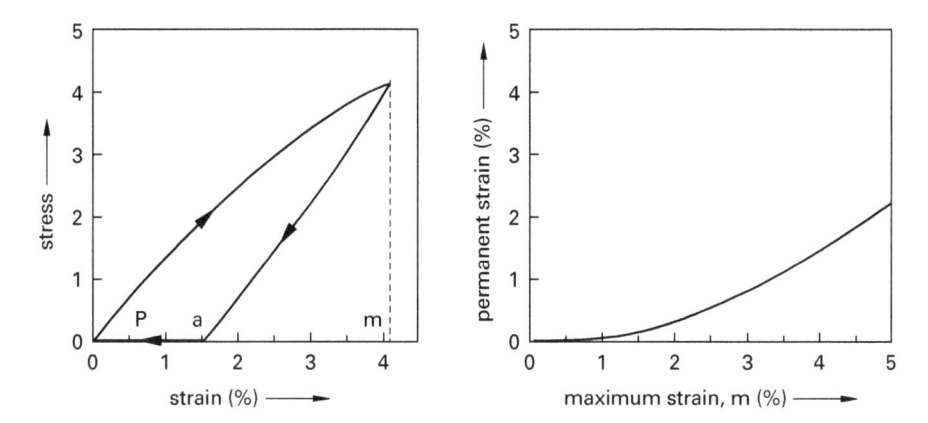

Figure 2.22. Left-hand picture: Force-elongation diagram up to a certain strain (m) and unloading. *P* indicates the permanent strain after a certain time at zero force. Right-hand picture: Permanent strain as a function of maximum strain.

Figure 2.22a shows a cyclic deformation of a paper at a constant strain rate. The paper is loaded up to a certain maximum elongation, m, and is unloaded. At a force of zero at point a, the paper is left unloaded. The elongation will therefore wish to return slowly to zero but, after a long time, it exhibits a permanent elongation, *p*. As shown in *Figure 2.22b*, the permanent change increases with increasing maximum elongation.

2.14 Creep Properties in Tension

If a material is subjected to a constant load, the elongation increases with time, *Fig. 2.23*. The phenomenon is a consequence of the viscoelasticity of the material. The same behaviour applies also for paper as shown in *Figure 2.24*, which show the strain as a function of time for a paper is subjected to different stresses.

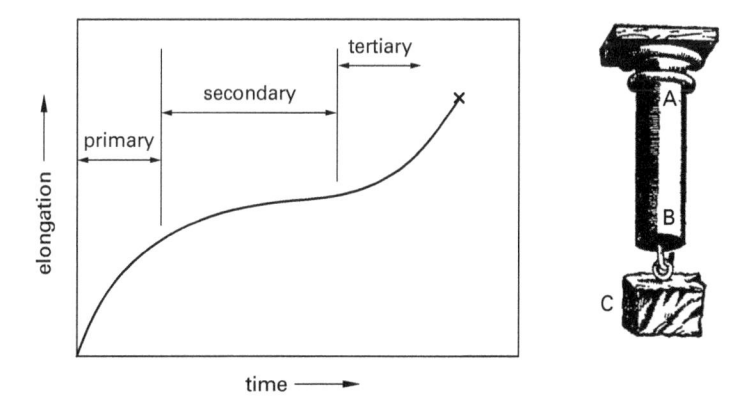

Figure 2.23. Creep of a material.

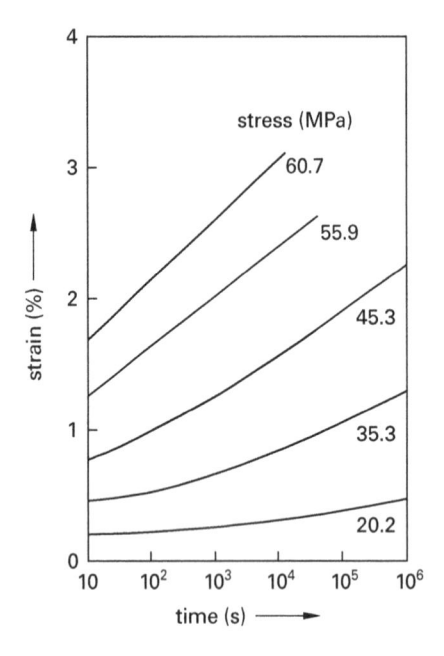

Figure 2.24. Creep of paper. The strain versus time at different stress.

In *Figure 2.25* the data are expressed as stress strain curves at different times. We obtain *isochronous curves*. The concept of isochronous curves will be further discussed later in relation to Section 2.16, viscoelastic properties in compression.

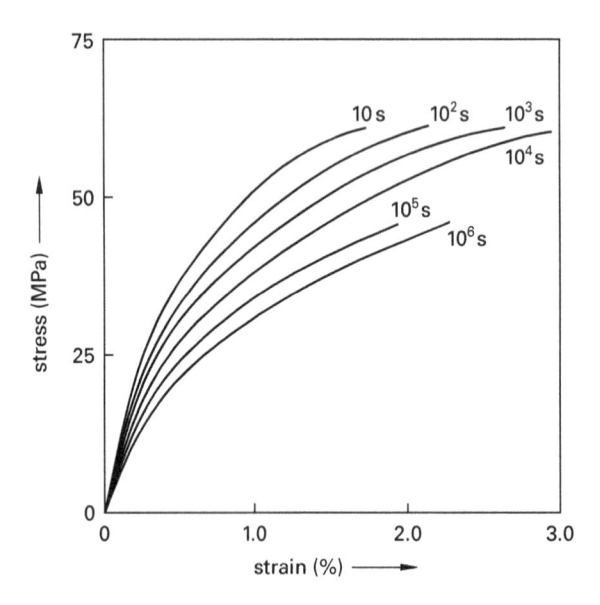

Figure 2.25. Stress strain curves at different times. We obtain isochronous curves.

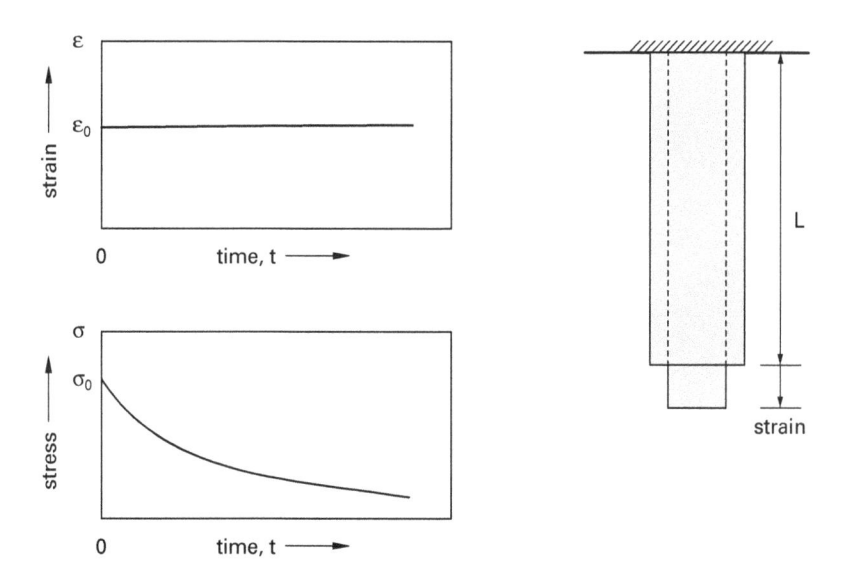

Figure 2.26. Principle of stress relaxation

If paper is strained to a given strain and kept at that strain, the stress decreases with time. The phenomenon is called *stress relaxation*. The principal behaviour is shown in *Figure 2.26*.

2.15 Compression Properties

When paper is compressed in the plane of the paper, the paper usually buckles, as shown in *Figure 2.27*.

Figure 2.27. Buckling of a paper strip under compression with a long clamping length.

The failure force per unit width and unit grammage in compression increases when the clamping length is decreased as illustrated for paper with different grammage in *Figure 2.28*. At clamping lengths of less than circa 0.7 mm, a plateau value is attained independent of the grammage of the paper.

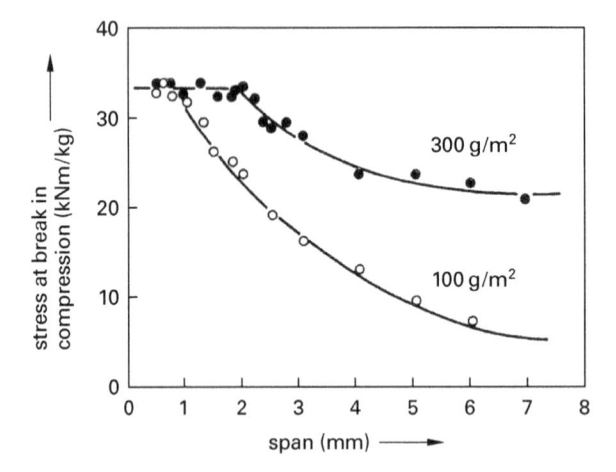

Figure 2.28. Stress at break in compression versus clamping length for paper with different grammage.

If buckling is prevented, for example by testing with a short clamping length, the pure compression properties and strength of the paper can be evaluated. According to SCAN-P 46, the compression strength can be evaluated by fixing the paper between two clamps with a clamping length of 0.7 mm, *Figure 2.29a*.

The paper can also be prevented from buckling by using columns which rest against the paper, *Figure 2.29b*. The columns must be used if the strain process is to be recorded, for example in studies of creep of paper in compression.

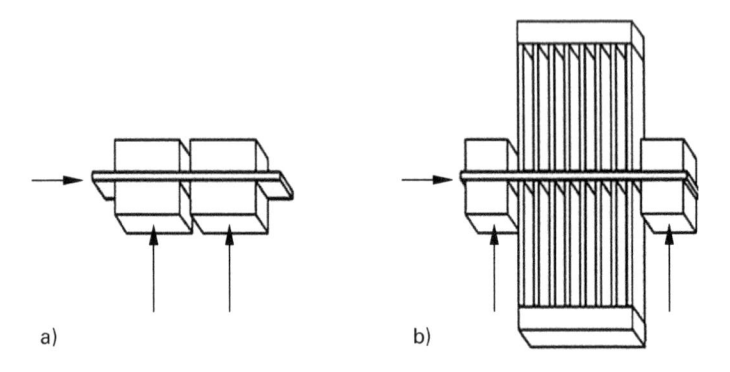

Figure 2.29. Two ways of evaluating the compression properties of paper. a) The SCAN-P 46 method with a short clamping length, short span compression test. b) Long clamping length with columns support.

Stress-strain curves for paper in compression are shown in *Figure 2.30*, where a comparison is also made with tensile properties. In a comparison of tensile and compression curves in MD and CD, the following can be observed:

- Paper is normally stronger in MD than in CD, both in tension and in compression.
- The strain at break is higher for MD than CD
- The slope of the curve is the same for small compressions in both tension and compression. This means that tensile stiffness = compressive stiffness.
- The strain at break in compression is lower than that in tension.

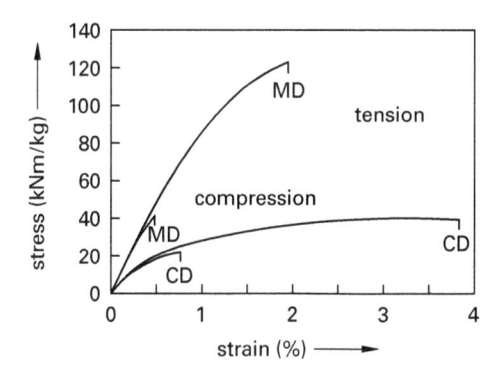

Figure 2.30. Stress-strain curves for paper. Comparison between tensile and compression properties in MD and CD. Positive stress and strain values are shown.

The compression strength of paper is measured primarily at high grammage, above ca. 100 g/m², and it is important primarily for the ability of corrugated board boxes and carton board packages to be stacked in storage.

Compression strength is the maximum stress which paper can support in compression. Compression strength σ_C^b is the maximum compression force per unit width.

$$\sigma_C^b = \frac{F_C}{b} \quad (\text{N/m}) \tag{2.21}$$

Figure 2.31. Compression index of paper dried under restraint plotted against density for a number of pulps. Different densities have been attained by varying the beating. OCC = recycled Old Corrugated Containers.

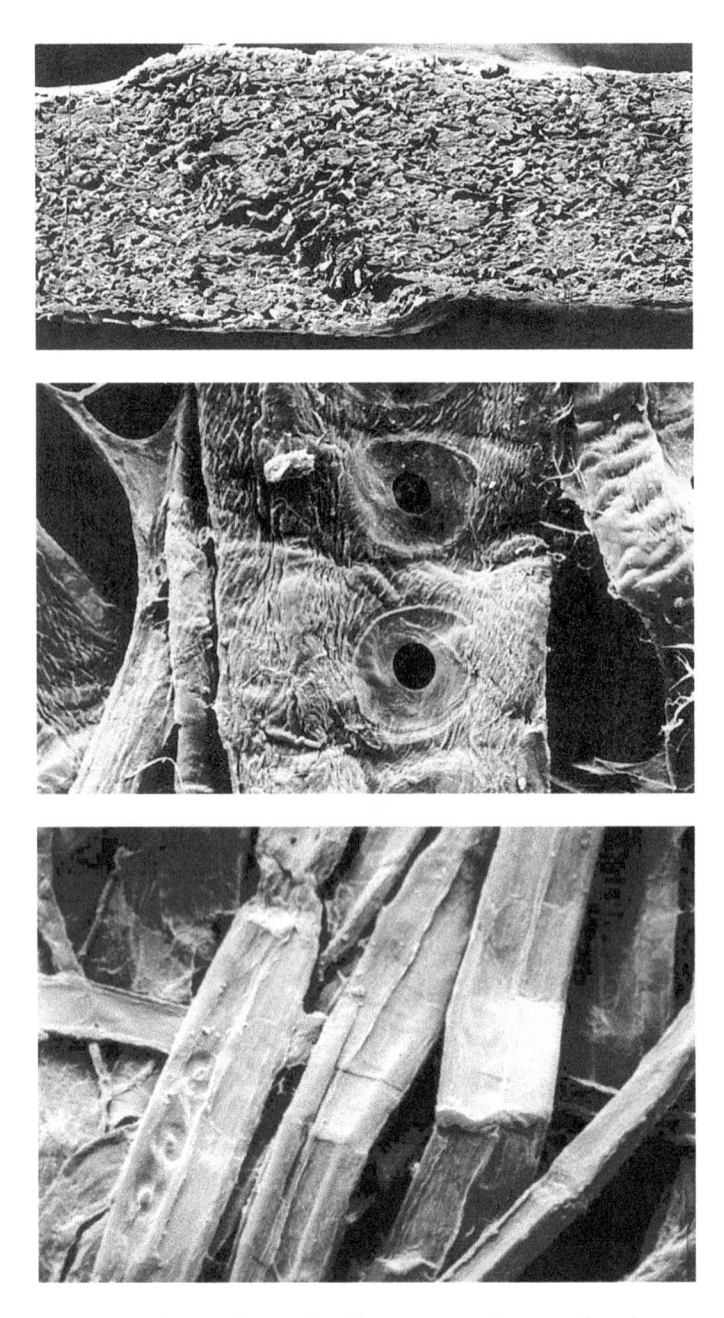

Figure 2.32. Electron micrographs of fracture zones of paper, subjected to compression loading.

The compression failure stress σ_T is the maximum compression force per unit width and unit thickness.

$$\sigma_T = \frac{F_T}{bt} \quad (N/m^2) \qquad (2.22)$$

Compression index σ_C^w is the maximum compression force per unit width and unit grammage.

$$\sigma_C^w = \frac{F_C}{b\,w} \quad (Nm/kg) \qquad (2.23)$$

Compression index varies between 20 and 40 kNm/kg for liner pulps, see *Figure 2.31*.

Figure 2.32 shows three electron micrographs of paper, one seen from the edge of the paper and two seen from above after compressive failure. The reason for the compressive failure is that the fibre wall collapses. For this reason, the compression strength is relatively insensitive to the length of the fibres.

There are many other methods of testing for the compression strength of paper. The most common is the ring crush test, where a paper strip is first formed into a ring and is then crushed edgeways in a special test apparatus, see *Figure 2.33*, where a flat plate is pressed against the ring-shaped sample. Technically, this type of measurement involves great difficulties such as buckling and difficulty to obtain sufficient parallelism of the loading plates.

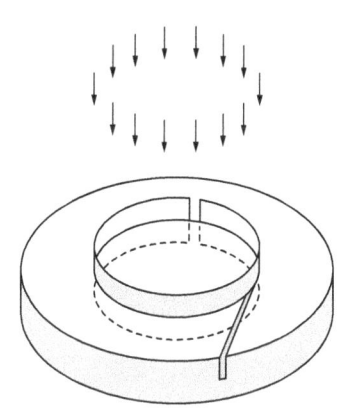

Figure 2.33. Ring Crush Test (RCT).

2.16 Viscoelastic Properties in Compression

The main use of carton board, liner and fluting is for packages of different kinds. An important property of these materials is the ability to protect the contents in the packages against external influences, especially in the long-term storage of the packages in stacks in a moist environment or in an environment where moisture or temperature vary with time. The material properties, which are important, are the creep properties of the materials in compressive loading.

In order to be able accurately to study the properties of paper in compressive loading, STFI has constructed the creep apparatus where stress, strain and time are recorded with a computer.

The following figures show in principle how the creep properties of paper can be recorded and evaluated. In the creep test, a constant stress is applied to the test piece, and the resulting strain is recorded as a function of time, *Figure 2.34*. It is possible to adapt the data in *Figure 2.34* to Equation 2.24 below, which relates stress, strain and time to each other.

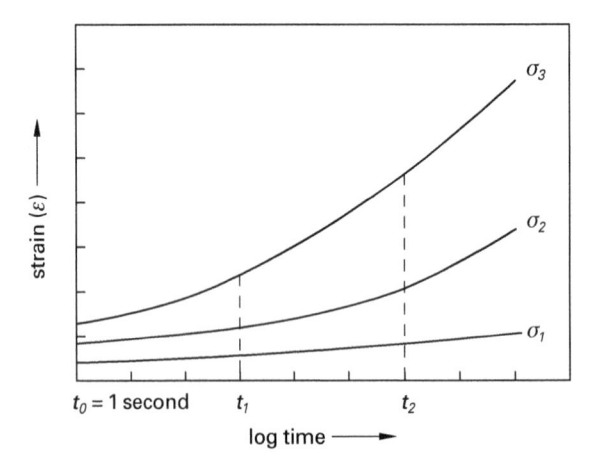

Figure 2.34. Creep curves at different stresses.

If the curves in *Figure 2.34* are intersected, corresponding stress and strain values are obtained. *Figure 2.35* shows such isochronous (from Greek iso = equal and chronos = time) curves at different times. Isochrones are often used for polymeric materials in strength dimensioning for structures, which must stand for a long time.

At small strains, the curves are linear and then deviate at high strains. The linear part of the curve corresponds to a linear viscoelastic behaviour of the paper while the non-linear part corresponds to a non-linear viscoelastic and plastic behaviour.

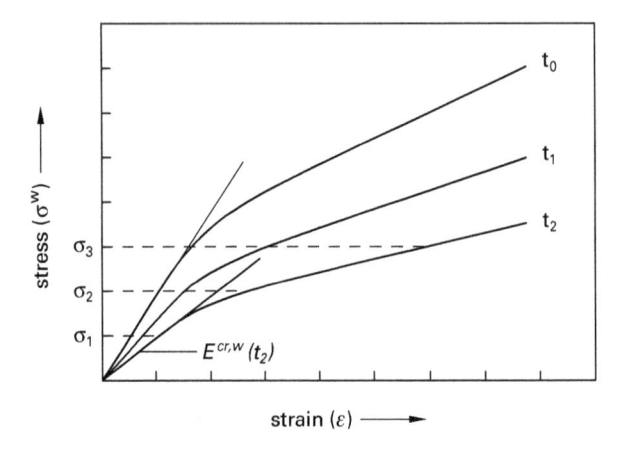

Figure 2.35. Isochronous stress-strain curves constructed from the creep curves in Figure 2.34.

To describe the isochronous stress - strain curves from creep curves, we use the following equation:

$$\sigma^w = \left[\alpha_1 \cdot \tanh\left(\frac{\alpha_2}{\alpha_1} \cdot \varepsilon\right) + \alpha_3 \cdot \varepsilon \cdot \tanh\left(\frac{\alpha_2}{\alpha_3} \cdot (100 \cdot \varepsilon)^2\right) \right] \cdot \left(\frac{t}{t_0}\right)^{-p} \tag{2.24}$$

where σ^w is the stress, α_1, α_2, α_3 and p are material parameters, ε is the strain at time t, and t_0 is the reference time, equal to one second.

The equation consists of two hyperbolic tangent functions to describe the shape and magnitude of a reference stress-strain curve and a power function to show the effect of time on the magnitude of the curve.

In tension, the stress-strain curve may become approximately linear at higher strains, which makes the second term in this equation necessary.

2.17 Tearing Resistance

The tearing resistance, which is a way of evaluating the crack sensitivity of the paper, is a strength dimension of central importance for paper.

A small cut is made in a pack of papers. In a pendulum apparatus, tearing is completed. The work is divided by the tearing distance and the average force is obtained, defined as tearing resistance.

Besides tensile index, tear index is the most important property in the characterization of pulp for paper. The tear apparatus that has been standardised internationally measures the tearing work with a pendulum tester of the Elmendorf type. *Figure 2.36* shows the principle. The tearing work is often so small that several layers of paper must be torn simultaneously. Most standards prescribe the tearing of four sheets.

Just for notification, there are other tearing methods in use. One that occurs in Europe is the Brecht-Imset method.

Figure 2.36. Tearing resistance according to Elmendorf.

Tearing resistance T^b is the mean force (N) during the tearing divided by the number of papers and the tear index T^w is calculated by dividing the tearing resistance by the grammage.

$$T^w = \frac{T^b}{w} \quad (\text{Nm}^2/\text{kg}) \tag{2.25}$$

Tear index lies in the range 10–30 (Nm²/kg).

2.18 Theory for Tearing Resistance

With increasing beating of chemical pulps, the tearing resistance usually passes through a maximum, *Figure 2.37*. The tear index of paper made from short-fibre hardwood pulp is considerably lower than that of softwood pulp, but it often increases even after a severe beating. Technical beating of hardwood pulps is seldom driven so far that the falling part of the curve is entered.

In spite of the central importance of the tearing resistance, the theory is weak. For unbeaten pulps with a low degree of bonding between the fibres, it is imagined that the tearing takes place by drawing the fibres out of the structure without breaking them, roughly as in *Figure 2.11A*. The tearing energy is then assumed to be determined by the length of the fibre and the number of bonding points. With increased beating, the fibres are bonded together in more and more places and the tearing energy increases. Finally, a situation is reached where the fibre breaks instead of being drawn out of the network, as in *Figure 2.11B*. The tearing energy required to create a fibre failure is then lower than if the fibre is pulled out of the network. The optimum tearing energy is obtained at a suitable distribution between pulled-out and broken fibres.

The tearing of machine-made papers and multi-layer structures follows complicated paths and delaminating phenomena are common, *Figure 2.38*. Different results are sometimes also obtained depending on the number of paper sheets tested.

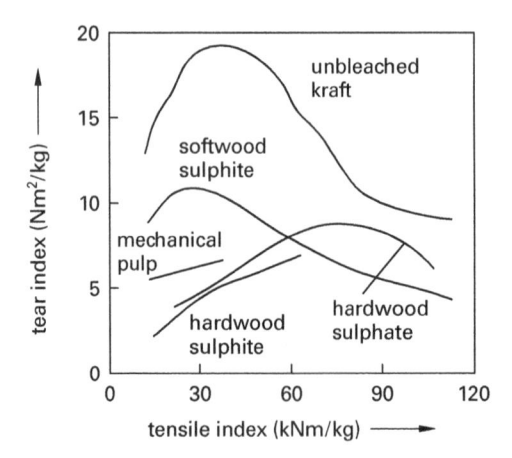

Figure 2.37. Tear index as a function of tensile index for some different pulps.

Figure 2.38. Tearing of thicker papers and highly oriented papers in the cross direction may be difficult to evaluate by the traditional tear method.

2.19 Bursting Strength

The bursting strength test was invented by Mullen about 100 years ago and is therefore often called the Mullen method. The Mullen test is thus one of the oldest methods for testing the strength of paper. In the testing, the paper is clamped in a ring with a centrally recessed rubber membrane, see *Figure 2.39*.

The membrane is pressed upwards by an increasing oil pressure and finally bursts the paper. The oil pressure at the moment of failure is read on a manometer and this defines the bursting strength. *Figure 2.40* shows the fracture appearance in the Mullen test.

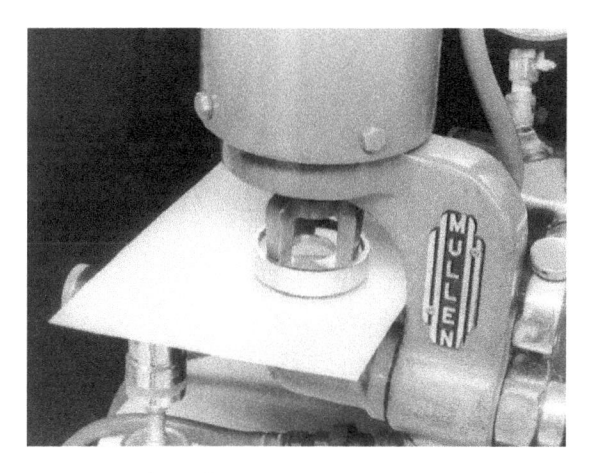

Figure 2.39. The Mullen test.

Figure 2.40. The fracture appearance in the Mullen test.

Bursting strength B is the pressure at break (Pa = N/m²). And the burst index is obtained by dividing the bursting strength by the grammage w.
Burst index is

$$B^w = \frac{B}{w} \quad (N/kg) \tag{2.26}$$

The burst index is in the range of 1–10 MN/kg. In spite of its age, and doubtful relevance, bursting strength is still used to characterize the properties of many paper grades. The bursting strength has been derived from the tensile properties as follows:

$$\text{Bursting strength} = k \left(\sigma_{T,MD}^b + \sigma_{T,CD}^b \left(\frac{\varepsilon_{MD}}{\varepsilon_{CD}} \right)^{0,5} \right) (\varepsilon_{MD})^{0,5} \tag{2.27}$$

where
σ_T^b i the tensile strength in MD and CD respectively
ε is the strain at break in MD and CD respectively

The bursting strength of the paper can thus be increased either by increasing the orientation of the fibres in the machine direction, which leads to an increase in the tensile strength in the machine direction, or by taking measures which increase the strain at break in the machine direction, for example by reducing the draws in the machine direction. Another measure is to reduce the strain at break in CD by preventing the shrinkage in the cross direction of the paper machine.

2.20 Bending Stiffness

The resistance of paper to bending, i.e. its bending stiffness, is one of the most important paper properties. High bending stiffness reduces the tendency for buckling of boxes when the contents press against the walls, prevents newspapers which we read from falling together under their own weight and gives runnability for example sack paper in sack manufacture.

Bending stiffness, per unit width b, of a strip of paper can be obtained if we know the bending moment and the curvature $(1/R)$ where R is the radius of curvature.

$$S^b = \frac{M}{\left(\frac{1}{R}\right)b} \quad (Nm) \tag{2.28}$$

It is often preferable to measure the bending indirectly by using mathematical relationships between downwards bending and curvature in the two-point method, which is most common for paper and carton board. The two-point method is shown in *Figure 2.41*. The bending stiffness is calculated by measuring the force required to bend a strip with a certain width and length through a certain angle.

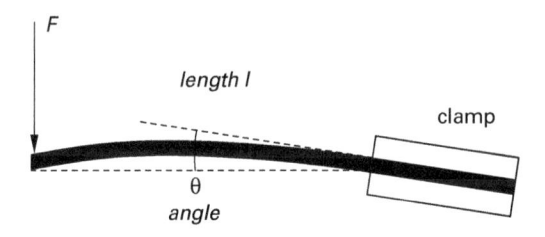

Figure 2.41. Bending stiffness determination by beam bending.

Bending stiffness can then be evaluated by bending a strip of paper, as in *Figure 2.41* in the elastic region. Bending stiffness per unit width b is as follows

$$S^b = \frac{F 60 l^2}{b \theta \pi} \quad (N/m) \tag{2.29}$$

Bending stiffness may be calculated from elastic modulus and thickness as follows

$$S^b = \frac{Et^3}{12} \tag{2.30}$$

As discussed previously, tensile stiffness index E^w is a better way of expressing in-plane stiffness of paper than the elastic modulus E.

Since $E^w = \frac{E}{\rho}$ and $\rho = \frac{w}{t}$ we get

$$S^b = \frac{1}{12} w^3 E^w \frac{1}{\rho^2} \qquad (2.31)$$

To be able to make quantitative calculations according to this equation, a true measure of the density of the paper is required, as mentioned earlier. We see that bending stiffness is proportional to grammage raised to the power 3 and we can define a property bending stiffness index.

$$S^w = \frac{S^b}{w^3} \quad (\text{Nm}^7/\text{kg}^3) \qquad (2.32)$$

Furthermore a low density, raised to the power 2, and a high tensile stiffness index are required.

A resonance method for the determination of bending stiffness based on the determination of the length at which a paper strip will come into resonance at a frequency of 25 Hz. The method is shown schematically in *Figure 2.42*.

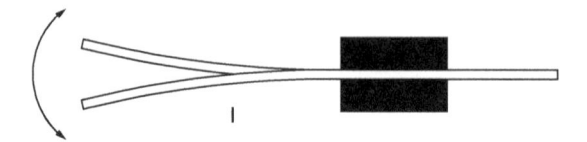

Figure 2.42. The bending stiffness for paper can be evaluated by determining the resonance length of the paper.

The bending stiffness is determined from Equation 2.33.

$$S^b = 2 \cdot 10^3 \, l^4 \, w \quad (\text{Nm}) \qquad (2.33)$$

where
l = resonance length
w = grammage
The bending stiffness index for paper lies in the range of 0.5–2 Nm/kg^3.

Paper and carton board are often manufactured in several layers. Through the use of laminate theory, it is possible to calculate and optimise the bending stiffness of multi-layer sheets theoretically. Usually pulps with a high tensile stiffness index are used in the outer layers and pulps with low density in the middle layers. Laminate theory will be discussed in another chapter.

2.21 Bending Resistance

In certain standards the bending resistance in terms of the maximum force, under standardised conditions, is measured with the two-point method according to *Figure 2.43*. An angle of 15°, a strip length of 10 or 50 mm and a strip width of 38 mm is often used in the standards. A bending curve for paper is shown schematically in *Figure 2.43*. One problem associated with bending to 15° is that the paper might be plasticised which has the consequence that the bending stiffness is underestimated.

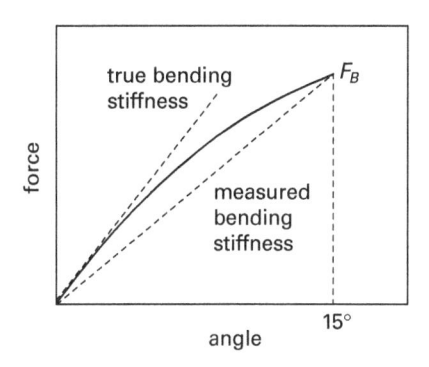

Figure 2.43. Schematic bending curve for paper. The force is plotted against the angle as indicated in *Figure 2.41*.

The relationship between bending stiffness and bending resistance can be calculated approximately according to the following equations.
For a strip length of 50 mm:

$$S^b = F_B \cdot 0{,}0837 \quad \text{(mNm)}$$

For a strip length of 10 mm:

$$S^b = F_B \cdot 0{,}00335 \quad \text{(mNm)} \tag{2.34}$$

Bending resistance can also be normalized approximately with respect to grammage and is then called bending resistance index.

$$F_B^w = \frac{F_B}{w^3} \quad (Nm^6 / kg^3) \tag{2.35}$$

2.22 Fold Number and Folding Endurance

According to the fold number method (SCAN-P:17), a paper strip is folded forwards and backwards at the same time as it is exposed to a constant tensile load. The folding endurance is the common logarithm of the number of double folds which a paper can endure before the strip breaks. There are some different measurement principles for fold number, for example Köhler-Molin, Schopper and MIT which all give slightly different results. The principle for the Köhler-Molin apparatus is shown in *Figure 2.44*.

Since the fold number can vary within very wide limits, from a few folds to above 10 000, it is often practical to work with the concept of folding endurance, the common logarithm for the fold number, instead of the number itself. A doubling of the common logarithm for the fold number is thus interpreted as a doubling of the folding endurance. It can be worth mentioning that folding endurance is the only existing fatigue test for paper.

The folding endurance is especially important for certain paper grades such as banknote paper, map paper, files and telephone catalogue covers. The folding endurance is also often used to evaluate ageing stability of paper used for archive purposes.

Paper can be folded and reopened many times like a hinge. In the folding of a package, folding marks are made in the carton board, so-called creases, without reducing the tensile strength noticeably. If the paper is wrinkled, permanent folds are formed with a rustling sound. It is not possible to bend paper to a double-curved surface without a fold or dent arising.

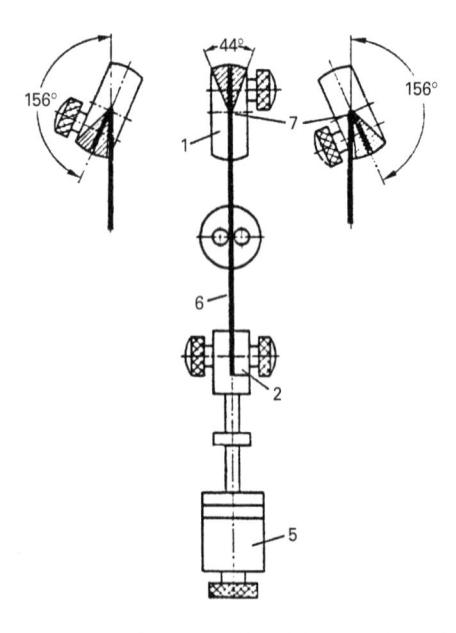

Figure 2.44. The principle for fold number determination according to Köhler-Molin.

2.23 Thickness Direction Strength and Delamination Resistance

In many contexts, the paper or board must have sufficient strength in the thickness direction of the sheet. A large number of methods are used in practice.

The *Z-direction tensile test* is the oldest method for measuring the internal bond strength. In this test, a paper is fastened between two specimen holders, and then pulled apart by the application of a tensile load perpendicular to the plane of the paper, *Figure 2.45*.

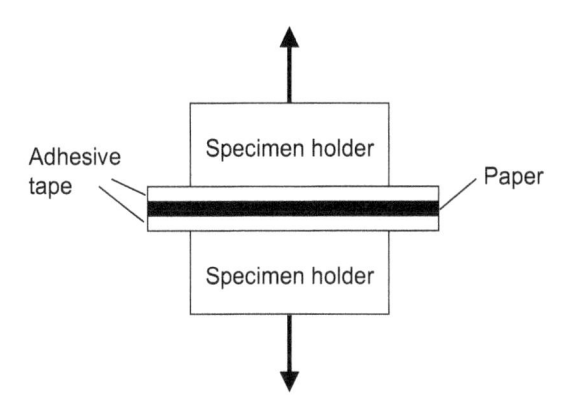

Figure 2.45. Z-directional tensile strength corresponds to the stress at failure of a test piece when pulling two solid blocks apart.

The Z-directional tensile strength is defined as the force required to produce unit area of fracture (kPa). A close relationship exists between Z-strength and apparent density for fully bleached chemical pulps, *Figure 2.46*.

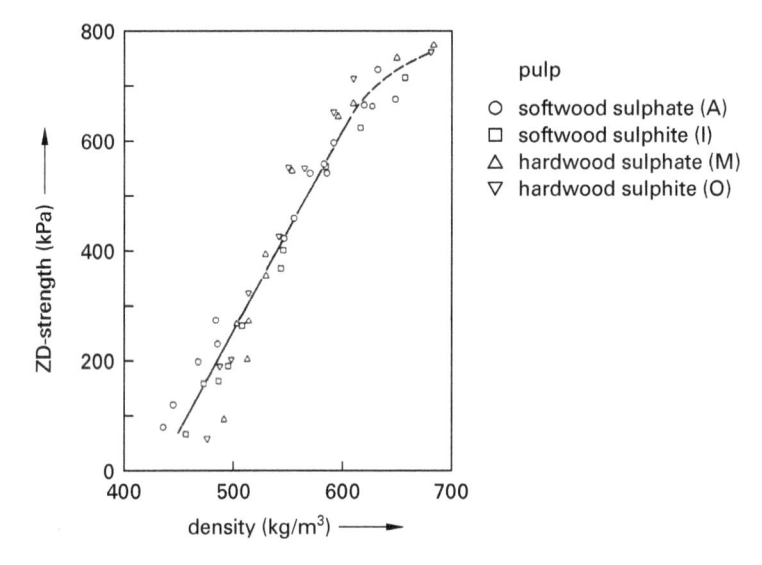

Figure 2.46. Z-directional tensile strength is closely related the density of the papers made from different chemical fibres.

The *Z-toughness method* is a rather new method for measuring the delamination resistance of paper materials It is based on fracture mechanics and it describes the crack propagation process in a material for the opening of a crack in mode I, normal loading. The Z-toughness value expresses the energy liberated per unit crack area formed, in fracture mechanics called the critical strain energy release rate.

In this method, the test piece is attached between two polycarbonate beams with double-sided adhesive tape to create a double cantilever beam (DCB). The shape of the beams is designed to give a constant force when the beams are pulled apart and the crack propagates through the test piece. If the force and the shape and bending stiffness of the plastic beam are known, the Z-toughness value can be calculated.

In this method, the paper samples were laminated on both sides with double-sided adhesive tape, and then cut into test pieces (25×150 mm) in CD. The test pieces were then fastened between the polycarbonate beams as indicated in *Figure 2.47*. After a pressing procedure, a force was applied so that a crack was initiated at the top end and propagated downwards. During the splitting of the paper, the force required for delamination was recorded as a function of beam deflection. The beams, with a constant width, were contoured to give a constant force during crack propagation. For practical reasons, the force at the maximum point in the delamination curve was used for the evaluation of Z-toughness, according to the equation:

$$G_{Ic} = c \cdot F^2 \quad (\text{J/m}^2) \tag{2.36}$$

where
c = constant depending on the elastic modulus of the polycarbonate beams
F = the force at the maximum point in the delamination curve

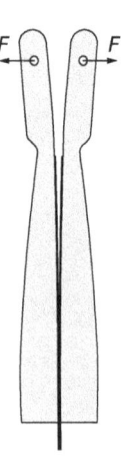

Figure 2.47. Shape of the contoured double cantilever beams. The position of the paper is shown and during the delamination the beams are in a vertical position. SCAN-P 90:03. Paper and board. Z-directional toughness.

2.24 Methods that Simulate Converting Operations

By simulating a printing or a converting process, it is possible to predict how a material will respond to the real situation, but this provides no knowledge or understanding of how the inner strength of the paper relates to the deformation or loading situation.

A number of methods for assessing the ability of paper to withstand forces involved in the splitting of ink films in printing processes have been developed. The *IGT tester* measures the resistance to picking and delamination of paper and board (International Standard 1980).

An oil or ink film of controlled thickness and controlled viscosity is applied on a printing disc. The printing disc is placed to form a nip with a semi-circular sector on which the test strip of paper or board is mounted *Figure 2.48*. The sector rotates with increasing velocity with a controlled constant printing force. The delamination resistance is defined as the minimum velocity at which delamination occurs.

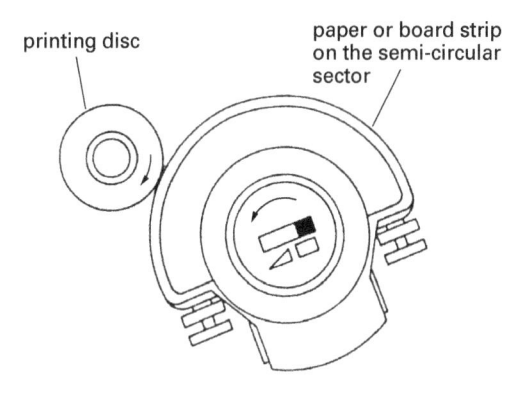

Figure 2.48. Principle of the IGT tester.

2.25 Methods for Production Control

The *Scott bond* test is a dynamic, multi-fracture test, which is commonly used for paperboard. The test specimen is laminated with double-sided adhesive tape and fastened between a rigid foundation and an angular-shaped piece of metal, as shown in *Figure 2.49*. A pendulum is released and hits the angle causing the test piece to split. A calibrated scale indicates the resistance to splitting in terms of the loss in potential energy during the swing of the pendulum. The Scott bond value is therefore expressed in energy units for a unit surface produced in this test (TAPPI 1971).

This test corresponds to a mixed mode loading situation which is hard to describe.

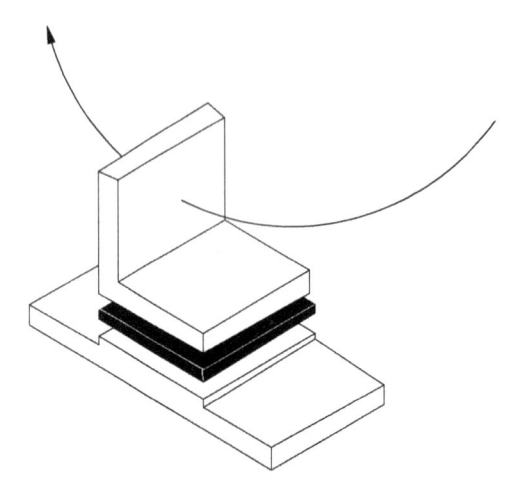

Figure 2.49. The Scott bond test gives the energy absorbed by the failure of the test piece, measured by the loss in potential energy when a pendulum hits an angle attached to the test piece by adhesive tape.

Different configurations of the *peel test* are shown in *Figure 2.50*. The T-peel test is commonly used for plastic materials.

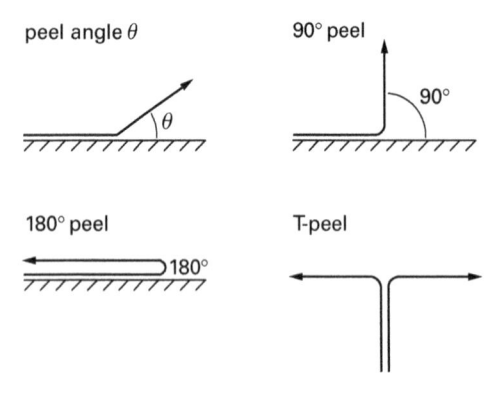

Figure 2.50. Several typical peel configurations.

The Brecht-Knittweis method (DIN1985) is similar to the Z-direction tensile test except that the loading is in the edge of the test piece instead of in the centre, *Figure 2.51*. A test piece (30x30 mm) is mounted between two blocks, the blocks are pulled apart, and the force at failure of the test piece is referred to as the delamination strength.

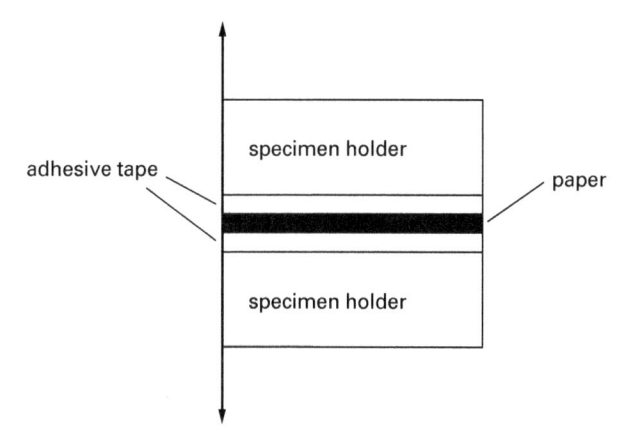

Figure 2.51. The Brecht-Knittweis method is similar to the Z-direction tensile test, but the force is applied at the edge of the test piece.

2.26 Fracture Mechanics

The fracture mechanics of paper are of great importance for the paper industry. The most obvious application of fracture mechanics to paper is to reduce the number of web breaks in converting operations. Fracture mechanics is further a powerful tool in the process of developing tougher, more fracture-resistant, paper qualities.

The fracture toughness of paper is a material property expressing the ability of paper to sustain mechanical loading in the presence of defects. The fracture toughness of paper can be evaluated by material testing.

The reason for using fracture mechanics in the paper industry is that paper very seldom is completely free from defects. In fact, paper may contain many different types of defects or cracks. The problem is that these defects occur very seldom but may cause a web break when the conditions are unfavourable, high force, large crack and low fracture toughness. *Figure 2.52* illustrates a few possible defects in paper.

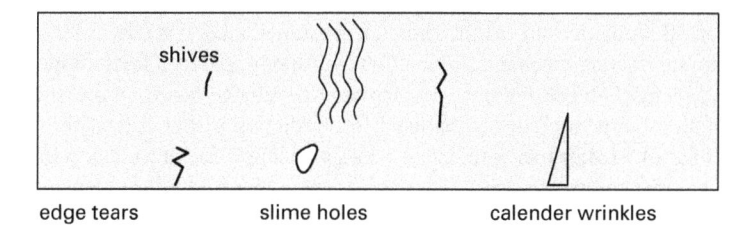

Figure 2.52. A few possible defects in paper.

The purpose of fracture mechanics theory is for instance to calculate the *critical force* F_{crit} (N) and the *critical elongation* δ_{crit} (m) of a paper web with a defect of known size. The fracture toughness of the paper material is a necessary input in such calculations.

A method for predictions of in-plane fracture of paper has been developed, based on non-linear fracture mechanics and the *J*-integral theory. This method, which has been standardized as SCAN-P77, is described further below.

Two different material tests are required for making predictions of fracture in paper possible, *Figure 2.53*. The first experiment is a tensile test on a rectangular paper test piece with the dimensions 100 by 15 mm. This material test is used to determine the mechanical behaviour of the paper material. The second experiment is a tensile test on a 100 by 50 mm rectangular test piece with a manufactured 20 mm centre crack. This second test, together with the known material behaviour from the first test, is used to determine the fracture toughness (J_{Ic}) of the paper material.

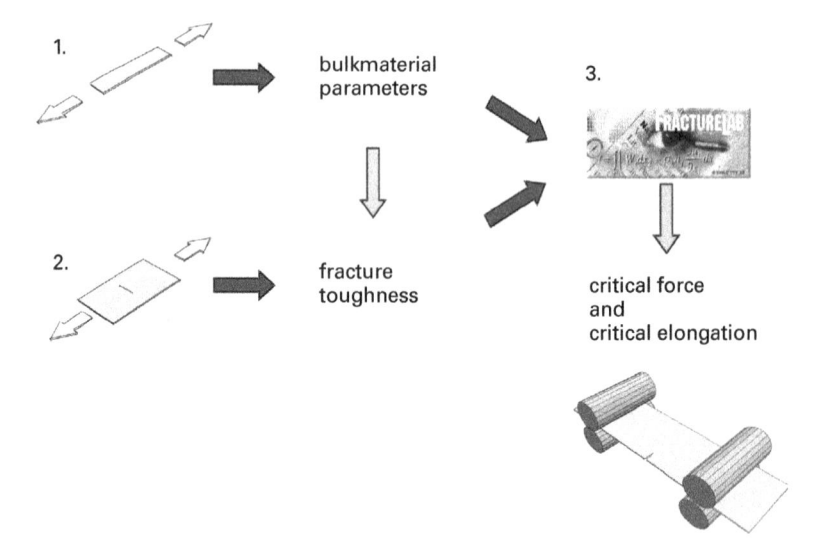

Figure 2.53. The evaluation principle for calculation the fracture toughness and the critical force and critical elongation for a given paper geometry and crack length.

The material behaviour and the fracture toughness of the material constitute the necessary data for making predictions of fracture possible. Furthermore, such predictions generally require the adoption of numerical methods, such as the finite element method (FEM). In order to make user-friendly and accurate fracture mechanics predictions in paper possible, a fracture mechanics software program, FractureLab see *Figure 2.54*, has been developed based on pre-calculated FEM reference solutions. FractureLab can be used for prediction of the critical force and the critical elongation of a full-scale paper web, based on a given crack size and crack position. FractureLab can also be used for predictions of the critical crack size in a full-scale paper web under a given loading situation.

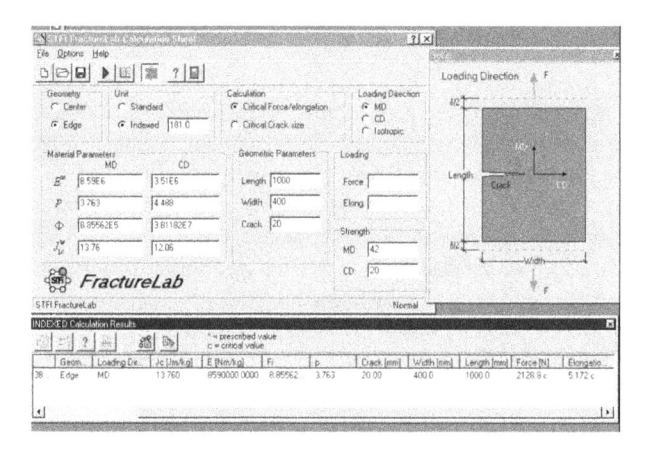

Figure 2.54. Fracture Lab calculates critical force and critical elongation for a given paper geometry and crack length.

Note: It is stressed that the fracture toughness should be used as a parameter in the fracture mechanics predictions of critical force and critical elongation only. It is not to be used for ranking of papers.

The following notations are used.

$$\text{Fracture toughness} = J_{IC}^b \ (\text{J/m}) \tag{2.37}$$

$$\text{Fracture toughness index} = J_{IC}^w = \frac{J_{IC}^b}{w} \ (\text{Jm/kg}) \tag{2.38}$$

The advantage with a predictive tool is that full-scale situations can be analysed at the laboratory. The fibre treatment and papermaking conditions can be optimised and expensive pre-trials minimised. *Figure 2.55* shows results from a trial where predicted and measured values for critical force are compared. The test piece dimensions are 100 mm long and 500 mm wide.

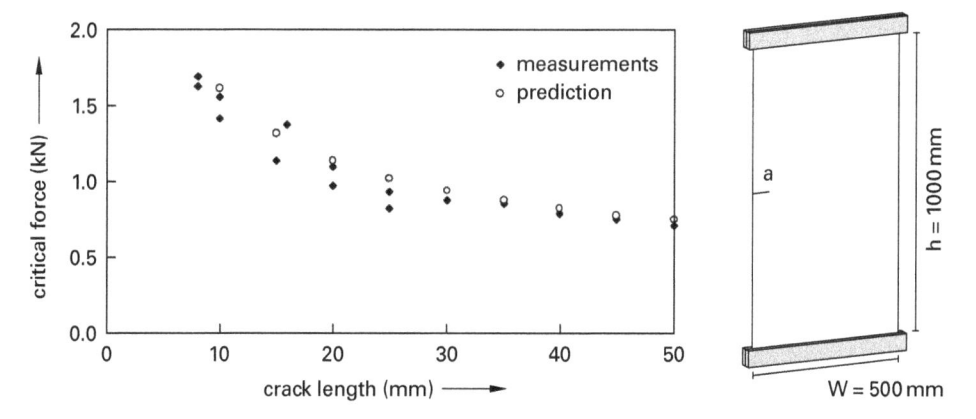

Figure 2.55. Results from a trial where predicted and measured values for critical force are compared. The crack is made at the edge, an edge crack.

2.27 Air Permeance

A paper property which is closely related to its structure is the ability of the paper to allow air to pass through, i.e. the air permeance. This property is especially important for example in the filling of cement in paper sacks. With increasing beating, the air permeance decreases.

In practice, air permeance measurements are carried out in the following way: Air is allowed to penetrate through a paper sample with given dimensions under standardized pressure, temperature and humidity, and the amount of air which passes through the paper per time unit is recorded. The air permeance P_a is calculated according to Equation 2.39:

$$P_a = \frac{Q}{A \Delta p} \text{ (m/s Pa)} \tag{2.39}$$

where
A = area, m^2
Δp = pressure difference across the paper, Pa
Q = the air flow through the paper, m^3/s

Air permeance according to Gurley (SCAN-P 19:78) is the most common method, *Figure 2.56*. In the Gurley-apparatus, the pressure difference is 1.21 kPa. The air permeance P_a can be evaluated as:

$$P_a = \frac{128}{t} \text{ (μm /s)/Pa} \tag{2.40}$$

Where t = the time in seconds required for 100 ml of air to pass through the test piece

Besides the Gurley-apparatus, there are also a number of other instruments for the measurement of air permeance, such as the Bendtsen, Bekk and Sheffield instruments.

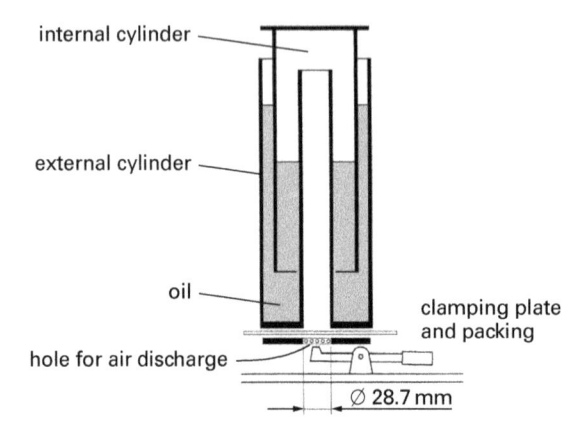

Figure 2.56. Sketch of Gurley-apparatus.

In the Sheffied-instrument, certain parts can be exchanged so that the measurement area of the paper can be changed. The openings are 1/2, 1, 2 and 4 inches in diameter and in theory; this means that the ratio of the areas of two successive openings is 4. This again means that the amount of air, which passes per unit time should vary in the same ratio, but a certain caution is advised here. An increase in the amount of air flowing through the apparatus can lead to a greater pressure drop in other parts of the apparatus and the real pressure drop across the paper is thus less. Leakage around the edge can also influence the measurement result. This may mean that the amount of air becomes slightly smaller than expected and the opening should not therefore be changed when a set of samples is being compared.

Both the Gurley-method and the method described in SCAN-9 26 for dense paper grades and carton board only consider penetrating pores and, with regard to gas transport, these are the most interesting. If it is also desirable to include the closed pores and to measure the total pore volume, other methods exist.

The air permeance is closely related to the pore structure of the paper. An illustration of this is given in *Figure 2.57*.

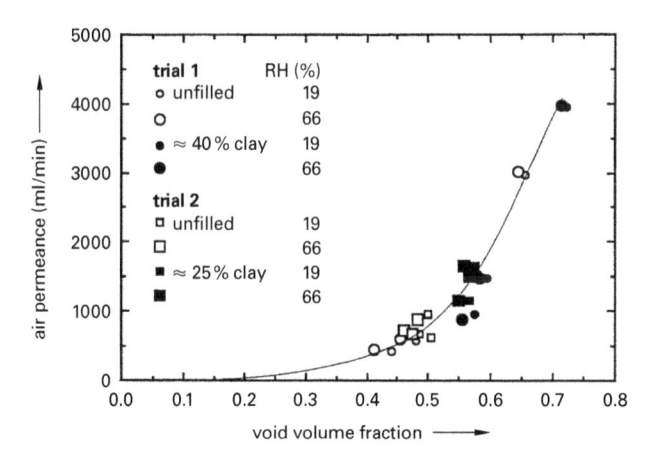

Air permeance versus void volume fraction for trial 1 and 2.
Trial 1. Different densities obtained by wet pressing.
Trial 2. Different shrinkage levels during drying.

Figure 2.57. Air permeance as a function of the void volume of the paper.

3 Development of Paper Properties during Drying

Torbjörn Wahlström
The Packaging Greenhouse AB

3.1 Introduction 69

3.1.1 Scope 69
3.1.2 Level of Observation 70
3.1.3 Drying of Paper 71

3.2 Nano Level – Forces Within or Between Fibres 73

3.2.1 Bonding Mechanisms 73
3.2.2 Network Forces 75
3.2.3 Capillary Forces 75
3.2.4 Chemical Forces 77

3.3 Micro Level – Fibre and Network 79

3.3.1 Single Fibre 79
3.3.2 Fibre Network 81

3.4 Macro Level – Paper Sheet 84

3.4.1 Shrinkage, Stretch and Paper Properties 84
3.4.2 Multi-Ply Board and Drying 89
3.4.3 Stock Preparation and Drying 90
3.4.4 Forming and Drying 92
3.4.5 Pressing and Drying 97

3.5 Machine Level – Moving Wide Web 97

3.5.1 Cylinder Drying 97
3.5.2 Drying Concepts 103

3.6 References 107

3.1 Introduction

3.1.1 Scope

This chapter deals mainly with the development of paper properties during drying and how they depend on the chosen drying concept. The influence of the unit processes preceding drying will also be dealt with to show how they influence the paper's behaviour during and after drying. The chapter aims at giving a conceptual understanding of fibre and network level and a ground for calculations and simulations on a homogeneous sheet level. Due to its usability in mechanical modelling and high sensibility to shrinkage and stretch tensile stiffness is used to demonstrate the drying related effects, but later in the chapter also influences on other paper properties are given.

3.1.2 Level of Observation

Depending of background of the researchers and their purposes drying has been studied on observation levels ranging from a nano to a macro level according to *Figure 3.1*.

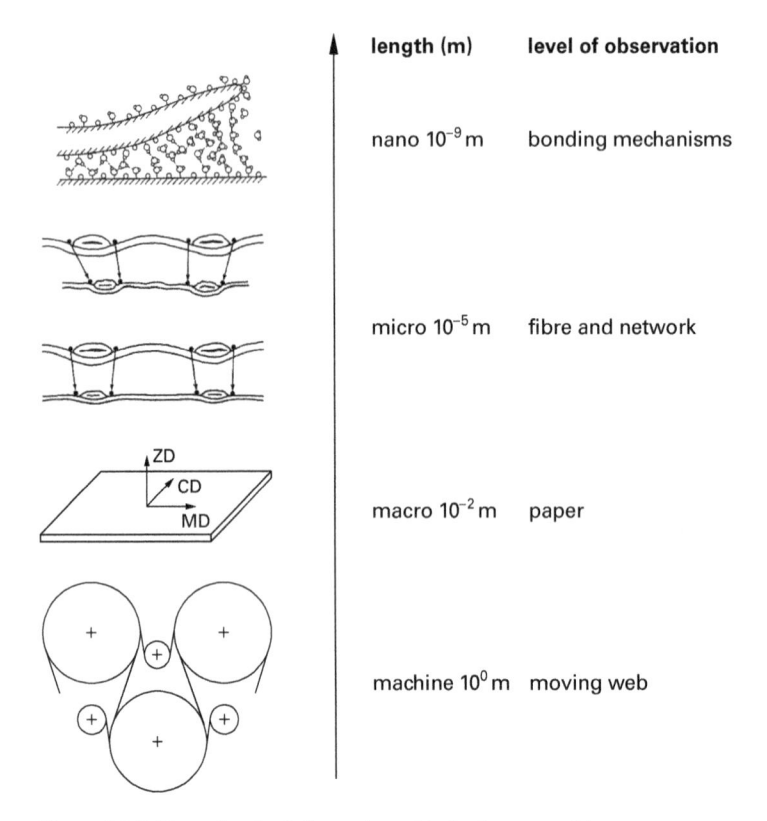

Figure 3.1. Different levels of observation of drying in papermaking.

The description in this chapter will start on the nano level with the bonding mechanisms within and between the fibres. On the micro level the behaviour of a single fibre and fibres forming a network is described. On the macro scale we treat the paper as a homogeneous material and describe its behaviour from different perspectives. Most emphasis is put on the macro scale since bonding and fibre behaviour are also dealt with in other chapters of this teaching material. Also the homogeneous perspective makes it possible to formulate models that can be directly applied on the next level of observation, the paper machine level. On this last level a wide and moving web is dealt with. Depending on paper machine concept the behaviour described on the macro level will develop in different ways.

3.1.3 Drying of Paper

Figure 3.2, shows schematically the amount of water in the different unit processes of paper making. Paper is normally formed from a suspension of fibres at a concentration of 0.1–1.0 % and in special cases at a concentration of up to 4 per cent. In the outflow on the wire, the suspension is air free until the so called dry line, where air begins to be sucked through the sheet. The dry content, or dryness, at the dry line is around 3–5 per cent. With air suction across dry suction boxes and the couch, the dry content can be increased to circa 20 per cent before the web enters the press section. In the press section, a further increase in dry content takes place by stages until a dry content of the order of 40 per cent has been attained. After pressing, water is present in the fibres and as free water between the fibres and in the lumen. The water-fibre system is now located in an atmosphere where the relative humidity is less than 100 per cent, which means that we are concerned with a drying process.

The chosen drying concept will have a large influence on the final paper properties but since paper is mainly dried by cylinder drying most emphasis will be put on this concept. In cylinder drying, the web is drawn through the dryer in the machine direction and heated on steam-heated drying cylinders. Evaporation of the water takes place mainly in the free draws between the cylinders. The paper properties develop with and is very sensitive to decreasing water content in the paper. The paper properties are also very dependent on the amount of shrinkage that the paper is allowed to develop during drying. In cylinder drying the paper can shrink in the cross direction in the free draws between dryer cylinders.

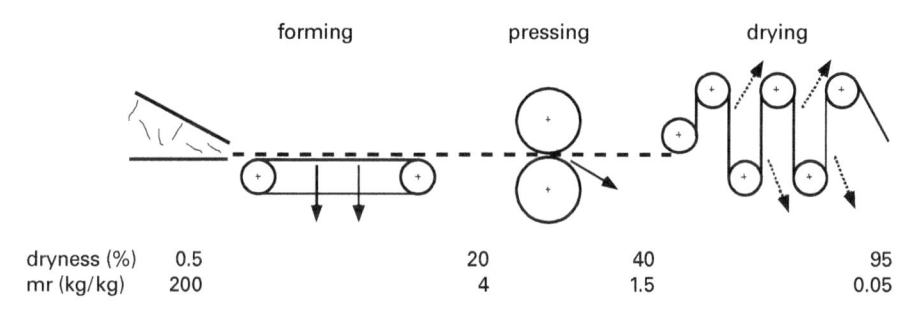

Figure 3.2. Water content in the papermaking process.

As a measure of water content in a paper sample the dry content, dryness or solids content, according to Equation 3.1, is close to an industry standard. However, dryness is not a linear measure of the water content in paper. Due to the close relationship between the water content and paper mechanical properties, a linear measure such as the moisture ratio, *mr*, defined according to Equation 3.2 is more useful. *Figure 3.3* shows the relation between moisture ratio and dryness. For example a 10 % increase in dryness from 40 to 50 % is equal to remove 0,5 kg/kg water, whereas an increase from 80 to 90 % equals only 0,15 kg/kg water.

$$\text{Dryness} = \frac{\text{Weight of dry material}}{\text{Total weight}} \qquad (3.1)$$

$$mr = \frac{\text{Weight of water}}{\text{Weight of dry material}} \qquad (3.2)$$

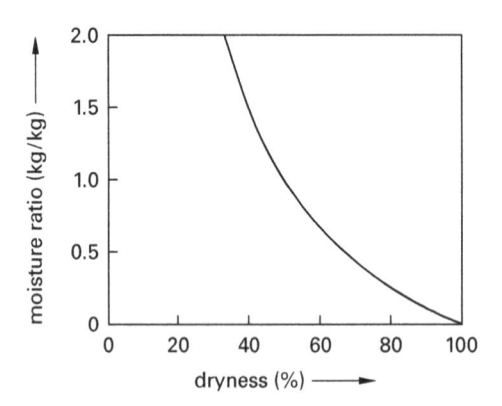

Figure 3.3. Relation between moisture ratio and dryness.

In the literature the water in wet pulps is classified in different ways relating to its location within or between the fibres but also to its properties. „Free water" has properties like normal bulk water and is found between the fibres and in large pores. Water that interacts and is bound to different degrees with the pulp has been named „Pore water". In *Figure 3.4* the „Pore water" is divided into semi bound or „Freezing" and bound or „Non-freezing" water, where the Non-freezing bound water is understood as directly bound water. First only free water is removed but below a water content of around 1.0 kg/kg all fractions are removed simultaneously until all Free and semi bound water has been removed. Finally only bound water remains. The different water fractions were measured with DCS, differential scanning calorimetry. Deodhar and Luner (1980) assumed that the largest pore that can carry only directly bounded water has a radius of 40 Å.

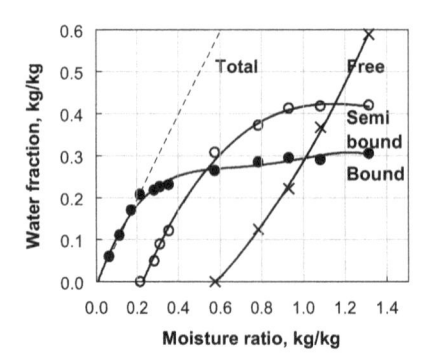

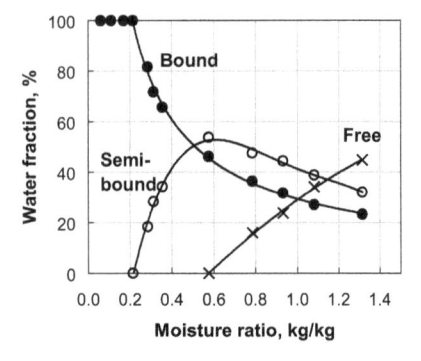

Figure 3.4. Absolute and relative water fractions in birch kraft pulp during drying (Maloney 2000).

3.2 Nano Level – Forces Within or Between Fibres

3.2.1 Bonding Mechanisms

In papermaking the water is removed gradually and the fibre surfaces are forced into contact with each other. *Figure 3.5* shows schematically how bonds are established between two fibre surfaces in forming, pressing and drying. Fibril threads, fines and hemicellulose form a swollen gel like layer which facilitates the creation of attraction forces. In a wet sheet, these bonds give a certain wet strength to the sheet.

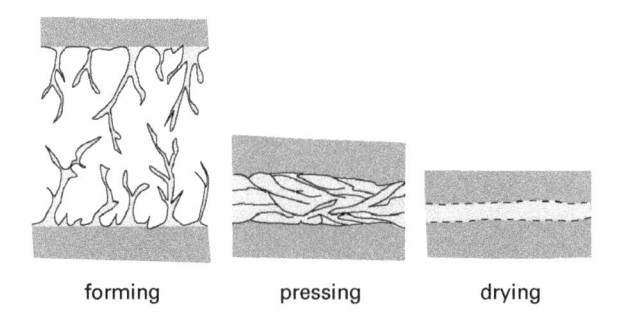

Figure 3.5. Schematic illustration of how bonds are formed between two fibres during papermaking. (Nanko and Wu, 1995).

The beating is very important for the size of the bonds. The beating makes the fibre more flexible and this facilitates bonding. The beating also creates a fine material which fills the cavities between the fibres. *Figiure 3.6* shows schematically the bonding structure between two fibre surfaces with increasing degree of beating from left to right. *Figure 3.7* shows an electron micrograph of a fibre structure which corresponds to the highly beaten structure in *Figure 3.6*. Note that the fines have filled the cavity between the fibres, and that this creates a denser sheet.

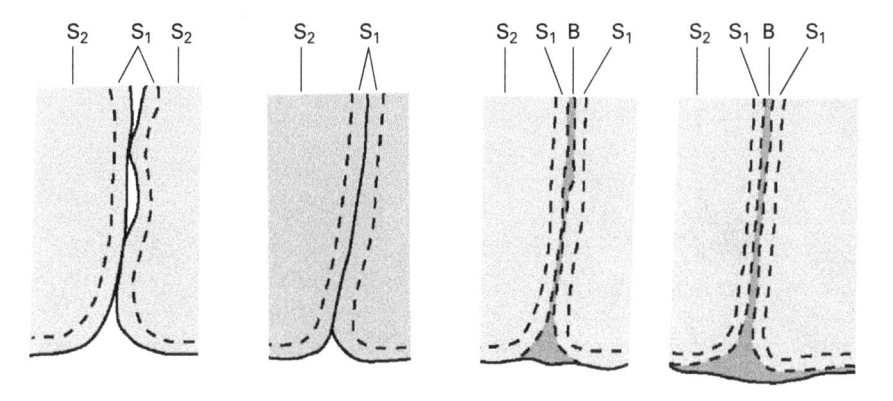

Figure 3.6. Schematic illustration of the bonding structure between two fibres with increasing degree of beating from left to right. S1 and S2 are layers in the secondary wall of the fibre, B is fines. (Nanko and Wu, 1995).

Water is a necessary component to create strong paper. Attempts have been made to form paper in a dry state but these papers have considerably lower strength than the wet-formed material. Through the different processes of forming, pressing and drying, the properties of the paper can be influenced within wide limits. *Figure 3.8* shows in principle how the tensile index of a paper increases as the dry content is increased. Three types of forces can be distinguished which act in different solids content ranges: mechanical or network forces, capillary forces (so-called Campbell forces) and chemical forces.

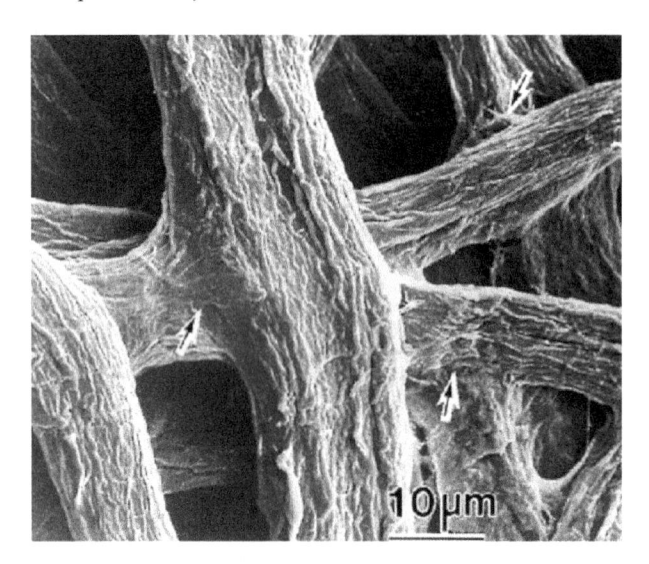

Figure 3.7. Bonds between two well-beaten fibres in a paper. (Nanko and Wu, 1995).

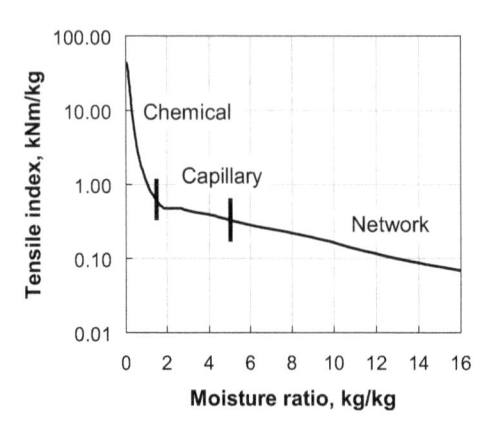

Figure 3.8. Development of tensile index for paper of sulphite pulp with decreasing moisture ratio. (Robertson).

3.2.2 Network Forces

A fibre in free rotation in a diluted fibre suspension covers a spherical volume with a diameter equal to the fibre length. The maximum concentration of totally free fibres with the average length L can then be represented by closely packed spheres with the diameter L. The fibre suspension flows almost like water and there are no network forces.

The sediment concentration is the lowest concentration where a network with an inherent strength is formed. If a known weight of fibres are carefully added to water in a beaker and allowed to sink by sedimentation, the height of the sediment gives the weight concentration. For a sulphate pine pulp the weight or sediment concentration is 2–3 g/l or 0,2–0,3 %.

If an amount of fibres exceeding the sediment concentration is carefully added to water the suspension does not automatically achieve network strength. It may be necessary to add turbulent energy by stirring. During the stirring operation the fibres are deformed from their natural shapes. After stirring the fibres tries to retain their original shapes but may be hindered and locked by other fibres, thus network strength has been created. As is shown in *Figure 3.9* three contact points are needed to lock a fibre in a given position.

If the mechanical locking of the fibres are crucial for the network strength it is required that the network strength is increased with increasing fibre concentration. Wahren has shown with measurements in a rheometer that the network strength is dependent on the fibre concentration. According to his findings the network strength is zero at the sediment concentration and thereafter increasing.

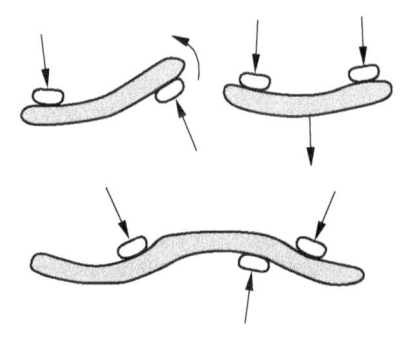

Figure 3.9. Three contact points are needed to lock a fibre in a given position.

3.2.3 Capillary Forces

Surface physics can be applied to the liquid meniscus in a fibre contact between parallel fibres according to *Figure 3.10*. The smaller the radius of curvature of the liquid meniscus during evaporation, the greater is the attraction between the fibres. The attractive force is proportional to the underpressure in the water, which is proportional to the surface tension and inversely proportional to the radius of curvature of the liquid meniscus. The underpressure and thus the attractive force theoretically approach infinity when the volume of water approaches zero. In reality, the increase in the attractive force gives rise to a local deformation of the fibres and the line contact is transformed to a surface contact, which limits the size of the force.

The existence of low pressures in a liquid with a small negative radius of curvature is the reason why fibres and fines attract each other when air is present, i.e. after the „dry line" in the paper machine. Capillary forces or „Campbell forces" is usually used as the collective name for these attraction forces. The fines created in the beating increase the capillary forces in the wet sheet, which is one of the reasons why a higher fines content gives a higher wet strength. Examples of capillary forces between parallel fibres in everyday life are paint brushes, hair locks and the well-known procedure of wetting a sewing thread before threading it through the needle. A wet thread is approximately 100 times stiffer than a dry thread. The attractive forces between perpendicular fibres are considerably lower than those between parallel fibres.

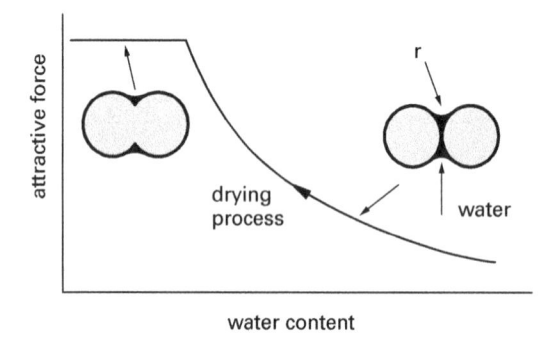

Figure 3.10. The negative pressure between two fibres increases with decreasing amount of water meaning that the attractive forces between the fibres increase up to an equilibrium level (Skelton).

Consider a paper after it has been beaten, formed and wet-pressed. The swollen fibres are then surrounded by thin water films. Gradually the water film around the fibres breaks, and the water is then collected where fibres have the closest contact. The capillary forces now attract the fibres to each other, as illustrated in *Figure 3.10*. The more softened and swelled the fibres are, from for example beating, the easier the fibres form themselves after each other when they are contracted by the surface tension forces. The fibres are also influenced by capillary forces which act in the lumen of the fibres, and this contributes to a flatter shape of the fibres after drying. The importance of the capillary forces for the development of fibre bonding can be illustrated in a comparison between laboratory sheets dried by freeze-drying and laboratory sheets dried in an oven, *Figure 3.11*. Freeze-drying means that the sheet is first frozen so that the water forms ice. The ice is then removed by sublimation (frozen water evaporates as vapour directly from the solid state). The freeze-dried sheet has much fewer collapsed fibres and free fibril threads can be seen in the structure. Nor is there any optical contact between the fibre walls, so that the light-scattering ability of the freeze-dried sheet is greater than that of the heat-dried sheet. The freeze-dried sheet also has a much lower strength than the heat-dried sheet.

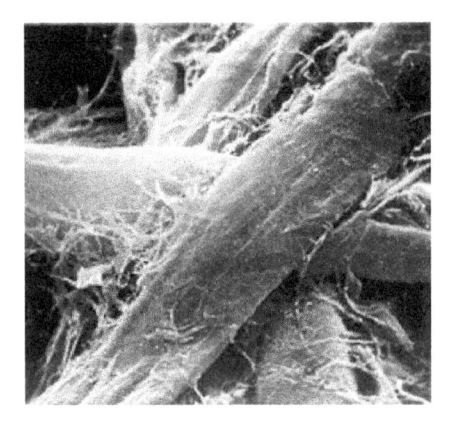

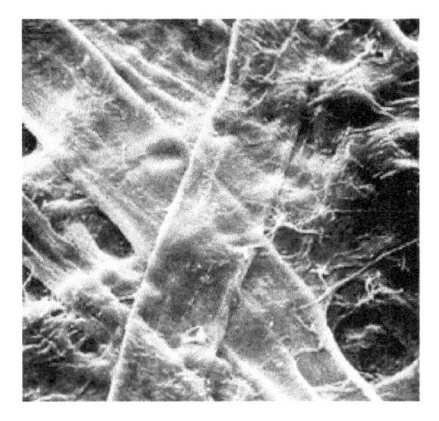

Figure 3.11. Left picture; Freeze-dried laboratory sheet of kraft pulp. Right picture; Heat-dried laboratory sheet of kraft pulp (Lobben).

3.2.4 Chemical Forces

When the water below the liquid menisci decreases and contracts further into smaller pores between and in the fibres, the radii of the menisci becomes smaller and the under pressure in the liquid becomes greater. The surface tension now very efficiently pulls the fibres into such intimate contact that chemical bonds can start to form between the fibres.

The atoms comprising the water molecule have a V-configuration with the oxygen atom at the angle and a hydrogen atom at the end of each arm according to *Figure 3.12*. The oxygen atom, which is strongly electro negative, attracts electrons away from the hydrogen atoms; this leaves it with a net negative charge and both hydrogen atoms with a net positive charge. This results in electrostatic attraction between the hydrogen atoms of one molecule and the oxygen atoms of a neighbouring molecule, an example of the so-called hydrogen bond.

Figure 3.12. Hydrogen bonding between two water molecules. (From Rance, Handbook of paper science).

The unique properties of water can be derived from the hydrogen bonds between the molecules. Hydrogen bonds are present in water to an unusually large extent, giving rise to a high degree of order. From the heat of fusion, it can be calculated that some 85 % of the hydrogen bonds present in ice remain unbroken in liquid water at 0 °C and even at 25 °C more than 80 % of the original number are still intact. Water, that locally may become almost crystalline, differs from a true crystal in that the hydrogen bonds are constantly broken and reformed. The resulting intermolecular attraction is acting through many hydrogen bonds and, although individually relatively weak, collectively requires considerable energy to overcome. This gives rise to the high heats of fusion and vaporisation, the high specific heat, the high surface tension and the fact that water is a liquid, and not a gas, at normal temperatures.

The strength of a hydrogen bond can vary slightly but it is approximately one tenth of that of the strong covalent forces, which hold organic molecules together. At short distances, the hydrogen bond is stronger than the dispersion forces, which lead to attraction between molecules. The condition for hydrogen bonds is that OH- or NH_2-groups exist in a molecule. If such groups lie close to each other, hydrogen bonds can be formed within the molecule, i.e. intramolecular hydrogen bonds. These can be found in the cellulose molecule and contribute to its stiffness. Hydrogen bonds can also be formed between different molecules, so-called intermolecular hydrogen bonds. In crystalline cellulose, such bonds can be identified between adjacent molecules. It is generally believed that the connecting force between cellulose chains is essentially due to the many hydrogen bonds.

The same mechanism of hydrogen bonding results in the attraction of water molecules to the hydroxyl, or other polar functional groups, of polysaccharides and of these latter groups one to another. *Figure 3.13A* shows a representation of two cellulose molecule's hydrogen bond to water molecules and, through these, in effect to one another. For convenience the V-configuration of the water molecules is shown in linear form. Considerable thermal effort is required to drive off these residual water molecules since their bonds with cellulose, even when ruptured, readily reform elsewhere. If carried far enough, however, the drying process reaches a stage at which the two cellulose molecules are bound through a monolayer of water molecules, as in *Figure 3.13B*, and ultimately directly one to another (*Figure 3.13C*) as in the crystalline zones of the cellulose micro fibril.

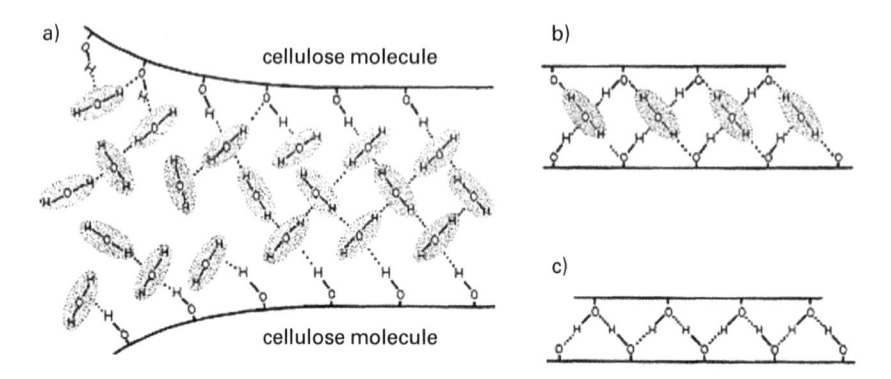

Figure 3.13. Hydrogen bonding between two cellulose molecules: (A) loosely through water molecules; (B) more tightly through a monolayer of water molecules; and (C) directly. (From Rance, Handbook of paper science).

Since bonds between cellulose chains within a fibril can be explained essentially by hydrogen bonds, it is not unreasonable to assume that hydrogen bonds can play an important role for the bonds between fibres. However, it should be pointed out immediately that the dispersion forces which exist between all objects of course also act between cellulose fibres. The fibres must come very close to each other for direct hydrogen bonds to be formed between the fibres.

Well beaten and swollen fibres develop good contact between the fibre surfaces, through the influence of capillary forces in the consolidation of the sheet, and this encourages the creation of hydrogen bonds. This is favoured for example by internal and external fibrillation and fines formation in beating and by the occurrence of hemicellulose on the fibre surface. Mechanical fibres are relatively stiff and do not give such well-developed contact surfaces as chemical fibres. Since the surfaces on mechanical pulp fibres consist partly of lignin with fewer OH-groups, there are fewer possibilities for hydrogen bonds to develop than on cellulose and hemicellulose. This is one of the reasons why mechanical pulps give a low tensile index. However, it should be remembered that the dispersion forces between the fibres can play an important role for the fibre bonds. For the dispersion forces, the geometrical requirements are not as strict as for hydrogen bonds. Also the dispersion forces do not decrease as rapidly with increasing distance.

The strength of a wet paper sheet is only a few per cent of the strength of the dry sheet. This can be understood as follows. When two fibres are to be separated by mechanical forces, hydrogen bonds must be broken and dispersion forces must be overcome. In a wet sheet, water molecules are always available which can penetrate the bonding zones and form hydrogen bonds between the exposed fibre surfaces. The dispersion forces are also much weaker when there is a medium between the objects which interacts with the dispersion forces.

3.3 Micro Level – Fibre and Network

3.3.1 Single Fibre

Shrinkage. Seen in a simplified way, crystalline micro fibrils lie at a small angle (5–15°) to the fibre axis. Between the highly crystalline cellulose fibrils lies a matrix consisting of an amorphous material, mainly hemicellulose and lignin. When the swollen fibres begin to dry, the matrix wishes to shrink. Since the crystalline micro fibrils lie in the length direction of the fibre, shrinkage is prevented in this direction. It is natural that the fibre instead shrinks most in its transverse direction. *Figure 3.14* shows how the cross-sectional area of a fibre decreases from the swollen to the dry state during drying. The authors reported cross section shrinkage of about 20 %, whereas the length or axial shrinkage of the fibre is in the order of only a few percent. The transverse shrinkage of the fibres in *Figure 3.14* is around 15 %, it will be shown later in this chapter that the shrinkage of the fibre has a large influence on the paper properties.

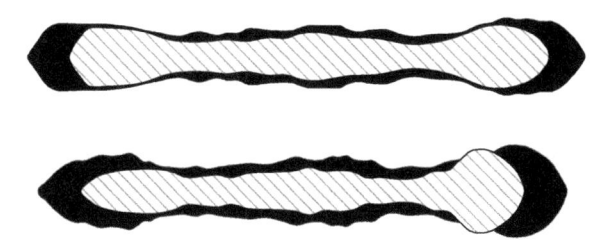

Figure 3.14. Example of how the cross-sectional area of a fibre is changed from a swollen state (black) to a dry state (lines). (Page and Tydeman 1966).

The left picture in *Figure 3.15* shows the development of the transverse shrinkage for a bleached kraft pulp fibre during drying. The transverse shrinkage was evaluated as the cross section width, D_{max}, reduction defined in *Figure 3.16*. The measurements were made with a confocal microscope on a surface fibre in a hand sheet. It is evident that the shrinkage of the fibre happens very late in the drying process. As a measure of the structural changes in the fibres the water retention value, WRV, was measured on the same fibres to characterise the hornification of the pores in the cell wall (right picture in *Figure 3.15*). The hornification takes place during the whole drying process. Down to a moisture ratio of 2.0 kg/kg (dryness up to 33 %) no hornification took place. Based on the results in *Figure 3.15* and visual observations a model of the shrinkage progress according to *Figure 3.16* has been proposed. The top row shows the shrinkage of a fibre cross section and the bottom row an element of the fibre fine structure. Phase A-C represents the large part of the drying process where almost no shrinkage of the cross section width occurs. Water is removed from large fibre pores and voids between the lamellae of the fibre wall are collapsed and causes a reduction in the WRV. The fact that almost no cross section width shrinkage can be observed in this region means that the shrinkage of the cross section area and the closure of large pores happen perpendicular to the lamellar layers. Note that phase A–C is the inverse of Scallan's well known swelling model. The late shrinkage observed in *Figure 3.15* happens during phase D in *Figure 3.16*, which shows the effect of drying of the fine structure. This last fraction of water is directly bound water dealt with in *Figure 3.4*.

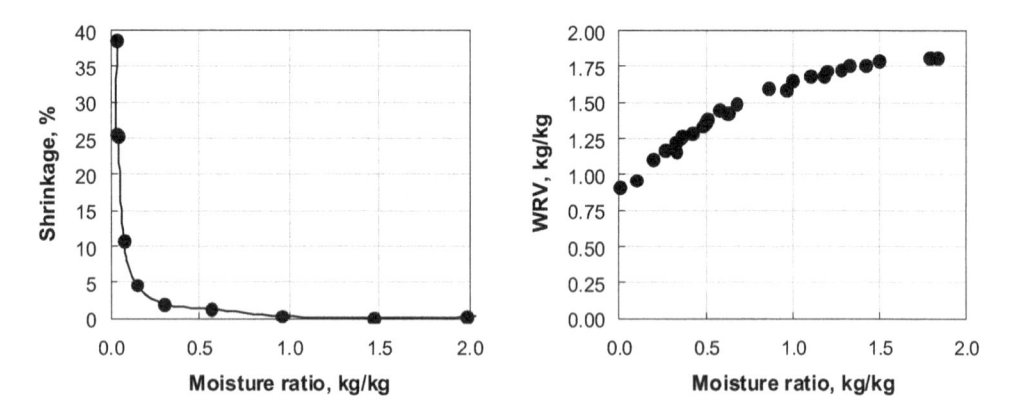

Figure 3.15. Transverse shrinkage and WRV development for bleached kraft pulp fibres during drying. (Weise et.al., 1996).

Properties and drying history. The left hand part of *Figure 3.17* shows the tensile stiffness and strain at break versus a large compression of a single fibre during drying. The fibre species was a holocellulose springwood fibre. The stiffness was reduced and the strain at break increased by the compression. The right hand part shows the elastic modulus and strain at break versus an applied load in grams on a single fibre during drying. The fibre species was a long leaf pine holocellulose springwood fibre. The stiffness was increased and the strain at break was reduced by the load. Note that applying a load on the fibre is equal to actually stretching the fibre, the amount of stretch that the load corresponded to, was however not measured. The effect of the applied load on the fibre properties is sometimes referred to as the „Jentzen effect".

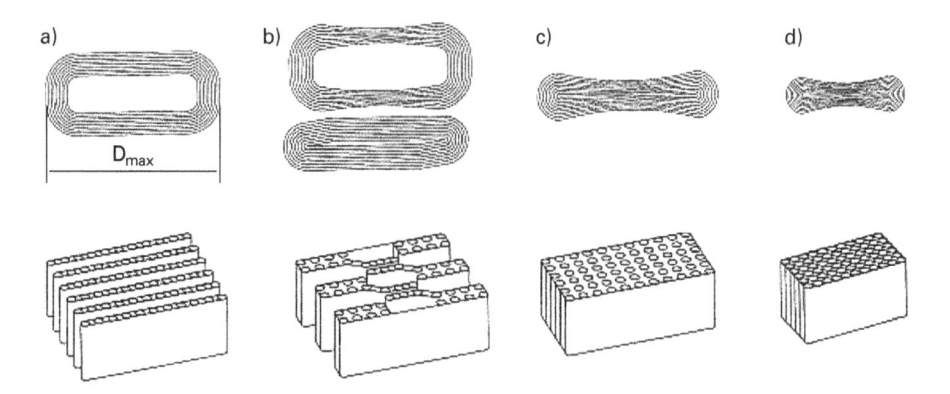

Figure 3.16. Shrinkage progress for a fibre cross section (top) and an element of the fibre fine structure (bottom). (Weise et al., 1995).

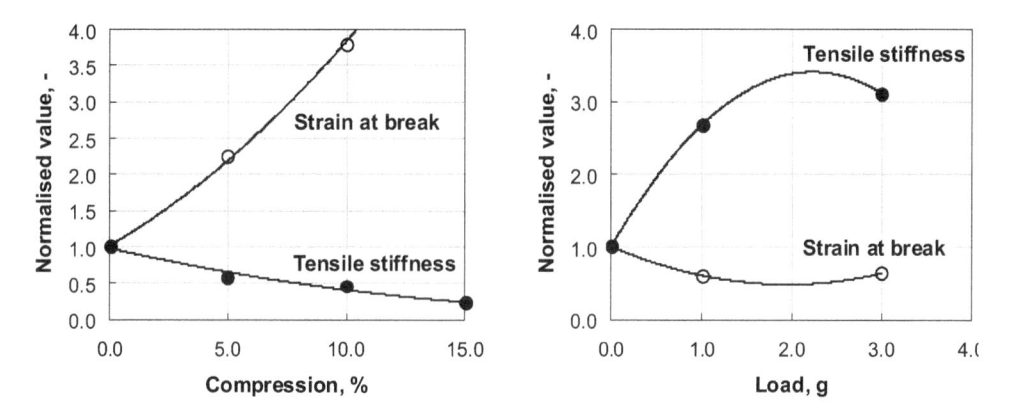

Figure 3.17. Tensile stiffness and strain at break of a single fibre during drying versus compression (Dumbleton) and load (Jentzen).

3.3.2 Fibre Network

Previous section dealt with the single fibres behaviour during drying. The question is now how their strength and the structural changes they undergo during drying is transferred to the network of fibres that we call a paper.

Since the fibres in their wet, swollen state have low compressive strength, the width shrinkage of the fibres is able to cause corresponding length shrinkage of each crossing fibre at the bonding sites. This shrinkage at the fibre bonding zones has been named „micro compressions" in the literature (Page and Tydeman). Note that this term is also used for a beating effect involving local damage of the fibre wall. The occurrence of micro compressions at the bonding sites requires that the bond develop first and thereafter the shrinkage. Considering the very late developed transverse shrinkage in *Figure 3.15* this seams to be reasonable.

It has also been shown experimentally that the shrinkage actually happens at the bonding sites. *Figure 3.18* shows the average longitudinal shrinkage of a sheet, an individual fibre in the sheet and the segments (bonding sites and free segments) of a fibre in a fibre network (sheet). The paper sheet was dried both free and restrained. Free drying means that there are no external forces which prevent the shrinkage. Note that the shrinkage for the free dried case happened only at the bonding sites. In the restrained dried case the sheet is clamped and the fibre as a whole cannot shrink. But the bonding sites were still shrinking and they were actually stretching the free segments. In the literature this is often referred to as activation of the fibre network. To the right in *Figure 3.18* is a schematic illustration of the measured results. *Figure 3.19* shows the distribution of the average segment shrinkage presented in Fig. 46.18. The distribution shows that also for free drying some of the free segments are stretched but the average result is zero.

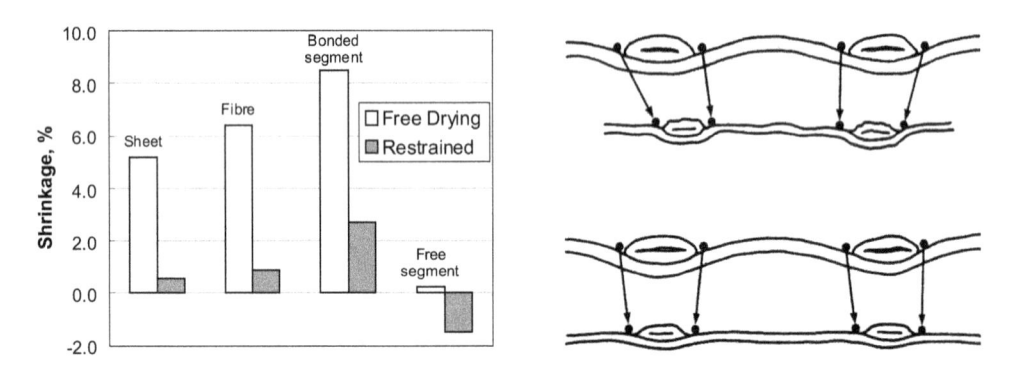

Figure 3.18. Shrinkage of the sheet, fibre and segments during free and restrained drying. (Nanko and Wu, 1995).

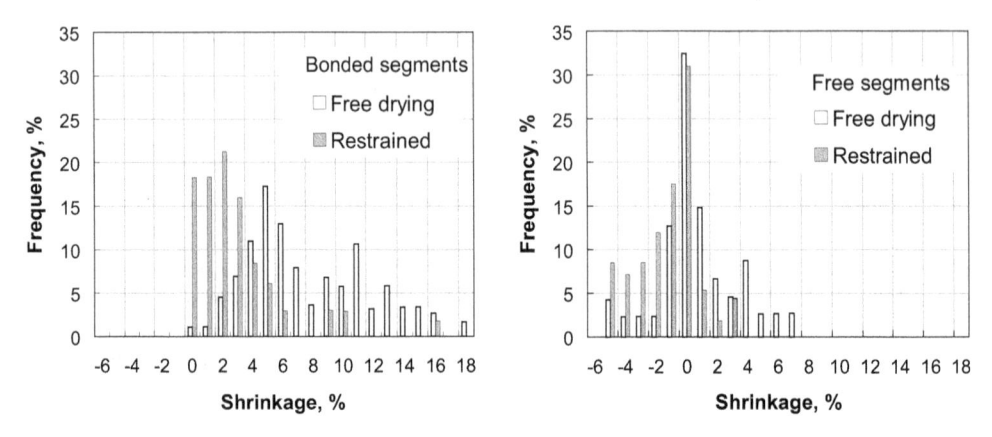

Figure 3.19. Distribution of shrinkage of bonded and free segments during free and restrained drying. (Nanko and Wu, 1995).

Figure 3.20 shows electron micrographs of paper after free and restrained drying. The micrographs confirm that the sheet dried under restraint has straighter fibres and fewer microcompressions than the freely dried sheet.

Figure 3.20. (Left)Electron micrographs of freely dried paper and (Right) paper dried under restraint. (Nanko and Wu, 1995).

The shrinkage is greater in CD than in MD for a free dried paper which is explained in the following way; in most papers, there is a larger number of fibres in MD than in CD. Each fibre crossing causes shrinkage with the development of micro compressions and each fibre has more bonding points per fibre in CD. Therefore, the sheet shrinks more in CD. Let's assume we have a unit cell of a schematic isotropic fibre network with ten fibres according to *Figure 3.21*. Isotropic means without direction and is in this case demonstrated with five fibres in each in-plane direction, MD and CD. The width of the unit cell is 300 µm and the fibre width is 30 µm. We know that the width shrinkage of a fibre is in the order of 20 % or 6 µm in this schematic case. The crossing fibre will however restrain a part of the fibres shrinkage potential. With reference to *Figure 3.18* it is assumed that the actual shrinkage of the bonding sites is 10 % or 3 µm in both directions of the bonding site. With these assumptions the shrinkage of the unit cell will be 5 % in both in-plane directions. If we now redistribute the ten fibres in the unit cell to achieve an anisotropic paper (a paper that has a direction or a MD and CD direction), we can for example get the networks and anisotropies according to *Figure 3.21*. The anisotropy, A, is calculated as the number of fibres in MD divided by the number of fibres in CD, giving the anisotropies 1.5; 2.3 and 4. The shrinkage in MD and CD for the four networks calculated with this simple model is presented in *Figure 3.22*.

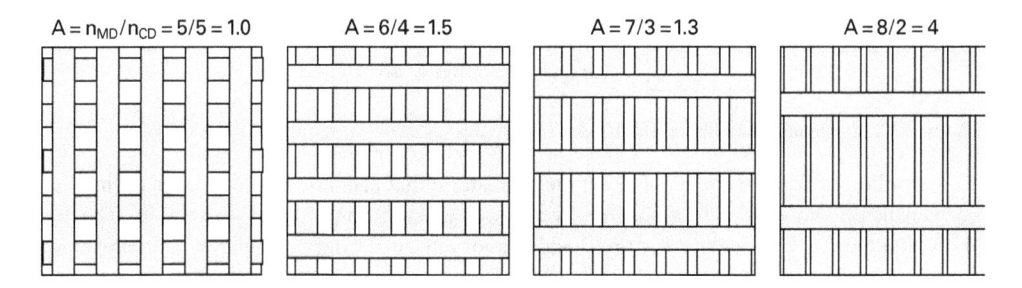

Figure 3.21. Schematic pictures of fibre networks with different anisotropies.

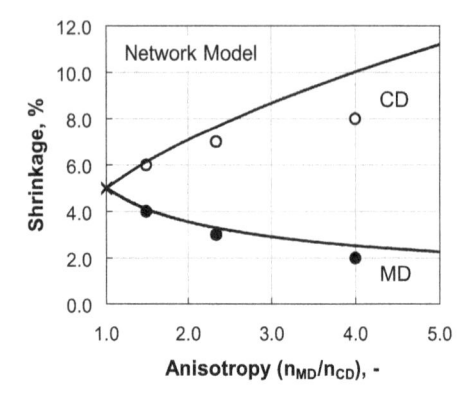

Figure 3.22. Free shrinkage in MD and CD for the fibre networks in *Figure 3.21*.

3.4 Macro Level – Paper Sheet

3.4.1 Shrinkage, Stretch and Paper Properties

Total strain accumulated during drying. If a paper is deformed, either stretched or allowed to shrink, during drying it will have a great effect on most of its paper properties. This deformation can be characterized with the relative elongation or strain, ε, according to Equation 3.3 and Figure 3.23 where δ is the elongation and L the original length of the sample. Note that a shrinkage or reduction in length from now on will be characterized by a negative strain even if it earlier in this chapter was defined as positive.

$$\varepsilon = \frac{\delta}{L} \tag{3.3}$$

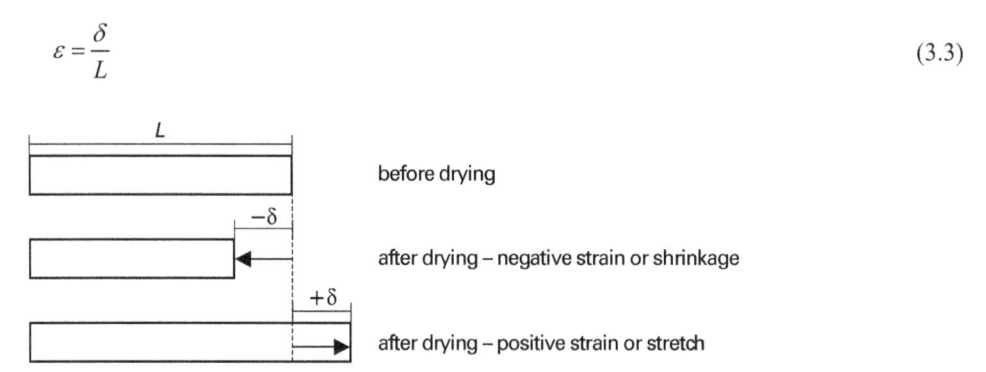

Figure 3.23. Illustration of shrinkage and stretch.

According to Wahlström et al (1999) the total strain that accumulates in a paper during drying can be decomposed into a sum of free shrinkage strain and mechanical strain. The free shrinkage strain occurs when moisture is added to or removed from the paper without any external forces acting on it. When moisture is added the paper expands and the free shrinkage strain is positive. Removal of water makes the paper shrink and the free shrinkage strain is negative.

Note that the definition is equal to the hygro expansion of paper with the difference that the main focus here is on removing, not adding, water.

The mechanical strain is induced in the paper by applying external forces. The mechanical strain is mainly a positive strain since paper is hard to compress in its in-plane directions. An exception with negative strain is however the so-called lateral contraction (Poisson's ratio) dealt with later in this chapter. A negative total strain is often referred to as shrinkage, a positive as stretch and a zero total strain after drying as restrained drying. Note that a standard hand sheet is dried restrained, which means that a mechanical strain equal to the free shrinkage strain is induced in the paper giving a zero total strain. Equation 3.4 shows the decomposition of total strain, ε, into free shrinkage strain, ε^{fs}, and mechanical strain, ε^{m}, for an isotropic hand sheet. Here, the paper will be treated as isotropic, i.e. without direction. In Paragraph 3.4.4, Forming and Drying, fibre anisotropy will be introduced. An isotropic paper has the same properties in MD and CD, which is typical for a laboratory made hand sheet.

$$\varepsilon = \varepsilon^{fs} + \varepsilon^{m} \tag{3.4}$$

Free Shrinkage Strain (Free drying). Drying without any external forces acting on the paper is often referred to as the drying strategy Free drying. The free shrinkage strain that occurs when a paper is dried with only free drying is often referred to in short as the „free shrinkage". The moisture range that the paper goes through during drying must then be defined. The starting point should always be before the shrinkage starts, which means around press dryness (mr = 1,5 kg/kg, dryness = 40 %), to some reference point in the dry end, for example bone dry or complete drying in a conditioned room at 23 °C, 50 % RH.

The free shrinkage strain for an isotropic hand sheet develops exponentially versus moisture ratio, mr, as shown in *Figure 3.24*. The solid line represents Equation 3.5 with the free shrinkage, $\varepsilon^{fs(mr=0)} = -4,6$ % (the reference point in the dry end is $mr = 0$) and $k = -2,6$.. Note that the free shrinkage strain does not develop as rapid as the shrinkage of the individual fibre did in *Figure 3.15*. The explanation for the difference may depend on the moisture gradient that develops in the thickness direction of a paper during drying. When the surface layer dries first it will cause some shrinkage of the sheet also at a higher moisture ratio of the whole paper.

$$\varepsilon^{fs} = \varepsilon^{fs(mr=0)} e^{k \cdot mr} \tag{3.5}$$

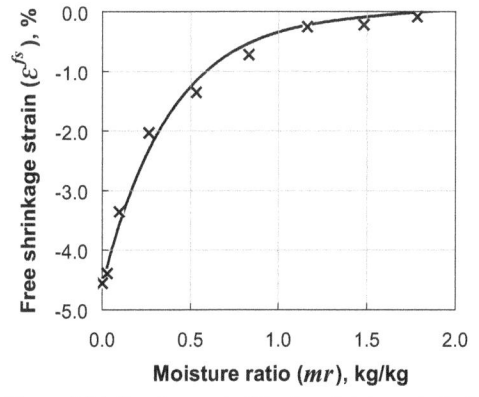

Figure 3.24. Development of the free shrinkage strain during drying for an isotropic hand sheet.

Drying stress (restrained drying). If the total strain according to Equation 3.4 is kept at zero during the entire drying process, a stress will develop in the paper. The drying strategy where no shrinkage or stretch is allowed during drying is often referred to as restrained drying. The drying stress for an isotropic hand sheet develops exponentially versus the moisture ratio, mr, The solid line in Figure 3.25 represents Equation 3.6 with the final drying stress, $\sigma^{mr=0} = 4$ N/m² and $k = -4$. The final drying stress is defined as the final point in the diagram at a defined final moisture content, 0 kg/kg in Figure 3.25.

There is no standard method for characterisation of drying stresses. Normally heated clamps or pre drying of the sample at the clamping area is used to avoid water leakage next to the clamps. The excess water causes a reduction of the maximum stress measured. Stress relaxation will take place during and after the drying process which makes the measurement depend on drying time and temperature. Restrained drying with the same temperature and time gives a linear relation between stiffness and final drying stress before relaxation (Htun and Fellers,1982).

Considering an anisotropic paper the drying stresses are greatest in MD due to the larger number of fibres in this direction, compare with *Figure 3.21*. A final drying stress in the sheet of the order of 10 kNm/kg corresponds to the stress obtained in tensile testing at an elongation of circa 0,1 per cent. 0,1 per cent is in the elastic range, failure occurs at 60–100 kNm/kg.

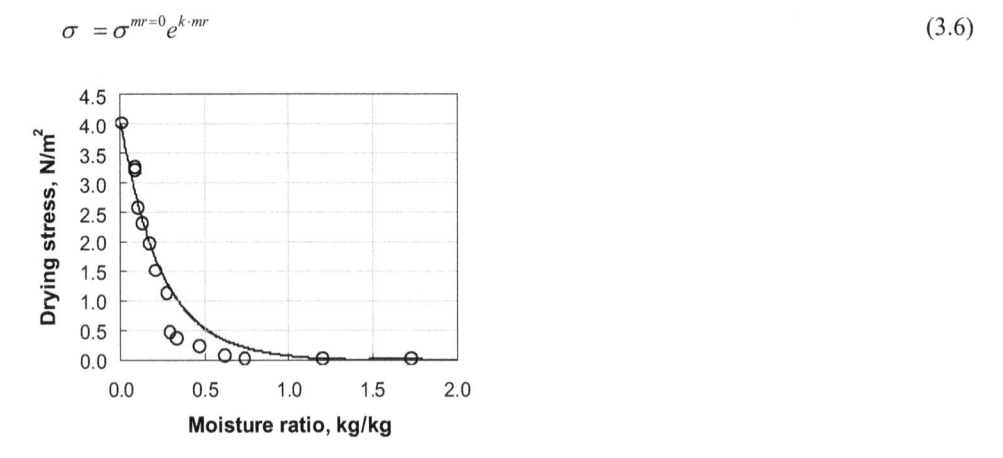

$$\sigma = \sigma^{mr=0} e^{k \cdot mr} \tag{3.6}$$

Figure 3.25. Development of drying stress during drying for an isotropic hand sheet.

Mechanical strain. When a mechanical strain, or stretch, is applied in one in-plane direction of a paper, for example MD as in *Figure 3.26*, it will contract in CD. *Figure 3.26* shows this contraction for straining of a wet strip of paper. The moisture ratio was 1,3 kg/kg which is about equal to a press dryness.

$$\varepsilon_{CD}^m = -\nu \cdot \varepsilon_{MD}^m \tag{3.7}$$

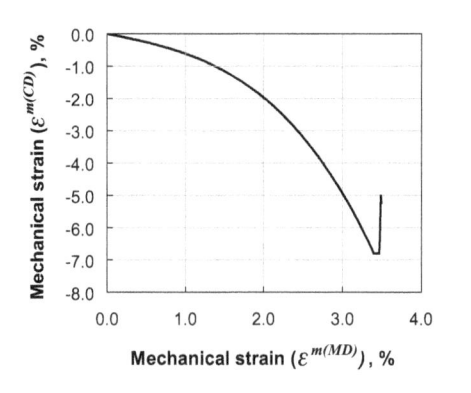

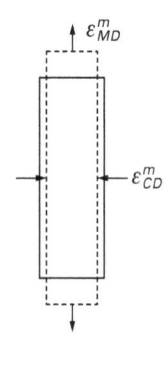

Figure 3.26. Contraction in CD due to a stretch in MD.

Development of paper properties during drying. Also the tensile stiffness index for an isotropic hand sheet develops exponentially versus moisture ratio. as shown in *Figure 3.27*. The solid line represents Equation 3.8 with the stiffness for a dry paper, $E^{mr=0} = 3,1$ MNm/kg for the free dried paper and 7,3 MNm/kg for the restrained dried paper. The exponential constant, $k = -4,8$ for both the restrained and free dried paper. Compared to the development of free shrinkage strain and drying stress an extra constant is needed to describe that the stiffness is separated from zero in the wet end of the relation. This constant, c, is equal to 0,13 for both the restrained and free dried paper in this case.

$$E = E^{mr=0} e^{k \cdot mr} + c \tag{3.8}$$

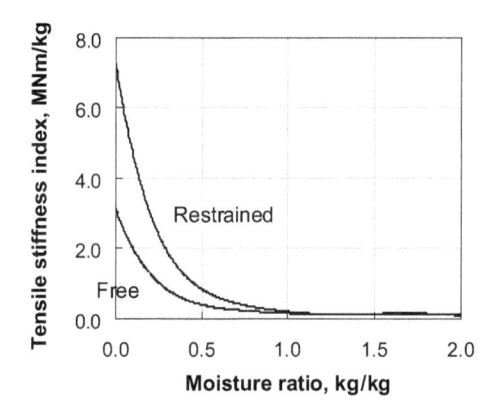

Figure 3.27. Development of tensile stiffness index during free and restrained drying for an isotropic hand sheet.

Relations between total strain and paper properties. There is a linear relation between tensile stiffness index and total strain accumulated during drying for isotropic hand sheets (Wahlström and Fellers 2000). With knowledge of the free shrinkage and the stiffness for a free and a restrained dried isotropic hand sheet the stiffness can be calculated for a given total strain with Equation 3.9. E^r represents the tensile stiffness index for a restrained dried paper and E^{fs} is measured on a free dried paper. ε^{fs} is the free shrinkage and ε is the total strain accumulated during drying. ε^r is by definition zero. The solid line in the left figure in Figure 3.28 shows this relation for isotropic hand sheets made of a bleached sulphate pulp with $E^r = 5,5$ MNm/kg, $E^{fs} = 2$ MNm/kg, $\varepsilon^{fs} = -6$ %. Note however that this relation is not valid for high positive strains where the stiffness is known to reach a maximum and thereafter decline.

There is also a linear relation between the other standard in-plane tensile properties and total strain accumulated during drying. The right picture in *Figure 3.28* shows relations between total strain and normalised stiffness, strain at break (stretch) and tensile index. Normalisation was made, with stiffness as an example, by dividing all values with the stiffness for restrained drying, $E^{norm} = E/E^r$. For restrained drying ($\varepsilon = 0$) the stiffness was 5,5 MNm/kg, stretch 3,3 % and tensile 44 kNm/kg. If Equation 3.9 is applied also here k is 0,11 for the stiffness, 0,03 for tensile and $-0,25$ for the stretch. Meaning that in this case stretch was most, and tensile least, sensitive to a change in total strain.

$$E = k\varepsilon + E^r \quad \text{where} \quad k = \frac{E^r - E^{fs}}{\varepsilon^r - \varepsilon^{fs}} \tag{3.9}$$

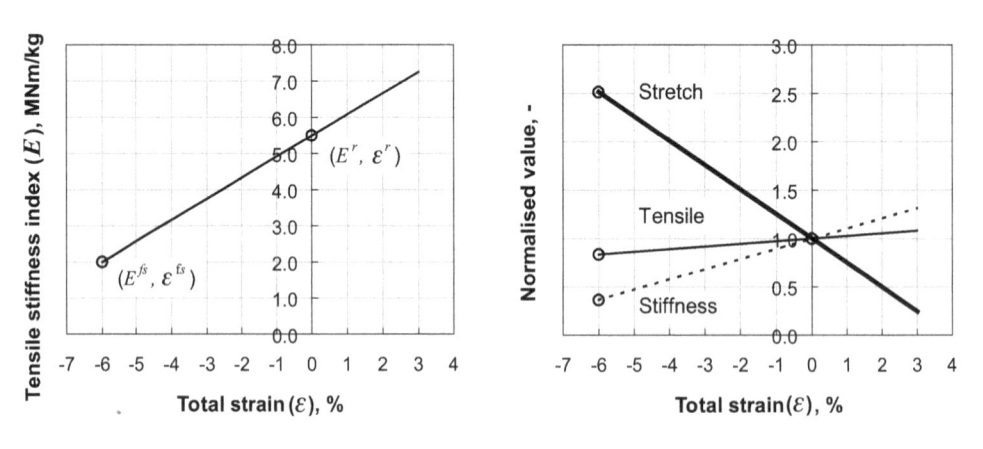

Figure 3.28. Left: Relation between tensile stiffness index and strain accumulated during drying for an isotropic hand sheet made of bleached sulphate. Right: Normalised results for stiffness, tensile and stretch (same paper).

Some paper properties that are not tested in a specified direction can be dependent on the change of the area strain rather than the strain in a certain direction. *Figure 3.29* shows an example for Bendtsen roughness and z-toughness plotted versus the total area strain accumulated during drying. The paper used were isotropic hand sheets made of a bleached sulphate pulp dried with a number of different drying strategies. The total area strain is the sum of the total strain in the two in-plane directions of the paper. The decrease in both roughness and z-toughness can be understood from *Figure 3.20*, which shows a more even surface but also more open structure for the restrained dried paper compared to the free dried.

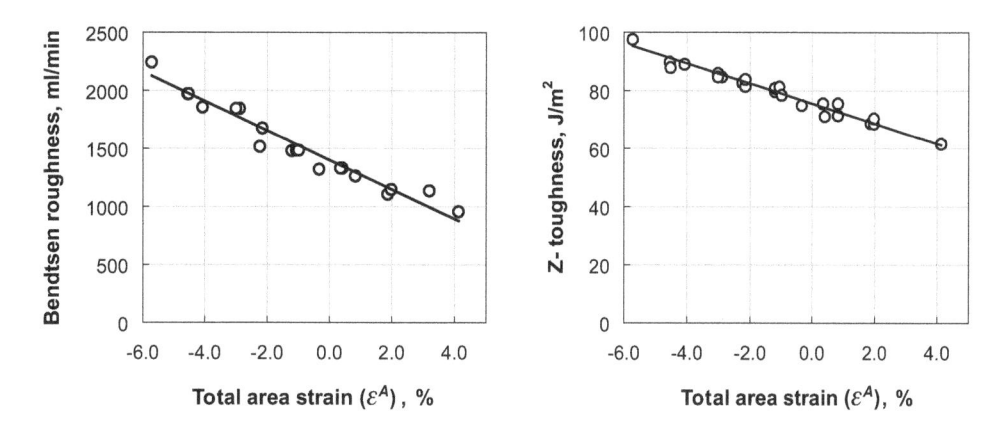

Figure 3.29. Roughness and z-toughness versus total area strain accumulated during drying for an isotropic hand sheet made of bleached sulphate.

3.4.2 Multi-Ply Board and Drying

Volume model. For modelling the free shrinkage strain, ε^{fs}, for a multi-ply laminate a volume model can be formulated according to Equation 3.10. The product of the grammage, w^r, stiffness, E^r, and free shrinkage strain for the laminate is assumed to be equal to the sum of the same products for the individual plies.

The stiffness for a restrained dried laminate, E^r, is calculated according to Equation 3.11, where n is the number of plies, E^r_i is the restrained tensile stiffness index, w the grammage and i denotes the individual plies. The grammage of the laminate, w^r, is calculated as the sum of the grammages of the individual plies.

$$w^r_{Lam} E^r_{Lam} \varepsilon^{fs}_{Lam} = \sum_{i=1}^{n} w^r_i E^r_i \varepsilon^{fs}_i \tag{3.10}$$

$$E^r_{Lam} = \frac{\sum_{i=1}^{n} w^r_i E^r_i}{\sum_{i=1}^{n} w^r_i} \tag{3.11}$$

Total strain and paper properties. Equation 3.9 can be applied also on the laminate and give a relation between the stiffness of the laminate, E, and the total strain accumulated during drying, ε, giving Equation 3.12. The stiffness for a freely dried laminate, E^{fs}, can be calculated according to Equation 3.13, where n, is the number of plies, w_i, the grammage and E_i, the stiffness that the individual ply has with the total strain equal to the free shrinkage of the laminate. E_i, is calculated for each ply using Equation 3.9 with $\varepsilon = \varepsilon^{fs}$.

$$E_{Lam} = E^r_{Lam} - \frac{E^r_{Lam} - E^{fs}_{Lam}}{\varepsilon^{fs}_{Lam}} \varepsilon \tag{3.12}$$

$$E_{\text{Lam}}^{\text{fs}} = \frac{\sum_{i=1}^{n} w_i^{\varepsilon_{\text{Lam}}} E_i^{\varepsilon_{\text{Lam}}}}{\sum_{i=1}^{n} w_i^{\varepsilon_{\text{Lam}}}} \tag{3.13}$$

Figure 3.30 shows relations between stiffness and total strain for a typical liquid packaging top, middle and bottom ply pulp. The top ply pulp is a bleached sulphate, the bottom ply an unbleached sulphate and the middle ply a sulphate and CTMP mix. The relations for each ply is calculated with Equation 3.9 and the input data in *Table 3.1*. Based on this the stiffness- strain relation is simulated with Equation 3.10–3.13.

Table 3.1. Input measured on hand sheets.

	Stiffness Restrained, MNm/kg	Stiffness Free, MNm/kg	Free shrinkage strain, %	Grammage Restrained, g/m²
Top ply (TP)	7,84	4,03	-5,56	126
Middle ply (MP)	5,62	2,82	-2,76	119
Bottom ply (BP)	8,51	3,77	-6,10	124

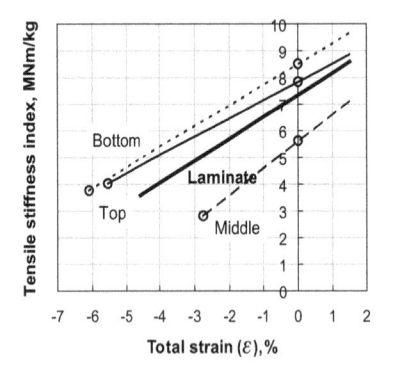

Figure 3.30. Measured input data on the plies and simulated laminate.

3.4.3 Stock Preparation and Drying

Raw material and beating. *Figure 3.31* shows a comparison in stiffness between paper made by an unbleached kraft pulp and an unbleached CTMP pulp. The pulps are unbeaten and beaten to two higher degrees and thereafter evaluated for restrained and free drying. The maximum beating energy was 200 kWh for both pulps. Generally it can be seen that the kraft paper has higher stiffness values at restrained drying compared to the CTMP paper. On the other hand the kraft paper shows a larger reduction with decreasing total strain. The kraft paper made of the most beaten pulp actually gets a lower stiffness than the CTMP when free dried. An interesting observation is what happens with the free dried papers when the beating is increased. For the CTMP both the stiffness and the free shrinkage strain is increased. For the kraft the stiffness first increases with increasing beating, but thereafter it is reduced again. Obviously the reduction in stiffness, caused by the free drying, overcomes the stiffness increase, caused by the beating, for high high beating levels.

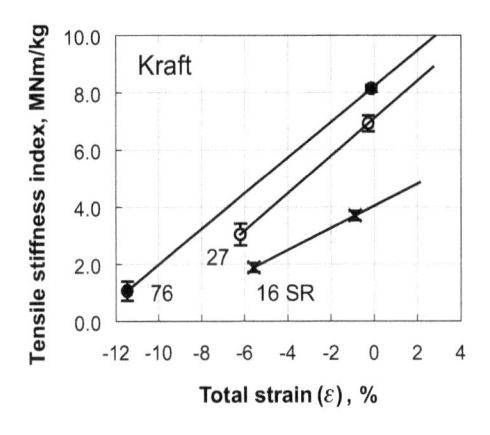

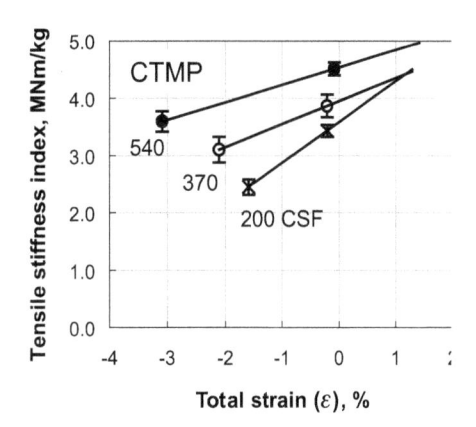

Figure 3.31. Stiffness for an unbleached kraft and an unbleached CTMP pulp, both unbeaten and beaten to different degrees versus total strain accumulated during drying.

Chemical addition. A dry strength additive with cationic charge functionality (Redibond 4300) was added to the 27 °SR kraft pulp in *Figure 3.31*. The stiffness was not affected by the dry strength addition. *Figure 3.32* shows the effect on tensile index versus total strain. The free shrinkage strain did not change as a result of the chemical addition, but the tensile index increased both for the restrained and free dried paper.

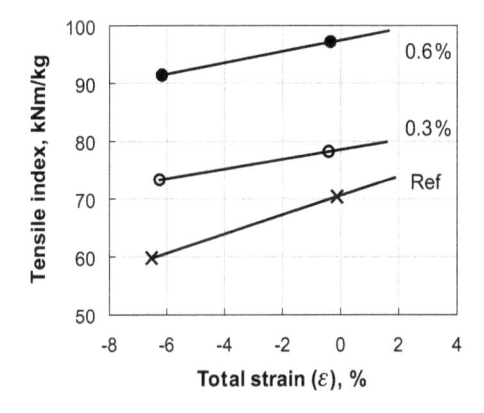

Figure 3.32. Tensile index for an unbleached kraft pulp (27 °SR) with an addition of a dry strength additive versus total strain accumulated during drying.

Fillers. A large particle size rhombohedral PCC filler was added to the 27 °SR unbleached kraft pulp used in *Figure 3.31* and *Figure 3.32*. Both the tensile stiffness and the free shrinkage are decreasing with increased filler amount (*Figure 3.33*). This is most likely due to that the interactions between the fibres are disturbed by the filler particles.

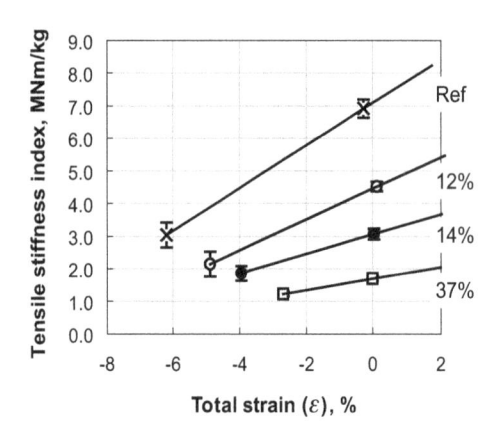

Figure 3.33. Stiffness for an unbleached kraft pulp (27 °SR) with different degrees of fillers added versus total strain accumulated during drying.

3.4.4 Forming and Drying

Fibre orientation and anisotropy. Effects at the headbox slice and forming section principally control fibre orientation. As the stock suspension is accelerated toward the slice, shear stresses in the flow cause the fibres to orient preferentially in the MD. When the jet impinges on the wire, a difference in the velocity between the fibres and the wire also creates shear stresses that cause an alignment. This effect is at a minimum when the jet and wire velocities are the same and becomes more significant as the velocity differential increases in either direction.

The amount of fibre orientation can be characterised by the anisotropy of the paper. An anisotropy is often defined as the ratio of some property in the two principal in-plane directions of paper, the machine or manufacturing direction, MD, and the cross machine direction, CD. Equation 3.14 defines the fibre anisotropy, or fibre orientation, where n is the relative number of fibres in the respective in-plane direction. The stiffness anisotropy for restrained dried paper is defined in Equation 3.15 and for freely dried in Equation 3.16. The anisotropy in free shrinkage strain is given in Equation 3.17. The anisotropy measured on machine made papers, as MD/CD for some property, will of course depend both on the fibre orientation and the shrinkage in CD and stretch in MD. This will be discussed in Section 3.5.

$$A_{Fiber} = \frac{n_{MD}}{n_{CD}} \tag{3.14}$$

$$A_{E^r} = \frac{E^r_{MD}}{E^r_{CD}} \tag{3.15}$$

$$A_{E^{fs}} = \frac{E^{fs}_{MD}}{E^{fs}_{CD}} \tag{3.16}$$

$$A_{\varepsilon^{fs}} = \frac{\varepsilon_{MD}^{fs}}{\varepsilon_{CD}^{fs}} \tag{3.17}$$

Free shrinkage strain – anisotropic. *Figure 3.24* showed the development of the free shrinkage strain during drying for an isotropic hand sheet. In *Figure 3.34*, the same relation is given for anisotropic hand sheets made of the same pulp. It shows that the exponential constant, k in Equation 3.5, has close to the same value regardless of anisotropy. Only the free shrinkage potential, $\varepsilon^{fs(mr=0)}$, varies. Consequently the development of the free shrinkage strain for anisotropic papers can be calculated with Equation 3.17 and Equation 3.23.

Paper properties – anisotropic. *Figure 3.27* showed the development of tensile stiffness index during free and restrained drying for an isotropic hand sheet. In *Figure 3.35*, the same type of relation is given for anisotropic hand sheets made of the same pulp. It shows that, as for free shrinkage strain, the exponential constant, k in Equation 3.8, has the same value regardless of anisotropy. The stiffness for a dry paper, $E^{mr=0}$, and the constant c varies according to the geometric mean assumption. Based on *Figure 3.27* the relations in *Figure 3.35* are given by Equation 3.15 and Equation 3.21 for restrained drying and Equation 3.16 and Equation 3.22 for free drying.

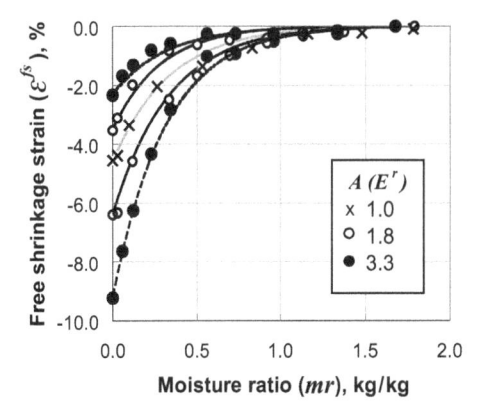

Figure 3.34. Development of the free shrinkage strain during drying.

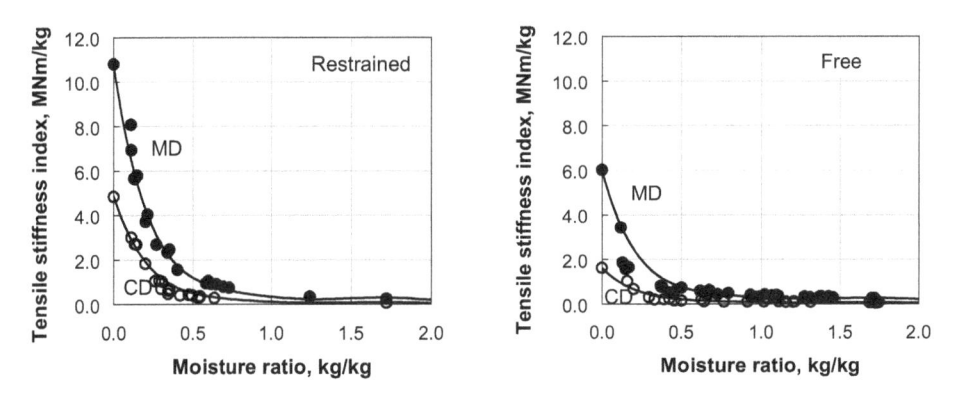

Figure 3.35. Development of tensile stiffness index during free and restrained drying for anisotropic hand sheets.

Geometric mean assumption. The fibre anisotropy is controlling the other defined anisotropies according to Equation 3.18–3.20. It is common practice to use the stiffness anisotropy for restrained dried paper as the fibre anisotropy according to Equation 3.18. The reason for this is that the fibre orientation is time consuming and not so straight forward to measure. The fibre anisotropy in machine made papers can have a value from close to 1 for certain board grades up to around 4 in newsprint made on high speed machines.

$$A_{E^r} = A_{Fiber} \tag{3.18}$$

$$A_{E^{fs}} = 2A_{Fiber} - 1 \tag{3.19}$$

$$A_{\varepsilon^{fs}} = \frac{1}{A_{Fiber}} \tag{3.20}$$

Schrier and Verseput (1967) found that the geometric mean of MD and CD taber stiffness was independent of changes in strength orientation. de Ruvo et al (1976) showed that the geometric mean of the elastic modulus in MD and CD of an oriented sheet coincided with the elstic modulus of an isotropic hand sheet made of the same raw material and pressed to the same density. The same relation was also true for the hygroexpansion. Htun and Fellers (1982) separated the effect of drying restraint and fibre orientation. They showed that, for a given drying condition, the geometric mean value of MD and CD properties is possible to use as an invariant measure of mechanical properties of paper. However if the value is to be regarded as an isotropic quality, the drying conditions must be the same in all sheet directions. Consequently the isotropic value can be assumed to be equal to the geometric mean of MD and CD according to Equation 3.21 and Equation 3.22 and also constant for different fibre anisotropies. It has lately been shown that the geometric mean assumption is valid also for the free shrinkage strain according to Equation 3.23.

$$E^r_{Iso} = \sqrt{E^r_{MD} \cdot E^r_{CD}} \tag{3.21}$$

$$E^{fs}_{Iso} = \sqrt{E^{fs}_{MD} \cdot E^{fs}_{CD}} \tag{3.22}$$

$$\varepsilon^{fs}_{Iso} = \sqrt{\varepsilon^{fs}_{MD} \cdot \varepsilon^{fs}_{CD}} \tag{3.23}$$

In *Figure 3.36 –3.38*, the validity of the geometric mean assumption is shown. Measurements on isotropic hand sheets (same as in *Figure 3.28*) were used as input to the simulations (solid lines). The restrained dried stiffness, E^r, was 5,5 MNm/kg, the free dried stiffness, E^{fs}, 2,0 MNm/kg and the free shrinkage strain, ε^{fs}, –6,0 %. The fibre anisotropy was varied between 1 and 4.

Figure 3.36 shows that the anisotropic behaviour of the tensile stiffness index for restrained dried sheets can be simulated based on the isotropic hand sheet data. The solid line represents the MD and CD tensile stiffness index, according to Equation 3.15 and Equation 3.21 with the anisotropy varied from 1 to 4. The x-mark represents the isotropic input and the open and closed circles the experimental results in MD and CD for anisotropic hand sheets made of the same pulp as the isotropic hand sheet.

Figure 3.37 shows that the anisotropic behaviour of the tensile stiffness index for free dried sheets can be simulated based on the isotropic hand sheet. The solid line represents the MD and CD tensile stiffness index, according to Equation 3.16, Equation 3.19 and Equation 3.22 with the anisotropy varied from 1 to 4. The anisotropy is the stiffness anisotropy for restrained dried sheets according to Equation 3.15. The x-mark represents the isotropic input and the open and closed circles the experimental results in MD and CD for anisotropic hand sheets made of the same pulp as the isotropic hand sheet.

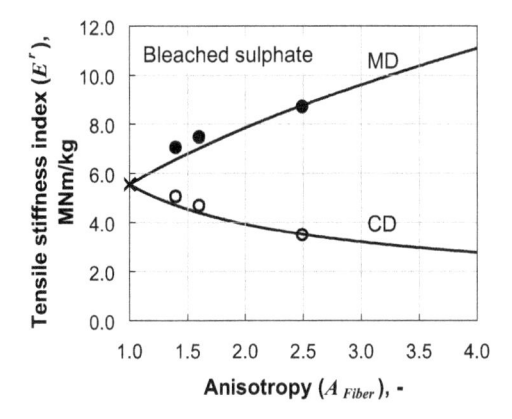

Figure 3.36. Experimental and simulated stiffness relations between restrained dried isotropic and anisotropic hand sheets.

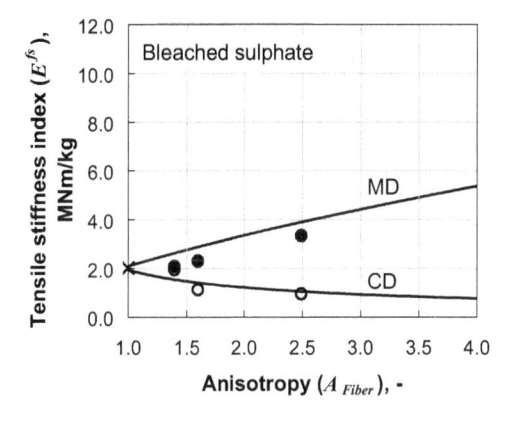

Figure 3.37. Experimental and simulated stiffness relations between free dried isotropic and anisotropic hand sheets.

Figure 3.38 shows that the anisotropic behaviour of the free shrinkage strain can be simulated based on the isotropic hand sheet data. The solid line represents the MD and CD free shrinkage strain, according to Equation 3.17 and Equation 3.23 with the anisotropy varied from 1 to 4. The anisotropy is the stiffness anisotropy for restrained dried sheets according to Equation 3.15. The x-mark represents the isotropic input and the open and closed circles the experimental results in MD and CD for anisotropic hand sheets made of the same pulp as the isotropic hand sheet.

Predicting anisotropic behaviour from isotropic hand-sheets. With isotropic hand-sheet data and Equations 3.9–3.23, the anisotropic behaviour for an arbitrary total strain accumulated during drying and anisotropy can be simulated based on isotropic hand sheet data. This means that the relations between stiffness and total strain accumulated during drying presented in many places in this chapter can be extended to the anisotropic case. Equation 3.9 was shown in *Figure 3.28* to be valid for isotropic hand sheets. Wahlström and Fellers (1999) and Mäkelä (2003) has shown a linear relation between total strain accumulated during drying and tensile stiffness index also for anisotropic paper, which makes Equation 3.9 valid also for MD and CD data.

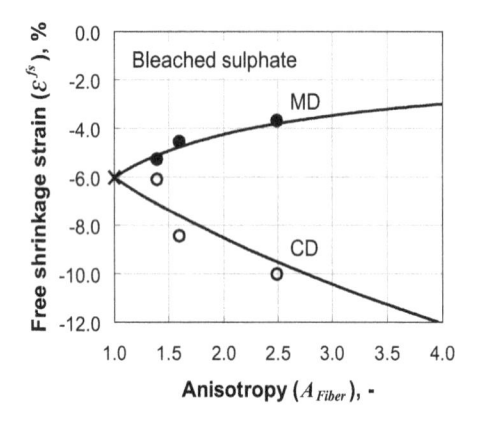

Figure 3.38. Experimental and simulated free shrinkage strain relations for isotropic and anisotropic hand sheets.

Figure 3.39 shows the simulated relations, using Equations 3.9–3.23, between tensile stiffness index in MD and CD versus total strain in the respective direction for the anisotropy 1, 2, 3 and 4. Input to the simulation was the isotropic data for the bleached sulphate hand-sheets that also was used as input to *Figure 3.28*. The restrained dried stiffness was 5.5 MNm/kg, the free

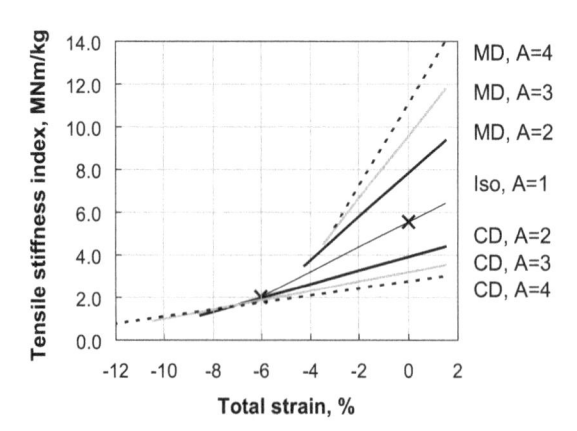

Figure 3.39. Simulated relations between tensile stiffness index in MD and CD and total strain in MD or CD for the anisotropy 1, 2, 3 and 4.

dried stiffness 2.0 MNm/kg and the free shrinkage strain −6.0 %, this data is also x-marked in *Figure 3.39*. The same result could also be retrieved from the relations in *Figures 3.36–3.38* by taking the stiffnesses and free shrinkage for the chosen anisotropies.

3.4.5 Pressing and Drying

Figure 3.40 shows a comparison in stiffness between hand sheets made by an unbleached kraft pulp (27 °SR) and an unbleached CTMP pulp (250 CSF). The two pulps were pressed in a laboratory platen press to different densities, as indicated in the figure, and dried free and restrained. An interesting difference between the behaviour of the kraft and CTMP made paper is that the free shrinkage of the kraft paper is unchanged, but increases for the CTMP paper. Probably the bondings between the fibres were favoured by the densification of the wet CTMP paper and the shrinkage of the fibres could better be transferred to the fibre network. The wet kraft paper already had bonding good enough for transfer of the fibre shrinkage, but still the stiffness was improved as expected. Note that the kraft pulp is not the same as in *Figure 3.31* although similar, but the CTMP pulp is (250 CSF).

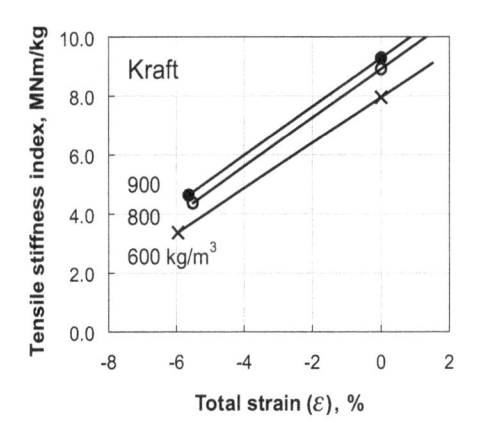

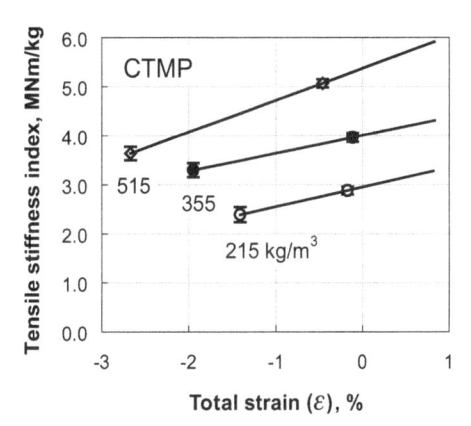

Figure 3.40. Stiffness for an unbleached kraft and an unbleached CTMP pulp, pressed to different density, versus total strain accumulated during drying.

3.5 Machine Level – Moving Wide Web

3.5.1 Cylinder Drying

Drying phases. The drying of paper is a dynamic process. Repeatedly and with very short time cycles, heat is transferred from the drying cylinder and water is evaporated in the free draw. The dynamic nature and its importance for the paper properties explain the difficulties in making use of data from static laboratory trials. The key to advancing the knowledge about drying, the last unexploited unit process in papermaking, is to study separately the sub-processes, the different drying phases.

A single felted, or single tier, dryer section with vacuum rolls, Valmet VacRolls, is here used as an example, to investigate the influence of the separate drying phases on paper properties. *Figure 3.41* shows one cycle of the dryer section divided into the different drying phases. From A to B, the paper is located *on the dryer cylinder* under the dryer fabric. The evaporation has been simulated with a physical model for heat and mass transfer developed by Wilhelmsson (1995).

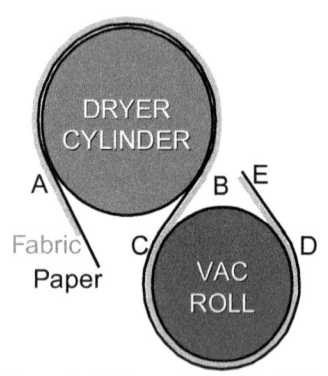

Figure 3.41. One cycle of a single felted dryer section divided into its different drying phases.

Figure 3.42 shows a simulation of how the temperature develops over the drying phases in one cycle. On the dryer cylinder, the temperature increases, particularly on the side of the paper towards the dryer cylinder. The temperature becomes high enough for evaporation to start, but the evaporation is reduced by the dryer fabric. *Figure 3.43* shows a simulation of the local evaporation rate, based on paper area, in one cycle. Evaporation is low but not negligible; about 20 % of the total evaporation occurs on the dryer cylinder. From B to C and from D to E the paper runs between the cylinders *in the free draw*. When the paper reaches the free draw the energy built up over the dryer cylinder leaves the paper as vapour. Initially the evaporation rate is very high but it decreases rapidly as the paper temperature decreases. About 40 % of the evaporation can take place in this phase. The remaining part of the evaporation takes place from C to D *on the VacRoll*. The dryer fabric is located between the paper and the VacRoll and does not therefore reduce the evaporation as it does on the dryer cylinder.

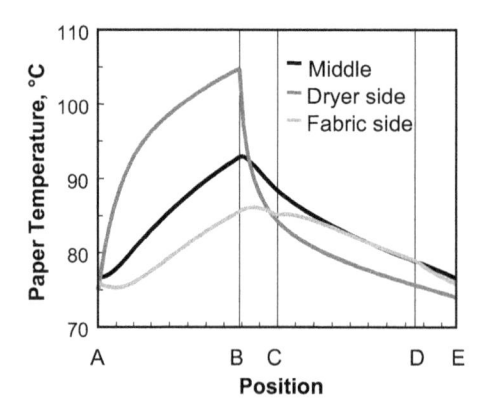

Figure 3.42. Simulated temperature in a paper passing through a drying cycle as in *Figure 3.41*.

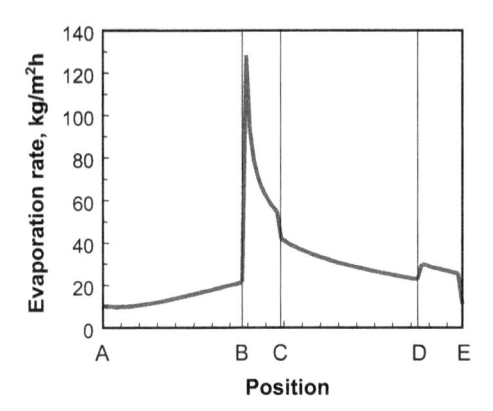

Figure 3.43. Simulated evaporation for a paper passing through a drying cycle as in *Figure 3.41*.

Phase 1; on the dryer cylinder. The first drying phase considered is when the paper is on the dryer cylinder and under the dryer fabric. If the paper is exposed only to this phase during drying, the relevance of this phase for the paper properties can be investigated. *Figure 3.44* and *Figure 3.45* shows results from a trial where 100 and 300 g/m^2 testliners were dried over a 1.83 m dryer cylinder in a single pass. The results for 0 kN/m were measured separately with the drying strategy free shrinkage. The pressure produced on the paper by the dryer fabric increases the heat transfer from cylinder to paper, but it also restrains the shrinkage of the paper. When the fabric tension is increased, the shrinkage is reduced and the final stiffness of the liner increases. Compared to free shrinkage, the impact on shrinkage and stiffness is high, but in the region of fabric tensions used on a production machine, 2–3 kN/m, the changes in shrinkage and stiffness were small. For grades that require high bulk, it is not possible to use a high fabric tension because of the densifying effect. Too low a fabric tension leads to a deterioration in surface properties and reduces the heat transfer.

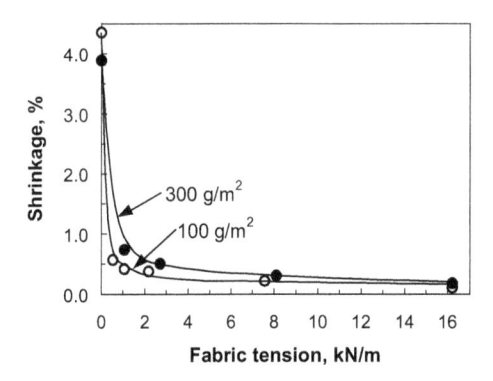

Figure 3.44. The shrinkage decreases with increasing fabric tension.

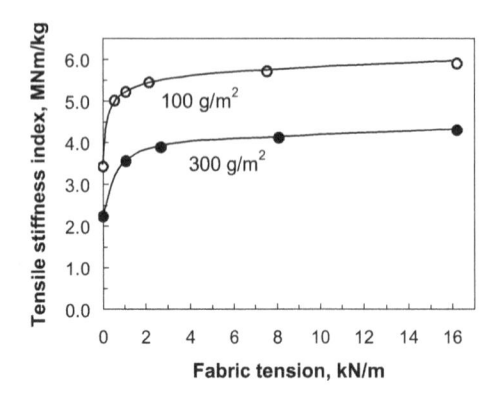

Figure 3.45. The stiffness increases with increasing fabric tension.

Phase 2; in the free draw. Hitherto, no attempts have been made to increase the understanding of the physics behind the shrinkage profile. With the finite element method, it has been possible to simulate stresses and strains in a paper web passing through a dryer section. The simulations have been aimed at investigating how the conditions in the free draw influence the shrinkage profile. The model was developed for a general orthotropic behaviour of the paper and the kinds of relations presented in *Figure 3.24*, *Figure 3.27*, *Figure 3.28* were used as input. The total strain in the paper was assumed to be the sum of a mechanical strain and a hygroscopic strain due to water evaporation. A mechanical strain can for example be due to a speed increase in MD between two dryer groups and the resulting contraction in CD. The hygroscopic strain is identical to the free shrinkage. It was assumed that when the paper web is on the cylinders it is restrained from shrinkage by the pressure from the dryer fabric. In the free draw between the dryer cylinders there are no forces acting on the edges of the web.

To validate the model, the shrinkage profile was measured on two paper machines and their respective free draw geometries simulated. *Figure 3.46* shows the measured shrinkage profiles. The dryer section with the double-felted configuration was 6.5 m wide and had a free draw

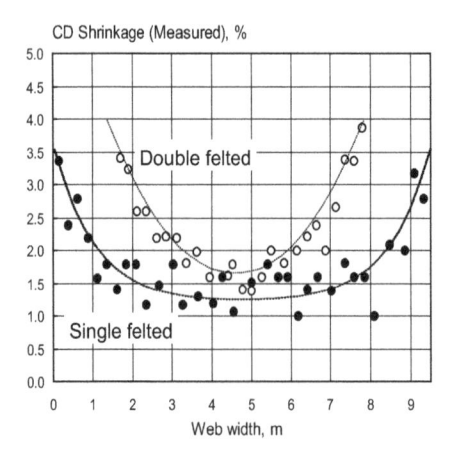

Figure 3.46. Measured shrinkage profiles for a single felted and a double felted dryer section.

length of 2.3 m. The free draw in double felting was defined as the length where the web is not in contact with the dryer fabric. In the single-felted case, the free draw was 0.9 m and the web width 9.5 m. *Figure 3.47* shows the results of the simulations of these dryer sections. There are some deviations in the absolute level of shrinkage and too low a prediction in the middle of the web. The model was, however, able to capture the general behaviour of the shrinkage and it gives a qualitative understanding of the influence of the studied variables.

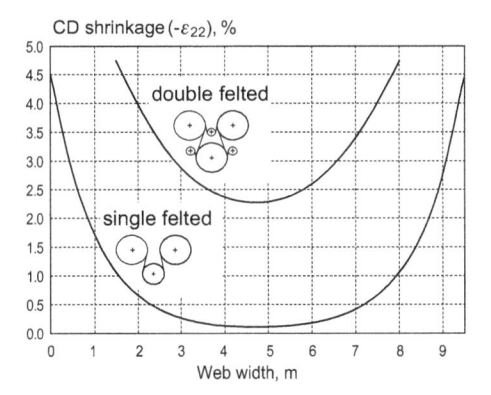

Figure 3.47. Simulated shrinkage profiles for a single felted and a double felted dryer section.

Figure 3.48 shows the results of a series of simulations where the length of the free draw was varied. When the length of the free draw was reduced, the shrinkage in the middle of the paper web decreased significantly. At the edges, the shrinkage also decreased but to a smaller extent. The region in the middle of the web with low shrinkage also became wider when the free draw was shortened, which means that the shrinkage gradient at the edges of the web became steeper. A shorter free draw reduced the total web width shrinkage, but the difference in shrinkage between the edge and the middle increased. This also means that the difference in most paper properties between the edge and the middle would have increased with a reduction in the length of the free draw.

The origin of the shrinkage profile can be explained by a difference in stresses. In CD, the stress became much higher in the middle of the web than at the edge where it was close to zero. The explanation of the higher stress was the boundary conditions, free edges in the free draw and a fixed situation when the web was under the fabric or on the VacRoll. The high stress in the middle of the web restrains the shrinkage more in the middle than at the edge. *Figure 3.49* shows the calculated CD stress fields in the final free draw for the simulated cases in *Figure 3.19*. The shortest free draw had the highest CD stresses and consequently also the lowest shrinkage.

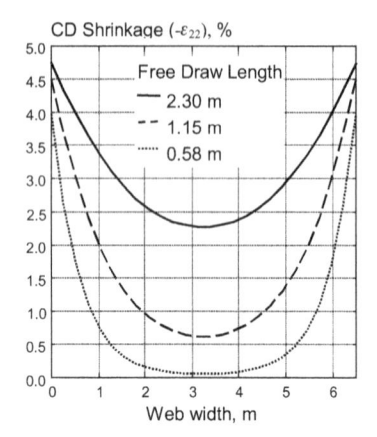

Figure 3.48. Simulated shrinkage profiles for three different lengths of the free draw.

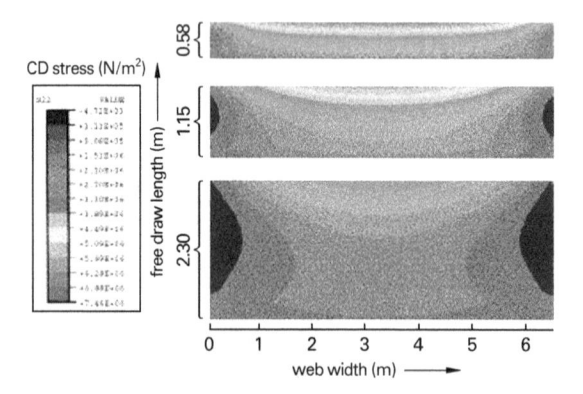

Figure 3.49. Simulated CD stresses in the final free draw for three different free draw lengths.

Phase 3; on the VacRoll. On the VacRoll, the web is pressed against the fabric by the vacuum inside the roll. This restrains the shrinkage but, as will be shown, not as effectively as the restraining effect of the dryer fabric over the dryer cylinder. To increase the understanding of this drying phase, a pilot trial was carried out. A pilot-machine-made 200 g/m² testliner was dried over a VacRoll in a single pass by blowing hot air onto the paper.

Figure 3.50 shows how the vacuum influences the development of the shrinkage during drying. The total web width shrinkage is highest without any vacuum in the VacRoll. The shrinkage starts at the same moisture ratio, around 0.9 kg/kg, in both cases. In the trial with vacuum, the paper was dried in a more restrained state and the final level of shrinkage was lower. The free shrinkage for this grade was 3.8 %. Since the shrinkage was about 2.6 % when the vacuum was zero, some restraint was obviously imposed on the paper even in this case.

If the restraint is caused by stress gradients, it can be suspected that this drying phase also gives rise to a shrinkage profile. *Figure 3.51* shows the shrinkage profiles measured on the dry 200 g/m² liner for three different levels of vacuum, 0, 1.1 and 4.8 kPa. With the highest vacuum, the shrinkage in the middle of the web is close to zero but it is as high as 2.6 % at the edge.

The profile for the paper dried without vacuum shows a more even profile than the high vacuum case but the total web width shrinkage in higher. The difference in shrinkage between the middle and edge of the web is greatest for the trial with the highest vacuum. As is indicated by laboratory data in this paper, this also means that the difference in paper properties between edge and middle was largest in this case.

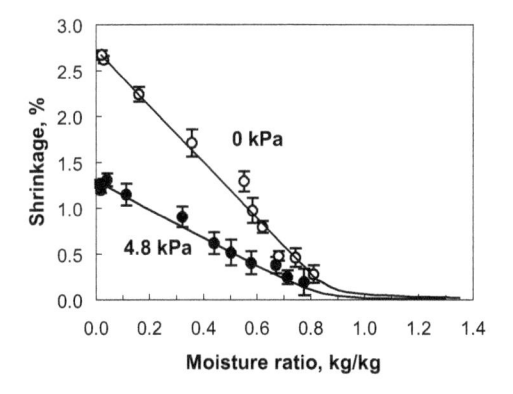

Figure 3.50. Development of shrinkage during drying with and without vacuum.

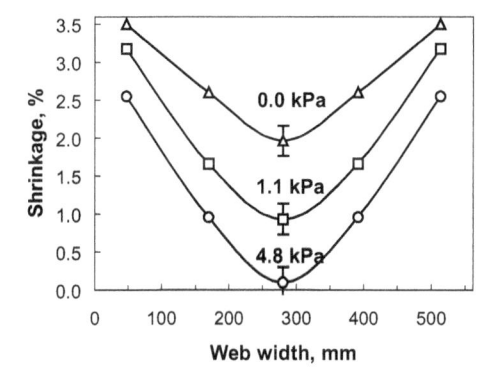

Figure 3.51. Shrinkage profiles for different levels of vacuum in the VacRoll.

3.5.2 Drying Concepts

Conventional double-felted drying section. The effect of cylinder drying on the mechanical properties of paper is very dependent on the shrinkage or stretch during the drying process. *Figure 3.52* shows a conventional double felted dryer section. *Figure 3.46* and *Figure 3.47* show the results from both a measurement and a simulation of the shrinkage in the cross machine direction of another double-felted dryer section. The free draw length in this case was 2.3 m and the web width 6.5 m. This and many other double-felted dryer sections in use have relatively long free draws, compared to new installations and in relation to the width of the machine.

The shrinkage in the cross direction of the paper machine is greater at the edges than at the middle of the web. This difference also creates a great difference in mechanical properties, as shown in the laboratory data presented in this paper. The simulations shows that the shrinkage profile is created by the situation with free edges of the web in the free draw and the restrained conditions under the dryer fabric.

Single-felted dryer section. In a single-felted dryer section, all the rolls at the top are steam-heated cylinders and all the rolls at the bottom are vacuum rolls (VacRolls), *Figure 3.53*. The sheet run is totally closed and supported through the entire dryer section for improved runnability. The total web width shrinkage is less than that in a conventional dryer section.

Historically, the speed and the width of the paper machine has increased to increase production, and the free draws have become shorter to improve runnability. *Figure 3.46* and *Figure 3.47* show how the shrinkage profile changes when the web width increases and the length of the free draw decreases. In the middle of the web, the shrinkage decreases significantly. Consequently a wider machine and shorter free draws reduce the total web width shrinkage and thereby increases the mean stiffness. The difference in shrinkage between the edges and the middle of the web increases however, which means that the differences in most paper properties between the edges and the middle also increase. The results presented in *Figure 3.50* show that a part of the shrinkage profile is also an effect of the conditions on the VacRolls.

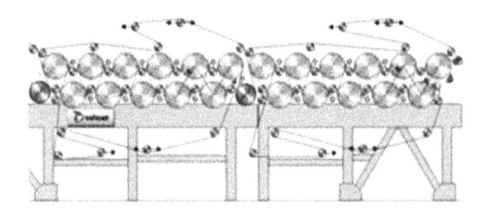

Figure 3.52. Conventional double-felted dryer section.

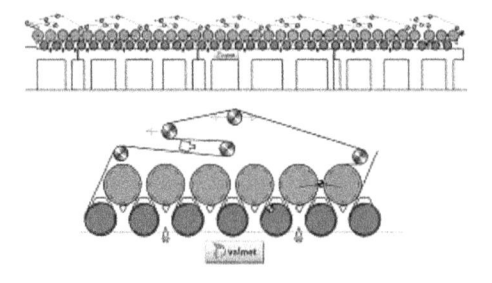

Figure 3.53. Single-felted dryer section.

Impingement drying. The most recently developed and commercialised drying process is the OptiDry impingement drying concept shown in *Figure 3.54*. The special features of the concept are the totally closed sheet transfer beginning in the press section, the use of two different forms of energy for heating, and a new high efficiency paper drying unit. The benefits are improved runnability, fast grade changes and a shorter dryer section. The concept consists typically of three air impingement modules added to a SymRun-type dryer section. On the impingement modules or OptiDry units the web is on top of the dryer fabric and is vacuum-supported by a large diameter VacRoll. The effect on paper properties does not differ significantly from that of the SymRun dryer section.

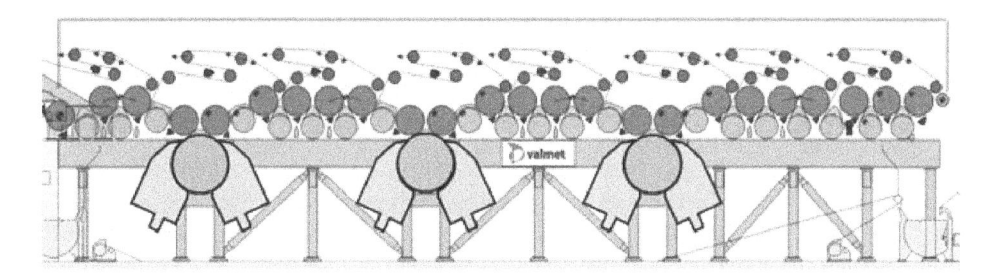

Figure 3.54. OptiDry impingement drying concept.

Condebelt drying. There are currently two board mills utilising the Condebelt process, the Pankakoski mill of Stora Enso in Finland, shown in *Figure 3.55*, and the Ansan Mill of Dong Il in South-Korea. In the Condebelt process, the wet web is carried on two permeable wires and fed between two long smooth steel belts. The web is dried by being under pressure (0.5 to 5 bar) and in contact with the upper steam-heated (111–159 °C) belt. As the moisture in the web evaporates, the vapour gene-rated passes through the wires and condenses on the lower cooler steel belt (Retulainen 1999).

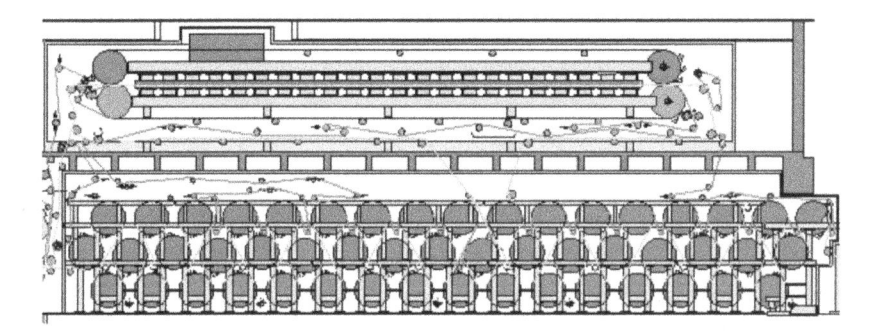

Figure 3.55. Condebelt

The z-directional pressure and the contact with the glossy, hot metal belt has two major consequences with regard to sheet properties. First, the pressure applied to the hot and moist web acts to plasticize fibres, improve bonding, increase density and create a smooth surface. Second, the pressure prevents all shrinkage during drying, which has earlier been shown to have a large effect on the paper properties in CD. The totally prevented shrinkage in the Condebelt unit, of course, also eliminates the shrinkage profile. *Figure 3.29* showed that a reduction in the shrinkage leads to a reduction in z-toughness. In the Condebelt process, this drawback is more than compensated for by the increase in density of the sheet.

Air-borne drying. In an Airborne dryer, the paper is dried by hot air blown towards the paper web by nozzles. *Figure 3.56* shows an Airborne dryer. The very long free draws extend from the left-hand to the right-hand side of the dryer. On each side, the web is turned on turning rolls. In the free draws, the air blown by the nozzles creates an air cushion that carries the web.

Figure 3.48 shows how the length of the free draw influenced the shrinkage profile in the paper. These results can be extrapolated towards very long free draws, which give a higher total shrinkage and a more even shrinkage profile. Only a small force is needed to pull the web through the dryer, and this gives minimum web tension. The very low web tension also allows for shrinkage in the machine direction. In a conventional cylinder dryer section, a small stretch is always necessary between the dryer groups to give sufficient web tension for good runnability.

Figure 3.28 showed that the tensile stiffness decreases when the shrinkage increases. This means that the strain at break, that is the extent to which it is possible to stretch the paper before it breaks, increases. This makes the airborne dryer suitable for the production of sack paper that requires a high strain at break. Most installations of this type of dryer section are however made for pulp drying.

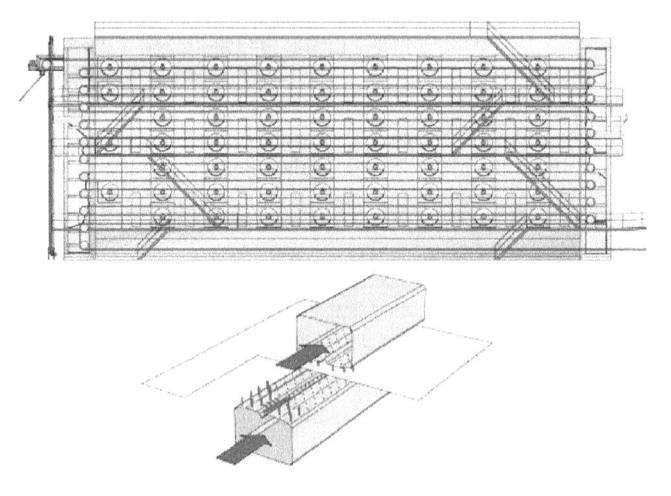

Figure 3.56. Airborne dryer.

Air Impingement Drying for Sack Paper. The increasing restraint in the dryer section due to shorter free draws and wider machines makes it harder to produce sack paper of good quality in a modern conventional dryer section. Sack paper requires high shrinkage, which is beneficial for producing a stretchable material. There are ways to increase the shrinkage in a conventional cylinder dryers section. Figure 3.57 shows an installation of impingement drying hoods in a conventional dryer.

Hot air is blown onto the web when it passes over a dryer cylinder. The web is not restrained from shrinkage by the fabric on these particular dryer cylinders, and this explains the increased shrinkage. The length of the free draw can then be considered as the length where the web is not restrained by a dryer fabric. Since the evaporation capacity is high in impingement drying, a large part of the drying takes place in this long „free draw".

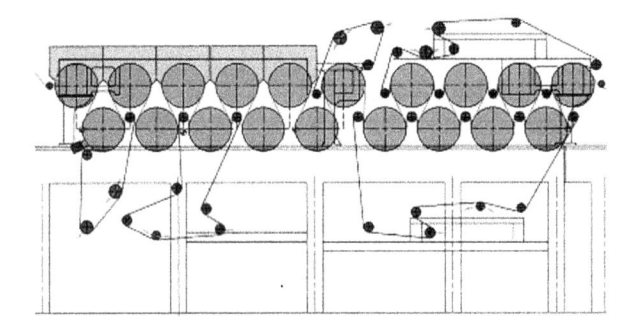

Figure 3.57. Air impingement drying for sack paper.

3.6 References

Htun, M., and C. Fellers. (1982) The invariant mechanical properties of oriented hand sheets. *Tappi Journal*, 65 (4): 113–117.

Maloney, T. C. (2000) How the Structure of the Cell Wall Influences Press and Dryer Dewatering. *Proceedings of the 12th International Drying Symposium IDS2000*, Elsevier Science, Amsterdam, Paper No.296.

Nanko, H., and J. Wu (1995). Mechanisms of paper shrinkage during drying. *International Paper Physics Conference*, Niagara-on-the-Lake, Canada, pp. 103–113.

Wahlström, T., and C. Fellers. (1999) Biaxial straining of paper during drying, relations between stresses, strains and properties. *TAPPI Engineering Conference, September 12–16, 1999*, Anaheim, CA, pp. 705–720.

Wahlström, T., K. Adolfsson, S. Östlund, and C. Fellers (1999) Numerical modelling of the cross direction shrinkage profile in a drying section, a first approach. *1999 International Paper Physics Conference, September 26–30*, San Diego, CA, pp. 517–531.

Wahlström, T., and C. Fellers (2000) Biaxial straining of hand sheets during drying- effect on in-plane mechanical properties. *Tappi Journal*, 83 (8): 91.

Weise, U., T. Maloney and H. Paulapuro (1996) Quantification of water in different states of interaction with wood pulp fibres. *Cellulose* 3(4): 189–202.

4 The Interaction of Paper with Water Vapour

Christer Fellers
STFI-Packforsk AB, Stockholm

4.1 Introduction 109

4.2 The Moisture Content of the Air 113

4.3 Climatology 114

4.3.1 Moisture Influences on Papermaking and Converting 115

4.4 Sorption Isotherms and Hysteresis 116

4.4.1 Desorption 119

4.5 Hygroexpansion 120

4.6 Permanent Dimensional Changes during Moisture Cycling 128

4.7 Measurement Methods 130

4.8 Curl and Twist 133

4.9 The Mechanical Properties at Different Moisture Contents 141

4.10 Mechanosorptive Creep 143

4.1 Introduction

The equilibrium moisture content of paper changes if the moisture content of the surrounding air changes and as a result the dimensions of the paper changes. Moisture can also be added in printing processes or other end-use situations. These changes can lead to great practical problems. For example, it cannot be tolerated that the paper changes its dimensions too much during the printing of the paper. A so-called register error occurs in multi-colour printing, where the colours are printed one at a time. Small errors can be adjusted during printing.

Another serious error is when a paper moves so that it deviates from the plane state. The paper exhibits curl and twist. Curl and twist are often the reason for failure in the sheet feeding of paper and carton board in printing presses and in the filling of packages in packing machines.

When adding moisture to paper, the result may also be thickness swelling, surface changes, and change in paper stiffness and strength.

Figure 4.1. When adsorbing moisture, paper deform in different ways.

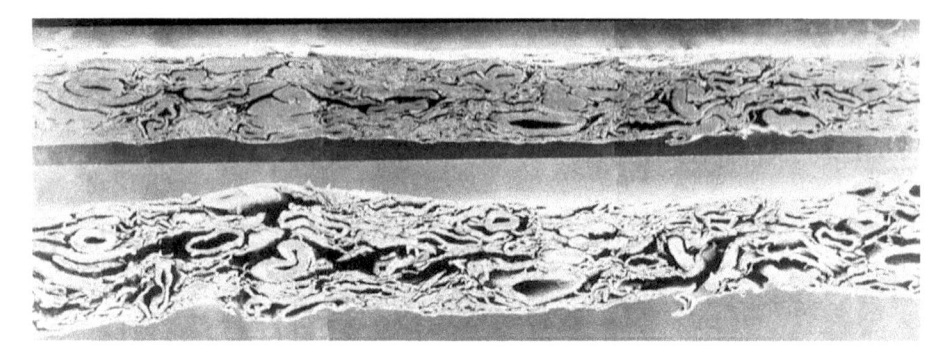

Figure 4.2. When adsorbing moisture, paper may experience thickness swelling and surface changes. Effect of wetting on calendered paper that contains mechanical pulp. Cross-sections of SC magazine paper (super calendered). Wetting by water 30 min and drying at room temperature.

When water vapour access to a paper-air system, the vapour diffuses into the structure. This process is called sorption. Throughout the papermaking process, the interaction between the paper fibres and water plays a major role. The amount of water vapour sorbed are expressed as moisture ratio, moisture content or dry solids content.

- Fibre concentration C
- Dry solids content DC
- Moisture ratio MR
- Moisture content MC

The concept of *fibre concentration* is used to indicate the amount of fibre in a suspension. In the paper machine, from the dry line in the wire section and forwards in the process, the quantity is usually called *dry solids content (DSC)*.

$$C(\%) = DSC(\%) = \frac{kg\ dry\ solids}{kg\ dry\ solids + kg\ water} \cdot 100 \tag{4.1}$$

The amount of moisture in the paper during its passage through the paper machine is often given as the *moisture ratio (MR)*, the relationship between the amount of water and the amount of dry substance. The moisture ratio is a linear measure of the water content in the sheet.

$$MR = \frac{kg\ water}{kg\ dry\ solids} \tag{4.2}$$

Moisture content (MC) is defined as the amount of water in relation to the amount of dry substance plus water. The concept of moisture content is appropriate to a finished paper.

$$MC(\%) = \frac{kg\ water}{kg\ dry\ solids + kg\ water} \cdot 100 \tag{4.3}$$

The relations between the different terms are given below.

$$DSC(\%) = 100 - MC(\%) \tag{4.4}$$

$$MR = \frac{MC(\%)}{100 - MC(\%)} \tag{4.5}$$

$$MC(\%) = \frac{MR}{MR+1} \cdot 100 \tag{4.6}$$

$$MR(\%) = \frac{100 - DSC(\%)}{DSC(\%)} \tag{4.7}$$

$$DSC(\%) = \frac{1}{MR+1} \cdot 100 \tag{4.8}$$

In general terms, diffusion is the tendency for a concentration gradient to be evened out by molecular movements. The rate of diffusion and the time to attain equilibrium depend on the initial moisture content of the paper, on the relative humidity of the air, on the thickness of the material and on the diffusion coefficient. *Figure 4.3* shows how fast different papers adsorb moisture.

If the relative humidity is cycled in a regular fashion as in *Figure 4.4a*, the corresponding moisture content versus RH-curve shows hysteresis (*Figure 4.4b*). This is a partly a consequence of the finite time for sorption to reach equilibrium.

In an ideal system, the properties of the mixtures are proportional to the concentrations; no heat release occurs and the processes are reversible. None of this is true for water vapour and fibres.

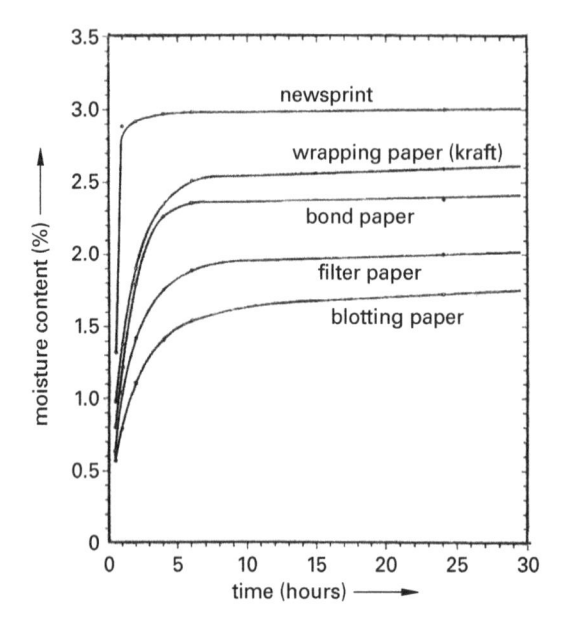

Figure 4.3. Rate of moisture adsorption from 35 % RH to 65 % RH.

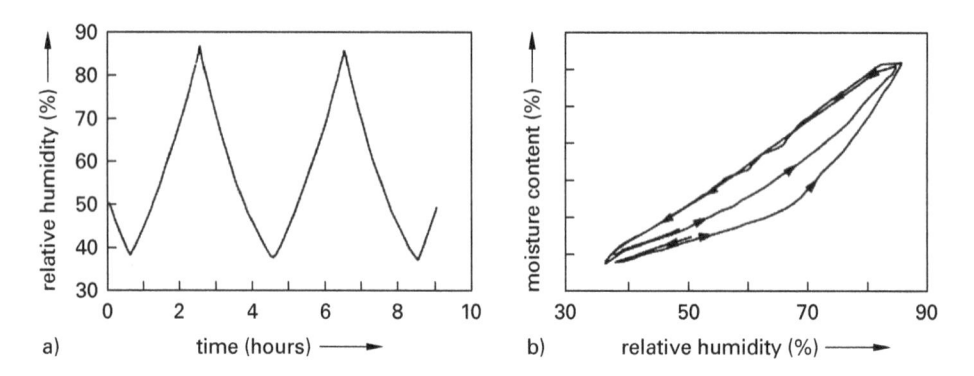

Figure 4.4. If the relative humidity is cycled in a regular fashion as in a), the corresponding moisture content versus RH-curve shows hysteresis as in b).

When moisture diffuses into dry paper fibres from wood, three things mainly happen:

- heat is released
- the weight increases
- the fibres swell

In this chapter, we shall describe more closely the phenomena that occur when paper fibres sorb water from air at different relative humidity.

4.2 The Moisture Content of the Air

The properties of paper depend to a large degree on the amount of moisture of the paper. The thermodynamic quantity determining the moisture content of paper is *the chemical activity of the vapour, which is equal to the relative humidity of the air*. The concept of Relative Humidity, usually abbreviated as RH is explained below.

Changes in moist air are often calculated with the help of a Mollier-diagram. A Mollier diagram has an angular coordinate system with axes representing enthalpy, h, and water content in the form of water vapour.

In this chapter, calculations relating to the interaction of paper with water vapour, a simplified Mollier-diagram will be used, *Figure 4.5*. An additional scale below in the diagram indicates the partial pressure of the vapour. Note that the temperature interval is chosen for room conditions.

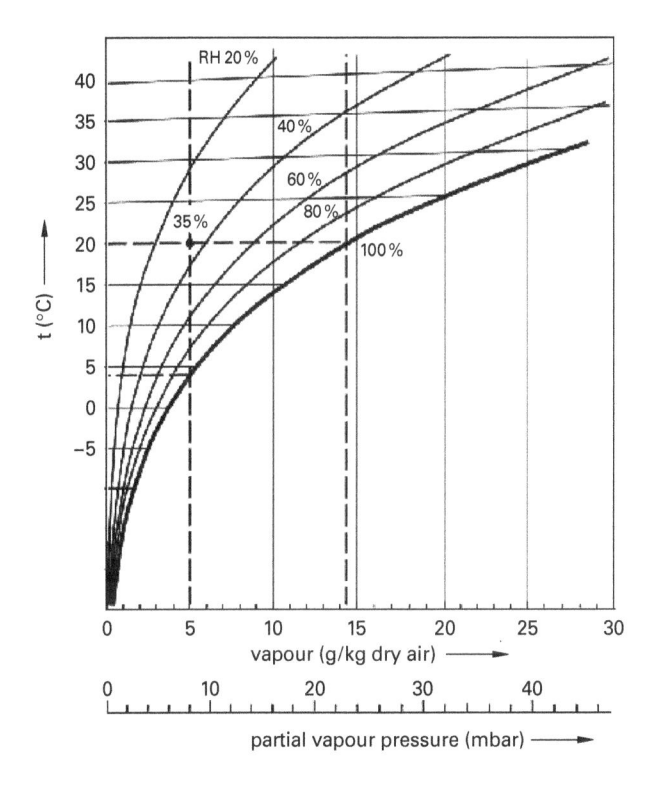

Figure 4.5. Mollier-diagram for moist air. (Ekroth, Granryd: Applied thermodynamics. Dep. for Energy Technology, KTH).

Air usually contains a certain amount of water, in practice 0–30 g water per kg dry air. The absolute amount of water at any temperature cannot exceed a maximum value, given by the saturation curve, which gives the maximum vapour pressure at each temperature.

For each temperature and water content, the relative humidity RH is calculated strictly with the help of Equation 4.9 as the partial vapour pressure divided by the maximum vapour pressure.

$$RH = \frac{partial\ vapour\ pressure}{maximum\ vapour\ pressure} \qquad (4.9)$$

The RH can also be expressed approximately as the amount of water that the air contains in relation to the maximum quantity that it could hold at the given temperature, according to the saturation curve.

$$RH \approx \frac{kg\ vapour\ /\ kg\ dry\ air}{kg\ vapour\ /\ kg\ saturated\ air} \qquad (4.10)$$

The following example shows the use of the Mollier-diagram:
Assume that air at 20 °C contains 5 g water per kg dry air. The partial pressure of the vapour is then 8.1 mbar. From the diagram, the maximum is 14.8 g water per kg dry air at this temperature. The partial pressure of the vapour is then 23.37 mbar. The relative humidity, according to Equation 4.11, is thus:

$$RH = \frac{8{,}1}{23{,}37} = 34{,}6\% \qquad (4.11)$$

The approximate Equation 4.10 gives:

$$RH \approx \frac{5{,}0}{14{,}8} = 33{,}8\% \qquad (4.12)$$

If the temperature of the air containing 5 g water/kg dry air decreases to 4.1 °C, the limit for the maximum possible atmospheric humidity is reached. If the temperature decreases further, condensed water is formed. This point is called the dew point.

4.3 Climatology

The dimensions and physical properties of paper such as electric resistance, heat conductivity and mechanical properties depend on the temperature and on the amount of moisture in the paper. As paper is a hygroscopic material, its moisture content depends on the RH and the temperature of the surrounding air according to theories of sorption isotherms and hysteresis, deeper described in Section 4.4 below.

The RH varies greatly outdoors from place to place. Across the oceans, the RH is largely constant, about 80 per cent, whereas the RH in desert areas seldom exceeds 40 %. Examples of RH-variations during the year at different places are shown in *Figure 4.6*.

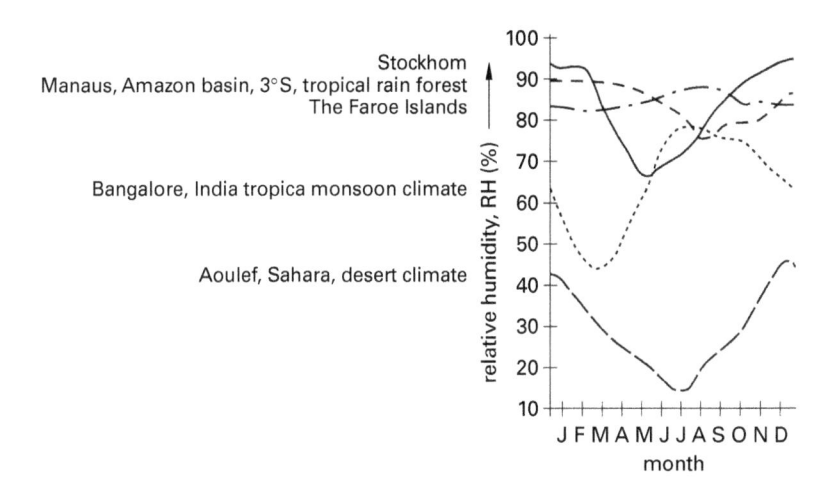

Figure 4.6. The annual variation in the relative humidity in different climate areas. Note that the climate indoors is completely unrelated to RH outdoors.

The relative atmospheric humidity also shows daily variations, due primarily to the temperature variations during the day. Note that RH indoors in each country is completely unrelated to RH outdoors.

4.3.1 Moisture Influences on Papermaking and Converting

Paper is normally manufactured with moisture content within the range of 6–9 %. The moisture content in paper is determined in practice by measuring the loss of weight of a paper when it is dried to constant weight at a temperature of 105 °C, in accordance with ISO 287 international standard. Usually papermakers want to supply paper with as high moisture content as possible to save drying heat. Paper is also sold according to weight and therefore the producers get paid even for the moisture in the paper.

However, it is easier to maintain an even dry content across the width of the paper machine if the dry content is high. For that reason over-drying is sometimes employed and the paper is then moistened again. As an example, over-drying is always done when producing qualities for supercalendering.

In printing houses, it is not desirable to have too high dryness in the paper, since the sheets can easily become charged with static electricity and stick together. Nor is too high moisture content in the paper beneficial, since the paper then becomes floppy and difficult to carry through the printing machines.

Indoors in paper mills, in storerooms, in transport vehicles etc, the RH will vary a lot as a consequence of changes in the humidity and temperature of the air. A printing house or a carton board factory must in their premises have an RH and temperature which are constant and they must work with paper that is in equilibrium with this atmosphere, therefore storage can be necessary before converting operations so that equilibrium will be reached.

For testing and classification of paper there is vital to take the climate variable in consideration. The standard conditions can vary from country to country. The practical problem, which the laboratories sometimes meet, is how to make it easy to control the climate in the room. It is expensive in warm countries to cool the air in the testing room. In countries with a warm and humid climate, it can also be difficult to dry the air.

4.4 Sorption Isotherms and Hysteresis

The usual way of studying moisture absorption in a material is to plot moisture content against RH at a given temperature, a sorption isotherm.

- The term *adsorption* and *desorption* designates water sorption on the surface of a material when moisture enters and leaving the material respectively.
- The term *absorption* is used for a more homogeneous water sorption inside the material.
- The limit between these two definitions is not sharp.
- *Sorption* is used as a general concept.

To measure the true isotherm of a material, the change in RH should theoretically take place in infinitely small steps and infinitely slowly over the whole interval of 0–100 % RH. This cannot, of course, be done in practice, so that each isotherm is a compromise depending on the choice of RH-interval and the time scale. It is therefore not immediately possible to compare isotherms of different origin without careful consideration.

It should be pointed out that experimentally, it is difficult to keep the RH stable at a high RH, so that sorption isotherms seldom include regions above 85–90 %.

When a porous material is exposed to water vapour, it can take a long time for the material to reach equilibrium through sorption. Air mixing plays a great role in reducing this time. This is probably the greatest reason why true isotherms are difficult to determine.

The diffusion of water into paper is an exothermal process, which means that heat is released. The differential heat of sorption is defined as the heat released when one gram of water is combined with an already partly water-filled fibre. The differential heat of sorption decreases with increasing moisture content.

For sorption of moisture to occur on cellulose fibres the free energy of sorption must be negative. The free energy is given by

$$\Delta G = \Delta H - T\Delta S \qquad (4.13)$$

where
H enthalpy (heat)
S entropy (order)
T absolute temperature.

The adsorption of moisture onto cellulose generally involves an ordering of the water molecules when they adsorb. Hence ΔS is positive and for adsorption to occur $-\Delta H$ must be larger than $T\Delta S$. This is indeed the case and sorption is therefore said to be enthalpy driven.

As mentioned earlier, the *RH is a measure of the chemical activity of the water* in the vapour phase. It is therefore the relative humidity of the air that determines the equilibrium moisture content of the paper at a given temperature, not the absolute moisture content of the surrounding air. This is shown with the adsorption isotherms for cotton in *Figure 4.7*.

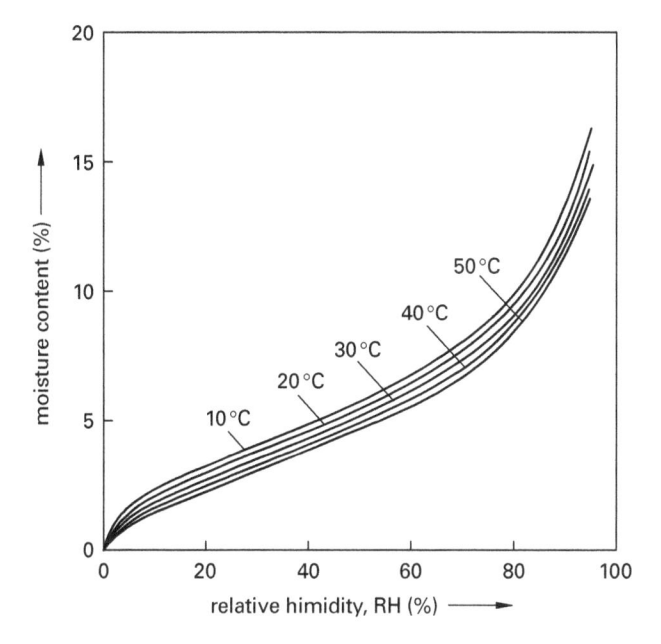

Figure 4.7. Adsorption isotherms at different temperatures for cotton based on relative humidity, RH, and based on the absolute moisture content of the air.

Although the absolute moisture content of the air varies by a power of ten between 10 and 50 °C at the same RH, the moisture content of the paper varies very little.

As a consequence of the fact that the adsorption of moisture into the cellulose fibre is an exothermal process, the cellulose will at a given RH have lower moisture content at a higher temperature. The reason why the sorption decreases with increasing temperature is that sorption is enthalpy driven. To maintain equilibrium it follows that if the temperature is higher the amount of adsorbed moisture must be lower.

(dry paper + water ⇔ paper in equilibrium + heat).

The curve shape in *Figure 4.7* is typical for the sorption of moisture in cellulose materials. In different parts of the curve, moisture sorption takes place by different mechanisms such as mono-layer sorption, multi-layer sorption and capillary condensation. These phenomena are treated below.

- *At low RH, surface adsorption* takes place as a mono-layer. The OH-groups of the cellulose molecule and of the hemicellulose molecule are considered to bind water through hydrogen bonds in a monomolecular layer on all available surfaces. As proof, it is usually stated that the first adsorbed water releases a large amount of heat (21 kJ/mole) and that this heat corresponds to the bonding energy of hydrogen bridges.

- *At a medium RH, surface adsorption* takes place as multi-layers. The behaviour of cellulose fibres with regard to sorption and swelling is similar to the behaviour of a gel. According to the gel swelling theory, absorption takes place in disordered areas in the cellulose, in the hemicellulose and in the lignin.
- *At very high RH levels,* moisture absorption takes place by *capillary condensation,* where larger capillaries in the fibre and paper structure are filled with liquid. The average size of the voids that are filled by water molecules can be described by the Kelvin Equation, *Figure 4.8* below.

$$\ln\left(\frac{P}{P_0}\right) = \frac{2\gamma M}{r\rho RT}$$

where γ = surface tension of water (0.072 N/m)
M = molecular mass of water ($18 \cdot 10^{-3}$ kg/mole)
r = radius of capillary (m)
ρ = density of the water (1000 kg/m³)

calculation based on the Kelvin equation at 25 °C

relative vapor pressure of water (p/p_0)	effective capillary radius (μm)
0.99999	106
0.9999	10.6
0.999	1.06
0.99	0.106
0.98	0.052
0.95	0.020
0.50	0.0015

Figure 4.8. The Kelvin Equation.

It must be pointed out that thermodynamic observations such as sorption and desorption isotherms can never explain how water is bound to the fibres. For this, other experiments are required e.g. with marked molecules or through the use of different spectroscopic methods.

The moisture content obviously depends on the chemical composition of the fibres in the paper, *Figure 4.9*.

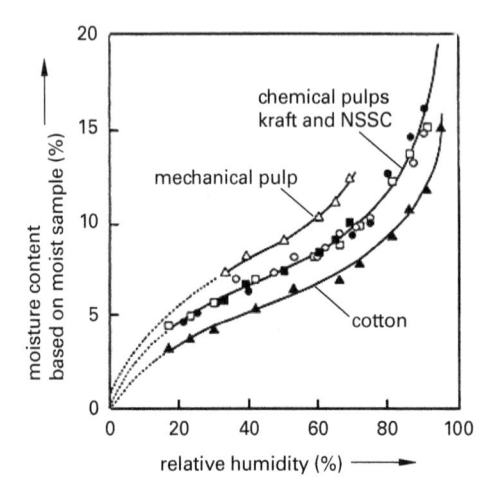

Figure 4.9. Adorption isotherms at 23 °C for papers made from different pulps.

4.4.1 Desorption

So far, the sorption process has been discussed. We shall now also discuss the reverse process, desorption. *Figure 4.10* shows three moisture sorption isotherms for paper. Curve 1 shows the desorption from a never-dried paper. Curve 2 shows the sorption of the paper up to 100 % RH. In the subsequent desorption according to curve 3, the paper has lower moisture content than in curve 1 at the same RH.

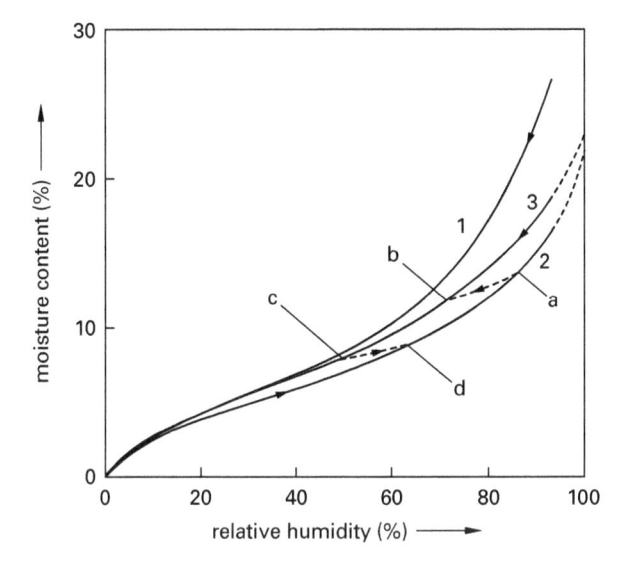

Figure 4.10. Typical moisture sorption isotherms for paper. The desorption curve (1) is for never-dried fibres, the sorption curve (2) is for the first moistening and the desorption curve (3) is for once-dried fibres.

It has been shown that a desorption curve always lies higher than the adsorption curve. The moisture absorption is thus not reversible. The process shows hysteresis.

The loop a-b-c-d shows approximately what happens if the adsorption stops at a, if the paper is then dried to b and c and then adsorbed to d and c again.

The inclination of the curve also becomes flatter, and it can be seen that the material becomes less and less hygroscopic. The physicist says that the material has been hornified. The effect can be considerable in recycled paper where the fibres have been dried many times. Porous fibreboard is made less sensitive to moisture and more dimensionally stable by being dried, hornified at a high temperature and thereafter carefully moistened. In paper testing, it is an advantage if the paper sample is conditioned from the dry side to the desired RH. All samples will then reach the given RH along the same isotherm.

The hysteresis effect occurs in other systems than cellulose-water, especially in those where long, more or less mobile molecules interact with the swelling medium.

There may be several reasons for sorption hysteresis. One important aspect is that during desorption the fibre gel is under tension and during adsorption the gel is under compression. From thermodynamics, it can be shown that a gel under tension has a higher vapour sorption and if the gel is under compression the sorption decreases.

4.5 Hygroexpansion

If the hygroexpansion is plotted against moisture ratio, moisture content or RH for a small change in the climate, linear relationships are obtained. It is then useful to define a property called hygroexpansion coefficient, which assumes that the increase in strain is linearly proportional to the RH change or change in moisture content, *Figure 4.11*.

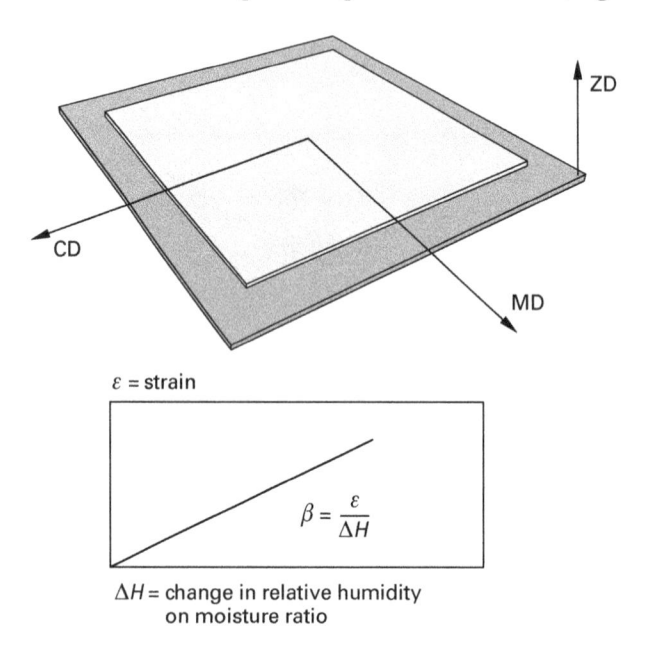

ε = strain

$$\beta = \frac{\varepsilon}{\Delta H}$$

ΔH = change in relative humidity on moisture ratio

Figure 4.11. Definition of the hygroexpansion coefficient β.

How can the moisture absorption affect the dimensions of a sheet? The reason is primarily the fibre's own swelling and the manner in which this swelling influences the shrinkage of the paper sheet on the paper machine. If the swelling of paper and wood during moistening is prevented by external limitations, large swelling forces arise. The effect can be used to blast stone through the use of wooden wedges that are moistened. When the paper shrinks, corresponding shrinkage forces occur. We shall below discuss the reasons for these effects.

Different pulps contain different amount of polymer components, *Table 4.1*, and are expected to swell differently due to the chemical composition.

Table 4.1. Relative chemical composition of some pulps.

	Relative composition		
Type of pulp	cellulose	hemicellulose	lignin
TMP	44.2	29.5	26.3
High yield kraft	59.5	19.1	21.4
Kraft	74.0	17.8	8.2
Bleached kraft	78.3	21.4	0.3

Figure 4.12 shows a comparison of hygroexpansion between different papers. No immediate conclusion if the chemical composition plays a role. It is however clear that drying conditions are important. In addition, beating and refining conditions as well as the chemical composition affect the shrinkage and consequently the fibre shape and the amount of fines.

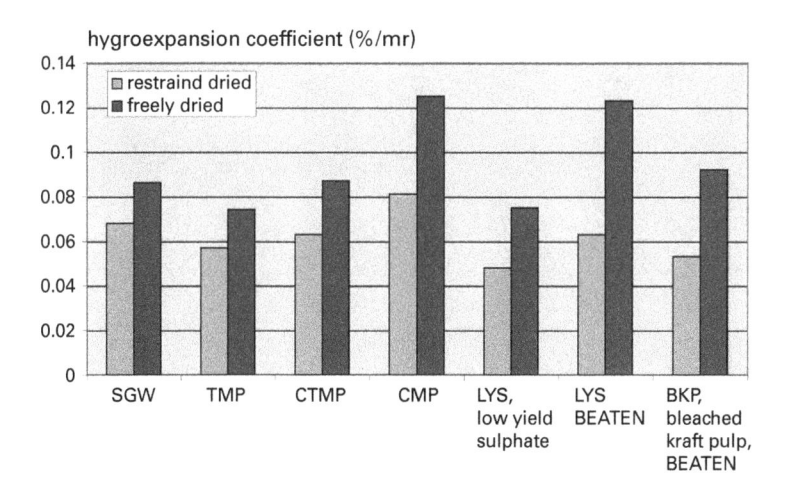

Figure 4.12. A comparison of hygroexpansion between different papers.

Swollen fibres shrink during the drying. They shrink approximately one per cent in the length direction while they shrink 20–30 per cent in their width and in the thickness direction. When they shrink, the fibres will attract each other in their bonding regions. *Figure 4.13* illustrates the changes in the sheet after restrained and free drying. If the paper is dried under restraint in a certain direction, fibre segments are stretched.

If the shrinkage takes place freely, the width shrinkage of the shrinking fibre contracts the crossing lengthwise fibre and micro-compressions are created.

A schematic representation of the development of micro-compressions during drying is shown in *Figure 4.14*.

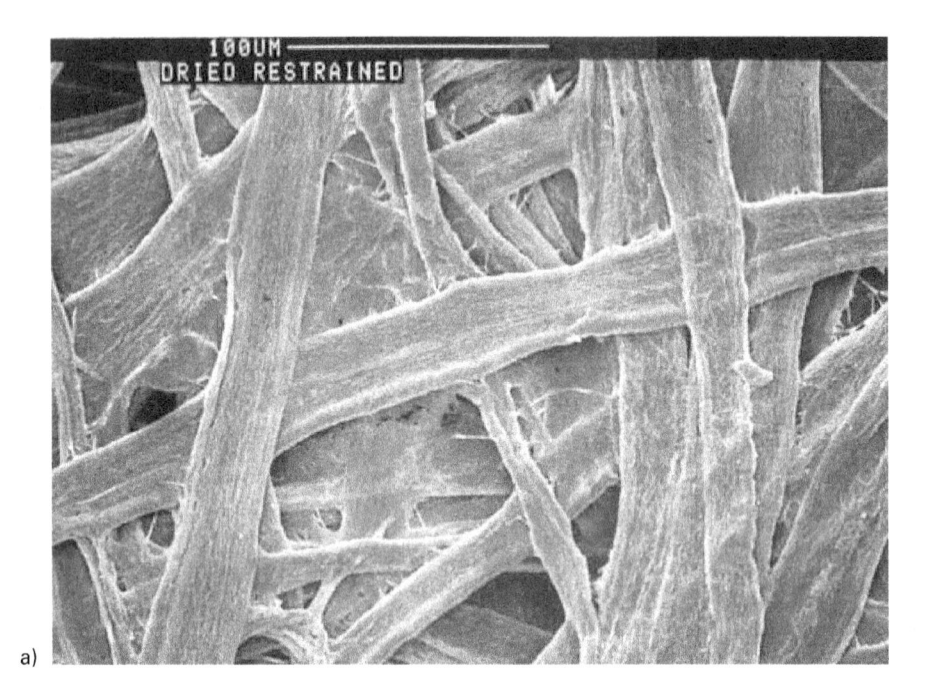

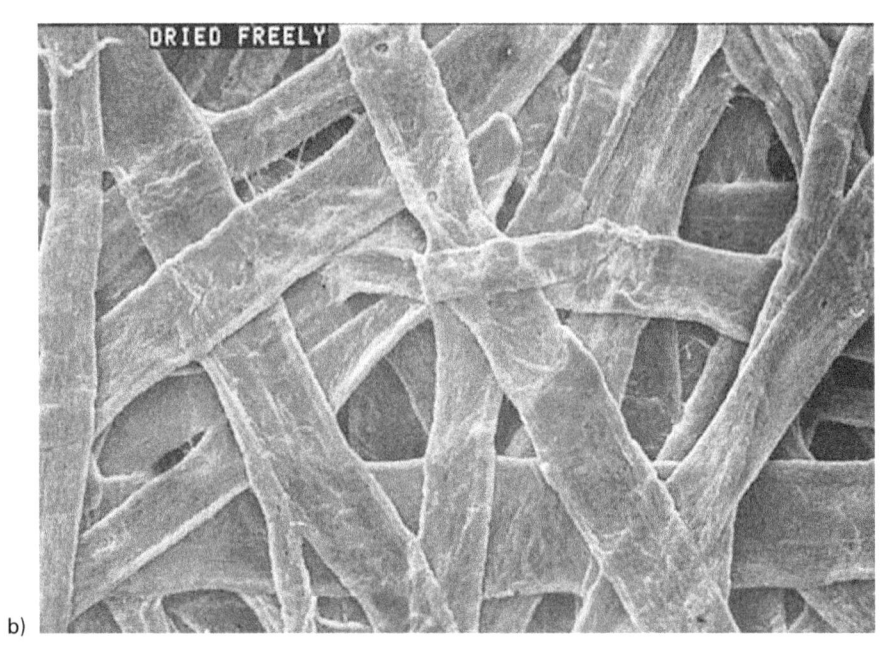

Figure 4.13. a) The structure of a paper, dried under restraint, b) The structure of a paper, dried freely.

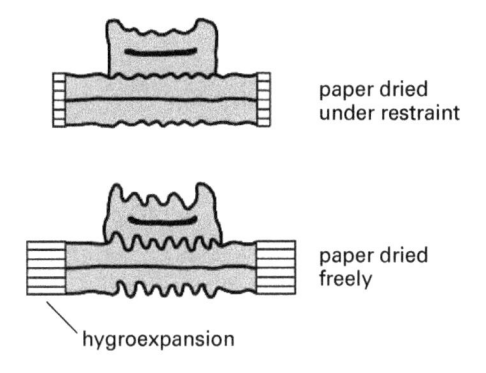

Figure 4.14. The structure in the bonding zone between two crossing fibres after different drying conditions. The hygroexpansion is higher for a freely dried paper than for a paper dried under restraint.

Let us assume that the dried sheet is exposed to a moisture increase so that the fibres swell. In the sheet dried under restraint, the fibres are mainly straight with small number of micro-compressions. The structure of the fibre now causes the swelling to take place in only a small part of the length direction of the fibre and this gives the paper a small hygroexpansion. In the freely dried sheet, the structure of the fibres is disturbed by many micro-compressions, and swelling easily stretches the lengthwise, folded fibre.

The most important paper technical factor which influences the dimensional change is thus the drying conditions. The more bonds formed and the greater the shrinkage during drying, the more micro-compressions are formed and the greater are the dimensional changes during subsequent moistening.

A more detailed mechanism for hygroexpansion may be viewed as follows. When a fibre in a paper adsorbs moisture, the thickness and width swells much more than the length. Any dislocation in the fibre will change its angle to the fibre direction and the fibre will be longer, *Figure 4.15*.

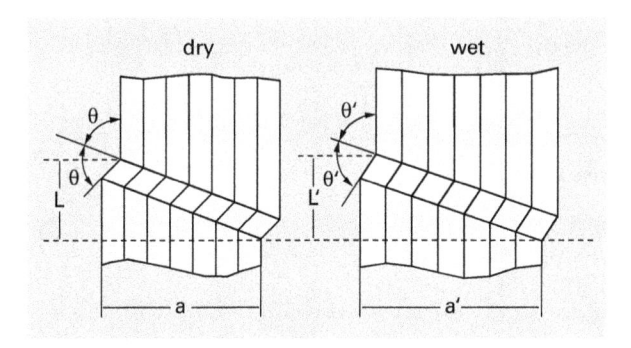

Figure 4.15. When a fibre in a paper adsorbs moisture, the thickness and width swells much more than the length. Any dislocation in the fibre will change its angle to the fibre direction and the fibre will be longer.

Shrinkage during drying gives rise to dislocations and there is a close relation between the hygroexpansion coefficient and the shrinkage during drying, *Figure 4.16*.

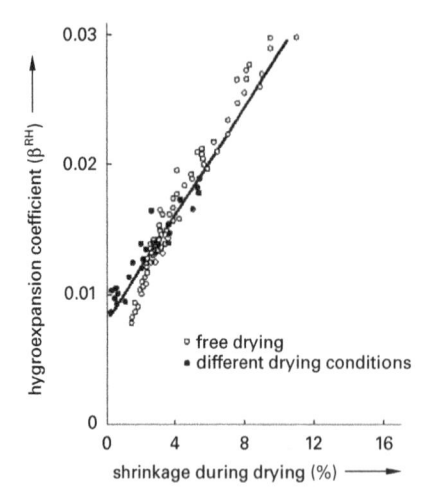

Figure 4.16. Shrinkage during drying gives rise to dislocations and there is a close relation between the hygroexpansion coefficient and the shrinkage during drying. The hygroexpansion coefficient β^{RH} is defined as the strain divided by the change in RH.

The hygroexpansion coefficient is dependent on the amount of fines and the density obtained by wet pressing, *Figure 4.17*. The higher the density the greater chance for fibre-to-fibre bonds and more micro-compressions for freely dried papers. For restrained dried papers a greater bonding results in less hygroexpansion, probably due to more straightening of fibre segments. More fines tend to increase the hygroexpansion.

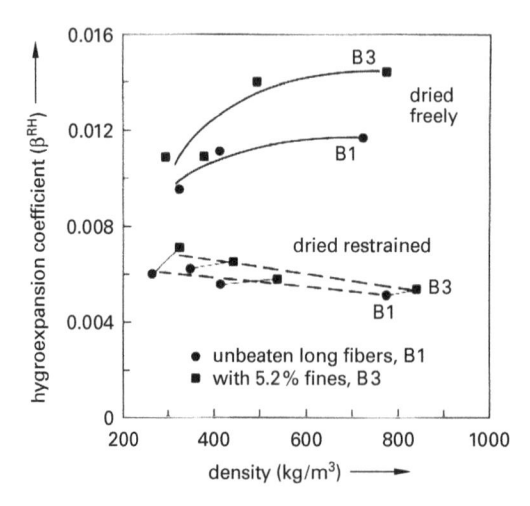

Figure 4.17. The hygroexpansion coefficient is dependent on the amount of fines and the density obtained by wet pressing.

Also the fibre curl affects the hygroexpansion, *Figure 4.18*. For freely dried papers the hygroexpansion increases with curler fibres. For restrained dried papers the fibre shape does not influence the expansion.

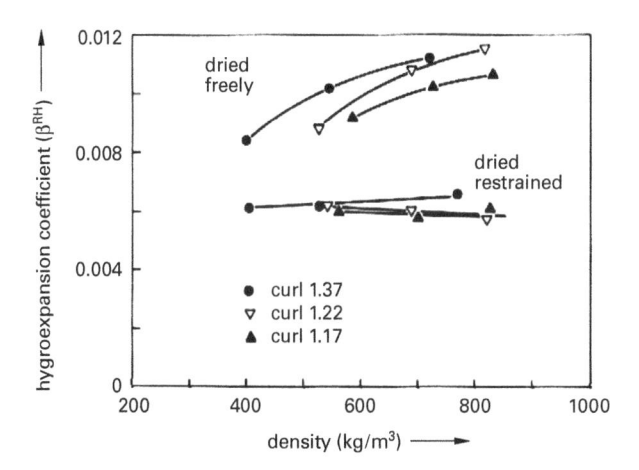

Figure 4.18. The effect of fibre shape on the hygroexpansion coefficient.

The more the fibres are oriented in the machine direction, the greater is the probability for bonded regions in series with CD, *Figure 4.19*. In MD we assume five fibres in this unit area and in CD three. In CD we have five chances for micro-compressions, in MD three. This naïve picture may explain the fact that the shrinkage and hygroexpansion increases the more the fibres are oriented in MD, in a paper machine. In *Figure 4.20* the same principle applies for restrained dried papers.

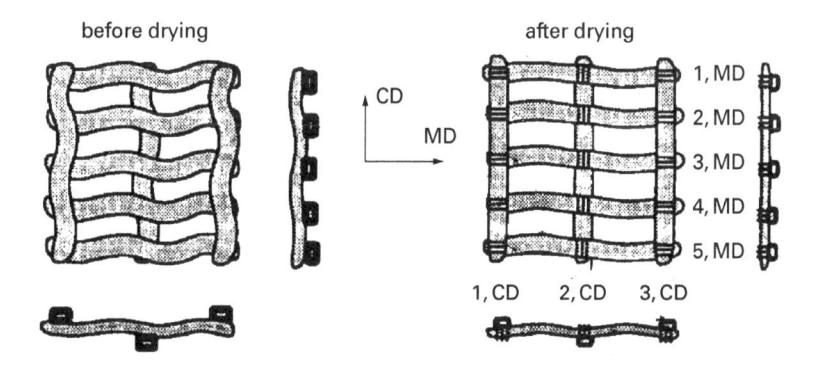

Figure 4.19. Micro-compression development as a function of fibre orientation.

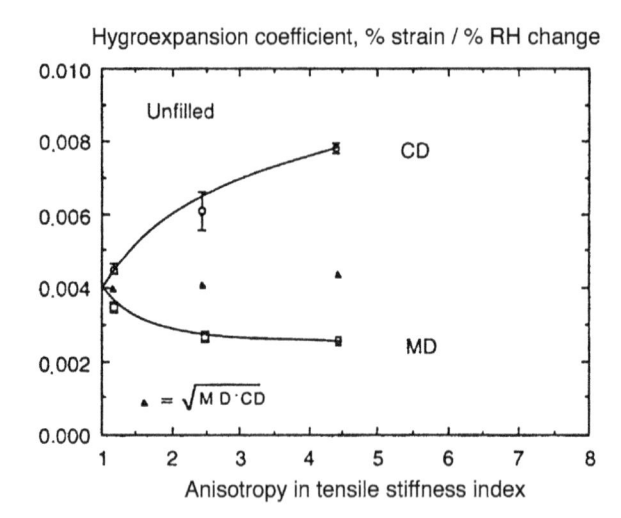

Figure 4.20. Hygroexpansion coefficient for restrained dried papers as a function of the fibre orientation.

Referring to *Figure 4.12* and *Figure 4.16* again, we can investigate if there is a correlation between hygroexpansion coefficient and shrinkage during drying for these pulps, *Figure 4.21*. While there is a general trend, it is apparent that the hygroexpansion did not show a particularly good correlation against shrinkage considering all kind of pulps.

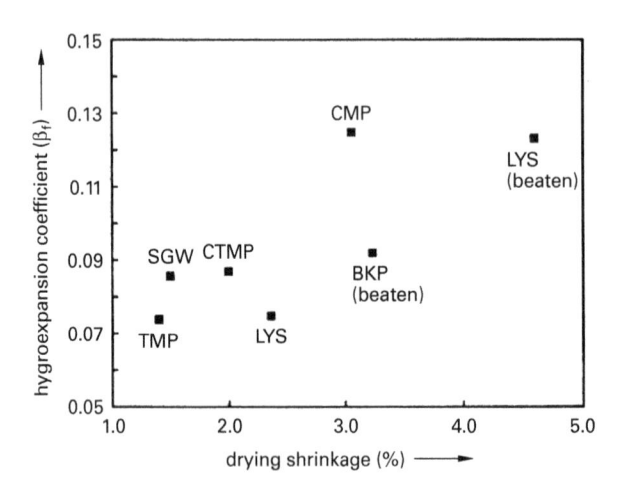

Figure 4.21. Hygroexpansion coefficient as a function of drying shrinkage for the pulps in *Figure 4.12*.

For a given pulp, it is evident that a reduction of the drying shrinkage would decrease the hygroexpansion. In *Figure 4.22* the effect of wet pressing on hygroexpansion is investigated for the pulps in *Figure 4.12*. In this figure it appears that the drying shrinkage is constant up to a certain threshold in density, beyond which drying shrinkage decreased.

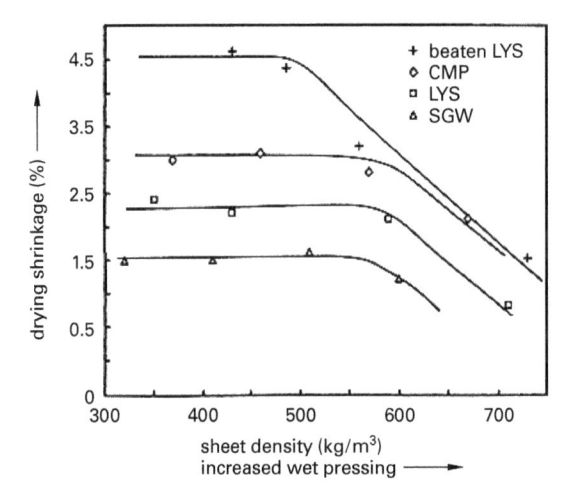

Figure 4.22. Hygroexpansion coefficient as a function of density obtained by wet-pressing, for the pulps in *Figure 4.12*.

The critical solids content at which the shrinkage starts falling is closely related to the WRV, Water Retention Value, *Figure 4.23*.

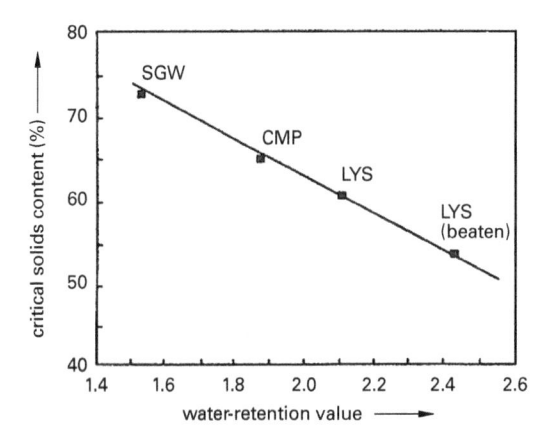

Figure 4.23. Critical solids content as a function of WRV, for the pulps in *Figure 4.12*.

The analysis so far has indicated that the hygroexpansion is related to the fibre orientation, the interaction between the fibres at the fibre bonds, the fibre shape and shrinkage. Furthermore the shrinkage is related to the WRV and the amount of moisture in the paper after wet pressing.

4.6 Permanent Dimensional Changes during Moisture Cycling

If a paper network is *dried freely* the fibres have reached the natural equilibrium state of deformation. If this network is adsorbing and desorbing moisture the network swells and contracts without any permanent deformation since the natural shrinkage state was already obtained during the drying phase.

If the network on the other hand is *dried under restraint*, the shrinkage of the network has been permanented in an un-natural state. Every time this network is adsorbing moisture the network swells but when the moisture enters the network it softens the fibres and makes the paper try to shrink part of the way to a freely dried state. When desorbing moisture the network deformation is permanented again. Every time the network passes a moisture cycle, it tries to shrink further until an equilibrium state is achieved.

Different terms for this phenomenon have been used in the literature. The term „release of built-in strains" seems to describe the situation best.

To achieve release of strains, RH of 65–70 % must be exceeded. The phenomenon is more marked in MD than in CD, due to the drying conditions.

The strain release phenomenon is illustrated in *Figure 4.24* where the RH is varied in the cycle 35-60-85-60-35 %. The upper curve shows the effect for a freely dried sheet, and it is evident that the dimensional changes are not changed with time.

The lower curve applies to sheets dried under restraint. The paper shrinks gradually with increasing number of moisture cycles.

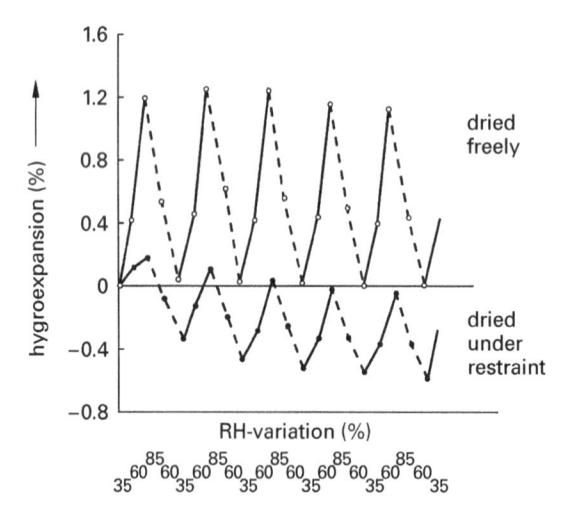

Figure 4.24. Dimensional changes in freely dried sheets (upper curve) and in sheets dried under restraint (lower curve).

The same phenomenon may be illustrated as strain versus moisture content as in *Figure 4.25* and *Figure 4.26* for a fine paper in CD where the paper has dried more or less freely and in MD where the paper has been dried under restraint or even strained a few percent.

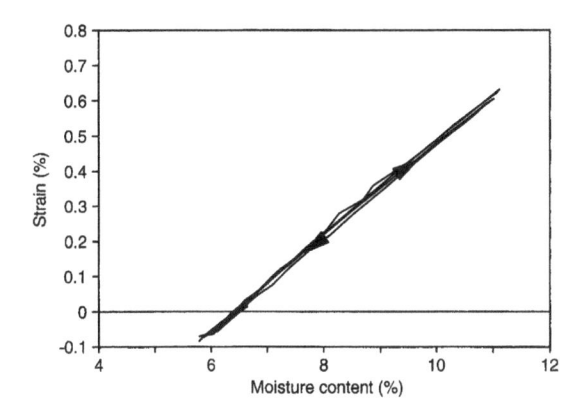

Figure 4.25. Hygroexpansion as a function of moisture content of fine paper (CD).

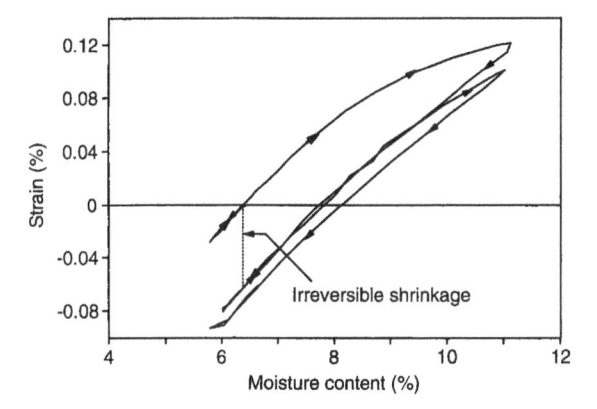

Figure 4.26. Hygroexpansion as a function of moisture content of fine paper (MD).

The dimensional interval in hygroexpansion is large for freely dried sheets. During repeated moisture cycling, on the other hand, the dimensional interval changes very little. In the case of sheets dried under restraint, the dimensional interval is relatively small, especially in the first cycle and increases with repeated moisture cycling. The sheet also shows a permanent shrinkage.

In the paper machine, built-in strains are built up more easily in the MD because the shrinkage is prevented. In CD, the sheet shrinks and the degree of built-in strains becomes small. The sheet then assumes its equilibrium position more rapidly in the CD than in the MD during repeated moisture cycling. The way in which a given paper shall be dried to give it the desired hygroexpansion properties must consequently mean that there must be a balance between whether a high hygroexpansion coefficient and a small change during moisture cycling are desired as in the CD or a small hygroexpansion coefficient but a large change during moisture cycling as in the MD.

The „release of built-in strains" sometimes results in extreme effects such as those described in the following two figures. An MD sample of newsprint was exposed to different amounts of moisture contents. *Figure 4.27* show that the paper first expands and then contacts again due to the release of the built-in strains. The consequence could be that the paper in a printing press becomes stiffer as the moisture increases.

The time effects of hygroexpansion strain with time after moisture exposure is shown in *Figure 4.28*. The starting atmosphere was 50 % RH.

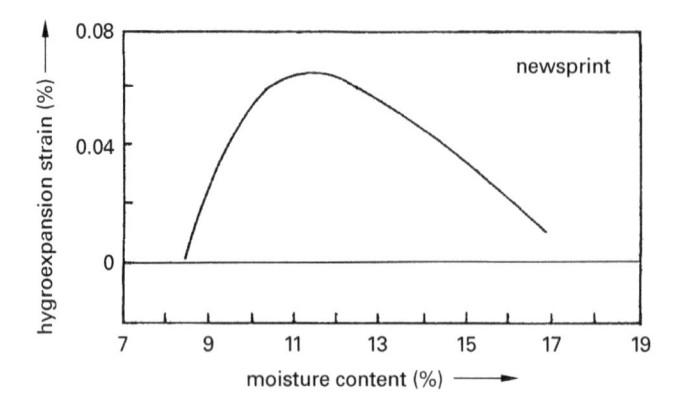

Figure 4.27. Hygroexpansion strain as a function of moisture content of newsprint in MD.

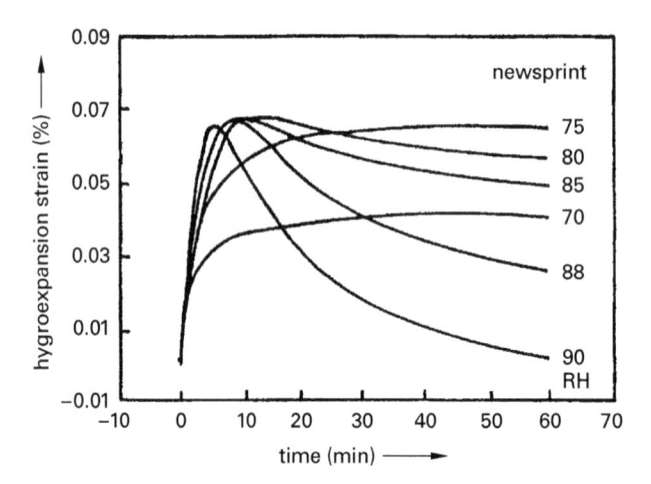

Figure 4.28. The time effects of hygroexpansion strain in MD with time after moisture exposure. The starting atmosphere was 50 % RH.

4.7 Measurement Methods

Many types of devices exist for recording the dimensional changes of paper during sorption. STFI uses a device shown in *Figure 4.29*. The feature of the principle is that the paper is flattened by a weight during measurement to avoid buckling of the test piece. The deformation is recorded by a deformation-recording device, LVDT.

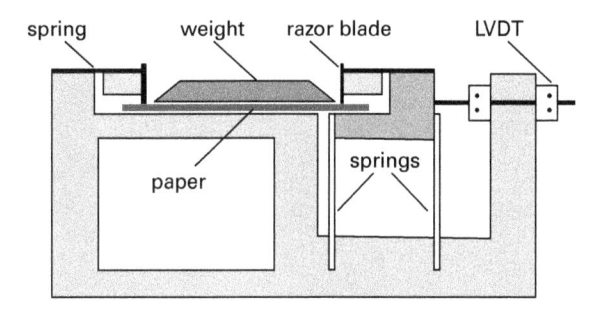

Figure 4.29. STFI deformation-recording device.

Speckle photography is a modern way to record two-dimensional deformations in a paper. In short, on the paper surface is applied small irregular dots. A camera detects the average movement of dotted areas and a very precise deformation recording can be achieved, *Figure 4.30*.

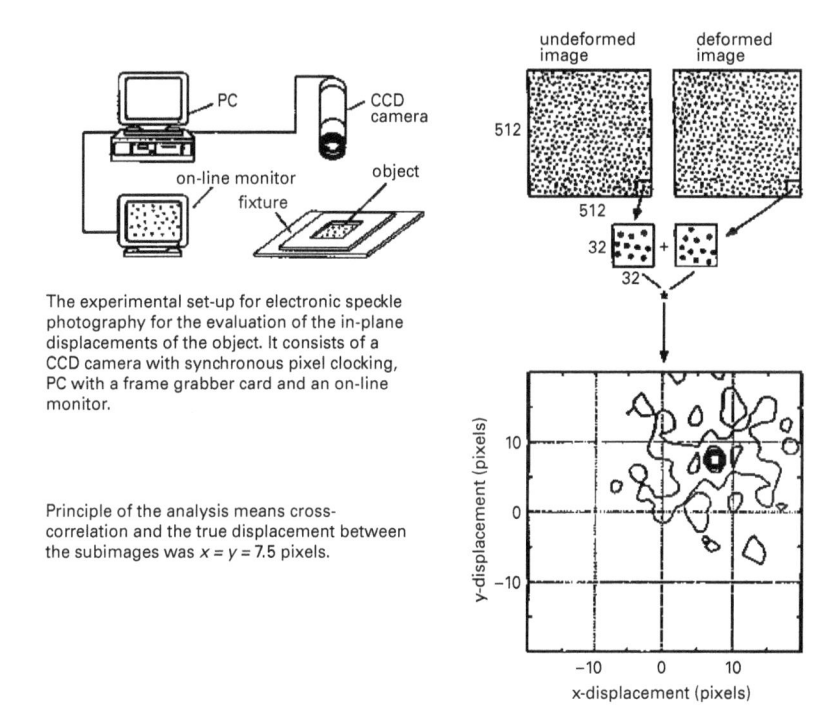

The experimental set-up for electronic speckle photography for the evaluation of the in-plane displacements of the object. It consists of a CCD camera with synchronous pixel clocking, PC with a frame grabber card and an on-line monitor.

Principle of the analysis means cross-correlation and the true displacement between the subimages was $x = y = 7.5$ pixels.

Figure 4.30. Principle of speckle photography.

The strains can be expressed in polar plots, *Figure 4.31*. In this case the symmetry of the hygroexpansion strain has an off-axis angle to MD. The same off-axis angle can also be obtained by ultrasonic techniques, *Figure 4.32*.

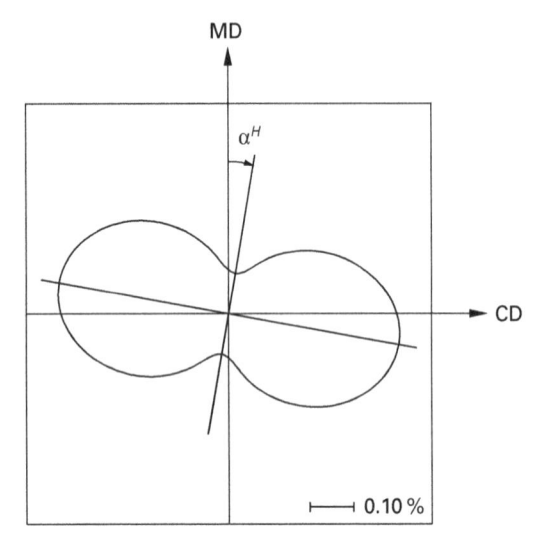

Figure 4.31. The strains can be expressed in polar plots. In this case the symmetry of the hygroexpansion strain has an off-axis angle to MD.

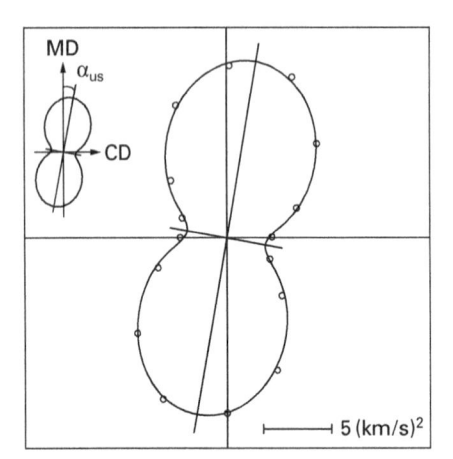

Figure 4.32. The stiffness index in a polar diagram where o) are measurement points. Note the tensor scale for stiffness index. The tensile stiffness orientation, α_{us} is defined. Copy paper 80 g/m².

4.8 Curl and Twist

Curl in paper or carton board is a situation where the sheet bends out of the plane because of a mechanical influence or because the temperature or moisture has influenced the sheet. Twist is a situation where the sheet is turned like a propeller.

Curl caused by moisture content changes depends on the fact that the paper has an asymmetrical structure in the thickness direction. One side then swells or shrinks more than the other and this causes the sheet to bend. The basic reason for curl problems is thus to be found in differences in the hygroexpansion of the different layers of the paper.

Curl due to moisture exposure can also have other causes. One such cause is the case where the sheet is moistened or dried from one side, *Figure 4.33*. The reasons are hygroexpansion and release of built-in strains as discussed before.

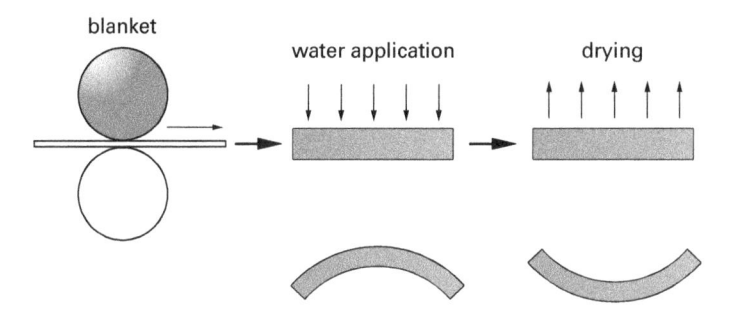

Figure 4.33. The sheet is moistened or dried from one side. The reason is hygroexpansion and release of built-in strains as discussed before.

Cockling, *Figure 4.34*, *Figure 4.35* and *Figure 4.36*, are out-of-plane movements locally over a paper.

Figure 4.34. Cockling is out-of-plane movements locally over a paper.

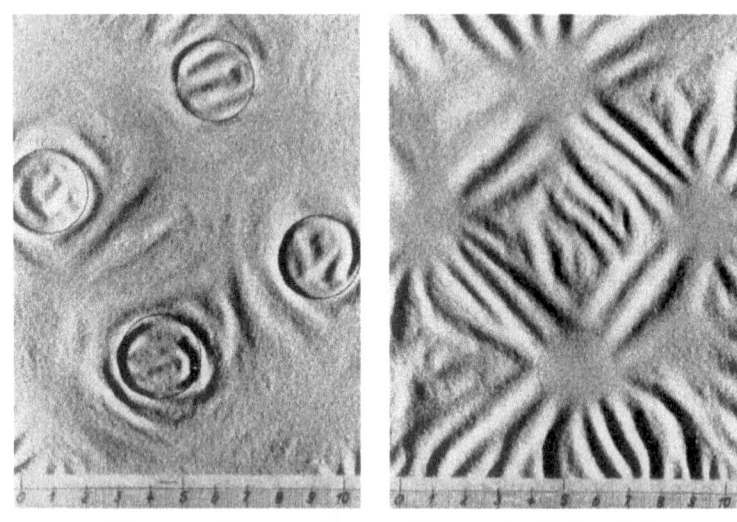

cockling caused by non-uniform moisture content before drying

cockled areas = dryer areas before drying

flat areas = wetter areas before drying

Figure 4.35. Cockling experiments.

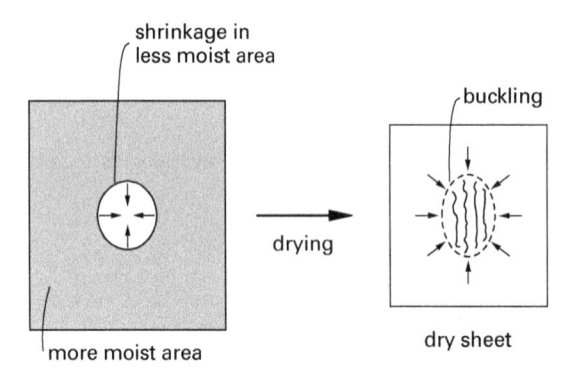

Figure 4.36. Cockling mechanism.

Cockling stems from its irregular distribution of strain over a paper surface. According to that theory, cockling takes place if paper inhibited in its shrinking from point to point.

Brecht, Müller and Weiss have tried to establish the correctness of this theory. Moist sheets of paper were prepared, which had an uneven distribution of moisture over its surface corresponding to a certain pattern. When the sheets were subjected to contact drying, there resulted the artificially produced cockling shown in *Figure 4.35*, which exactly reproduced the pattern of moisture distribution just mentioned. There were circular patches on the sheets, which before were less moist than the remaining regions in the two left-hand pictures and more moist in the two right-hand pictures. The cockles always coincided with the initially less moist spots, whereas the initially moister patches remained flat. The explanation is that the initially drier patches enter that stage sooner during the drying in which fairly large shrinkage is connected with relatively small evaporation. These spots thus tend to shrink more than the others. Consequently, there arise tensions on their margins that hamper shrinking. If the sheet was flat until then, the other parts of the sheet shrink as a whole towards the sheet centre in the further course of drying. The patches already dried cannot shrink any more; they are warped out o the plane and form cockles.

Since cockling comes from uneven shrinkage conditions, one may expect that it will manifest itself more markedly, on the whole, where greater shrinking is to be anticipated than in the case of lesser tendency to shrink.

An out-of-plane shape of a paper or corrugated board may be subdivided into three components, *Figure 4.37*.

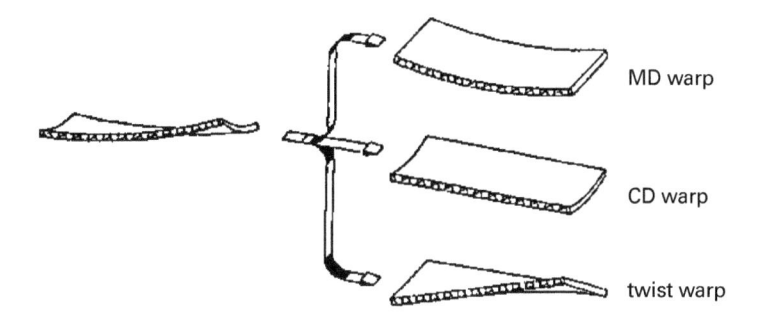

Figure 4.37. An out-of-plane shape of a corrugated board may be subdivided into three components, curl in MD, curl in CD and twist.

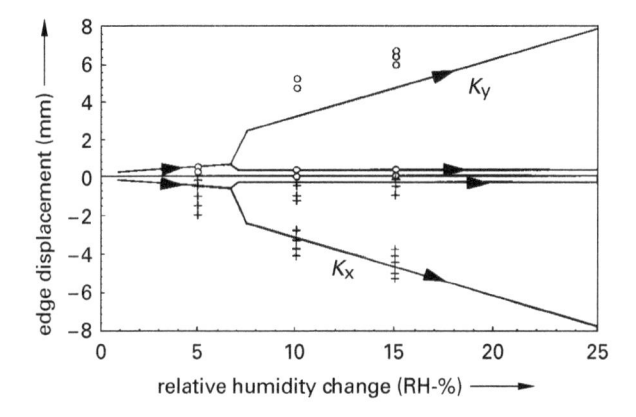

Figure 4.38. Measured curl components, K_x and K_y (crosses and circles, respectively) of a two-ply, 0°/90° laminate of two copy paper sheets vs. a change in relative humidity, ΔRH. The lines show a finite element calculation. The bifurcation point at $\Delta RH = 7\%$ indicates the buckling transition. After that, only one curl component is non-zero.

$$w(x,y) = \frac{1}{2}K_x \cdot x^2 + \frac{1}{2}K_y \cdot y^2 + \frac{1}{2}K_{xy} \cdot xy$$

Curl in two directions cannot exist at the same time, *Figure 4.38*. At a given RH or moisture content one of the directions dominates and the other returns to zero. The point is called the bifurcation point.

Figure 4.39 and *Figure 4.40* show typical behaviours for curl when the adsorption starts by adsorption and desorption respectively.

It is possible by laminate theory to calculate curl and twist if the properties of the plies of the paper are known. Simplified equations have been derived in special cases, when the paper is modelled as two plies with equal thickness, *Figure 4.41*.

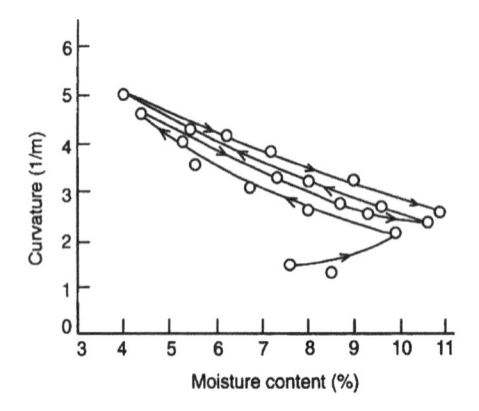

Figure 4.39. Bending curvature response of fine paper to cyclical humidity changes when the process starts from the adsorption stage.

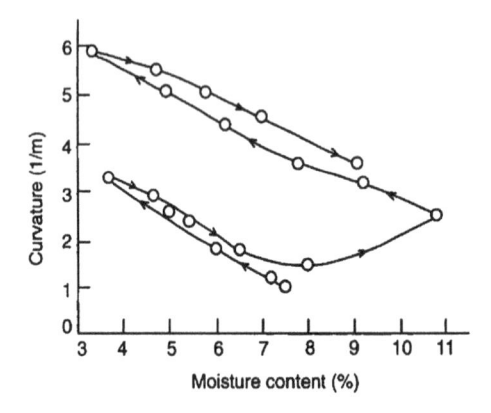

Figure 4.40. Bending curvature response of fine paper to cyclical humidity changes when the process starts from the desorption stage.

$$curl = \frac{1}{R} = \frac{3}{2t}\left(\beta^B \cdot H^B - \beta^T \cdot H^T\right) \quad (4.14)$$

β is the hygroexpansion coefficient for the bottom ply B and the top ply T respectively. H is the change in moisture.

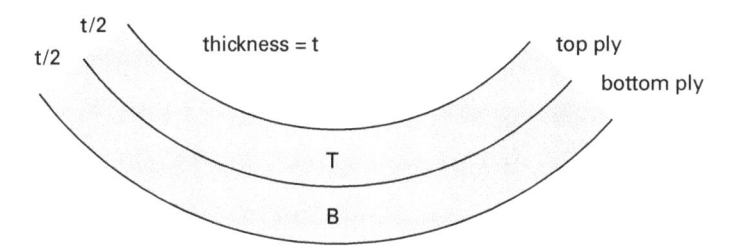

Figure 4.41. Paper is modelled as two plies with equal thickness.

Curl thus depends on the difference in hygroexpansion, and on the strain on the bottom and top plies. Curl decreases with increasing thickness. It can be shown that the tensile stiffness index of the layers plays a very small role.

The main requirement for a paper sheet to remain flat during moisture exposure is thus that the sheet is symmetrical with regard to its middle line in the thickness direction in both MD and CD. For the different layers, the symmetry includes both thickness and hygroexpansion.

Whether an anisotropic machine-made sheet curls in the MD or the CD depends on the dimensions of the sheet. A thin strip always curls in its length direction regardless of whether this is MD or CD. Symmetrical and round sheets tend to be dominated by curl in CD. Even if moisture diffuses into the sheet from both sides, curl can occur if the hygroexpansion is different in different layers. Curl is always balanced in practice by adjustment of the steam pressure in the upper and lower cylinders in the dryer section on the paper machine.

Reasons for unequal sidedness in hygroexpansion, which causes curl in a single-layer sheet, can also be different fibre orientation and the content of fines on the upper and bottom side of the sheet. For a multi-layer sheet, the layer structure itself can be asymmetrical.

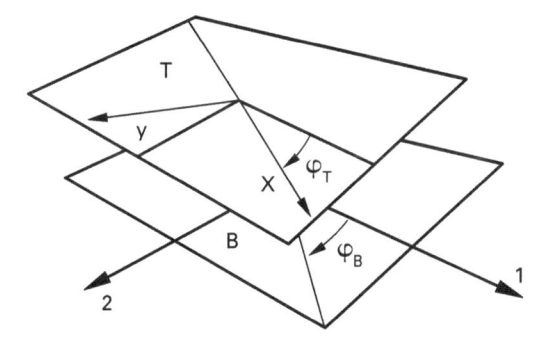

Figure 4.42. Twist occurs, for a two-ply paper when one or two plies have an off-axis symmetry. The definition of the off-axis angle is indicated.

Twist occurs if some layer in the paper is asymmetrical in relation to the MD, *Figure 4.42*. During exposure to moisture, such crookedness leads to a turning moment which twists the sheet.

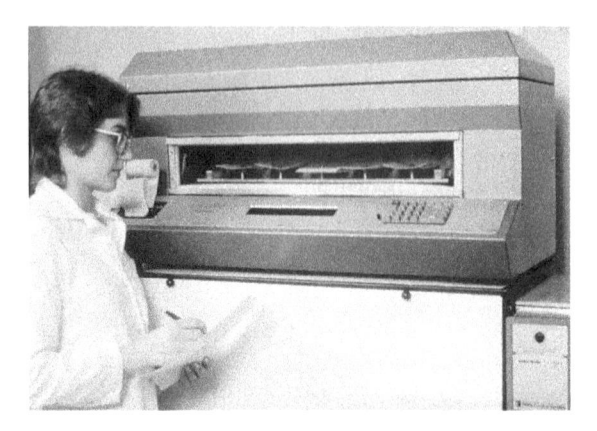

Figure 4.43. In the instrument the 100 mm by 100 mm samples are scanned by a laser gage and the three curl and twist components calculated.

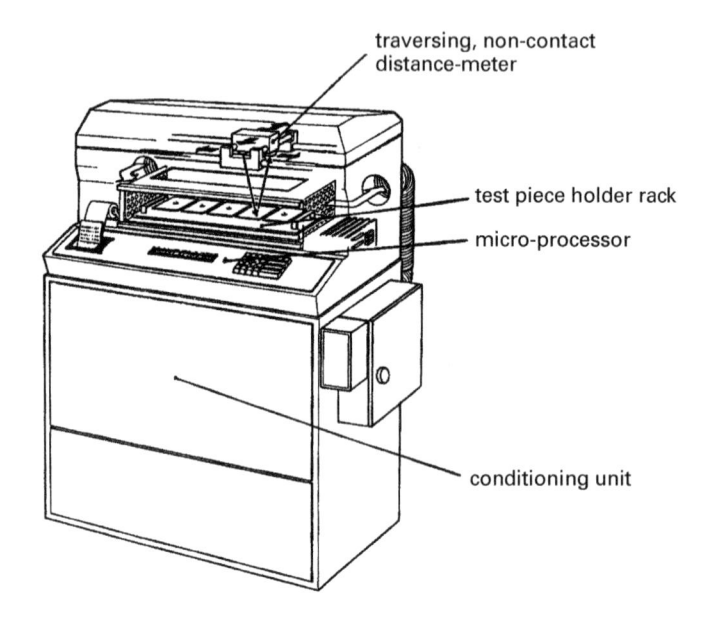

Figure 4.44. The curl and twist tester.

Twist may be expressed in a simple equation as follows for the paper shown in *Figure 4.42*.

$$k_6 = \frac{3}{t}\left[\phi^T H^T \left(\beta_y^T - \beta_x^T\right) - \phi^B H^B \left(\beta_y^B - \beta_x^B\right)\right] \tag{4.15}$$

β is the hygroexpansion coefficient for the bottom ply B and the top ply T respectively. H is the change in moisture and x and y are two perpendicular directions in the paper where 1 is MD and 2 is CD.

Twist may be measured in several ways. In the instrument shown in *Figure 4.43* and *Figure 4.44*, the 100 mm by 100 mm samples are scanned by a laser gage and the three curl and twist components calculated. Another technique involves scanning the shape of a circular test piece.

Some other techniques have been proposed. *Figure 4.45* shows the cross-cut method.

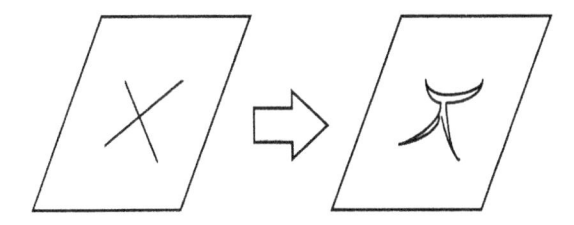

Figure 4.45. The cross-cut method for curl measurements

Working with narrow strips, the curl is evaluated by comparing the curvature against known curvatures or calculating the curvature by measuring the bending of the strip, *Figure 4.46*.

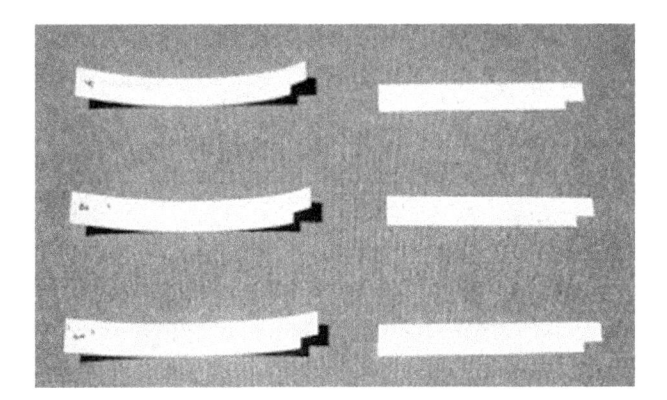

Figure 4.46. Curl measurements by using strips.

Any two-sidedness will influence curl and twist. The two-sidedness of some machine parameters have been investigated, *Figure 4.47–Figure 4.49*.

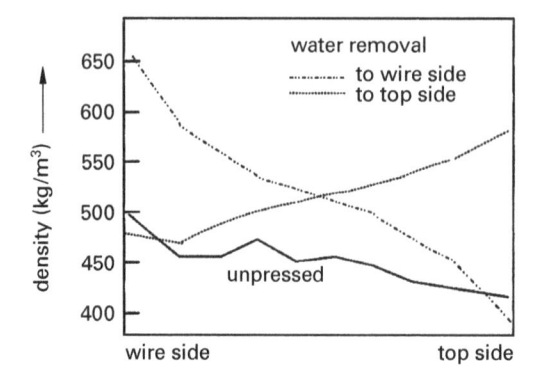

Figure 4.47. Two-sidedness in density created by wet pressing.

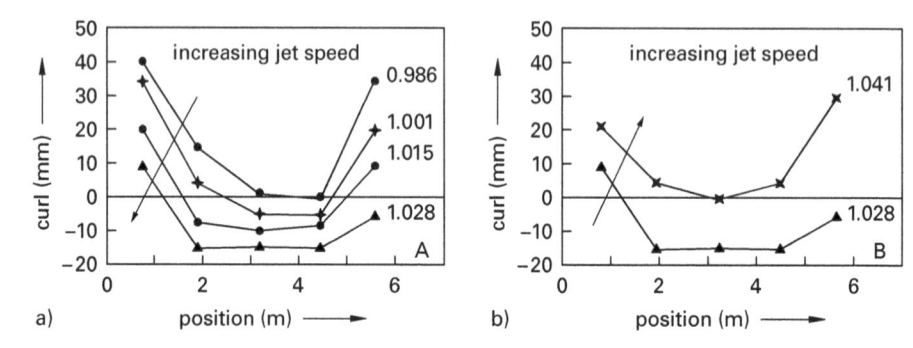

Figure 4.48. Curl due to different jet/wire ratios.

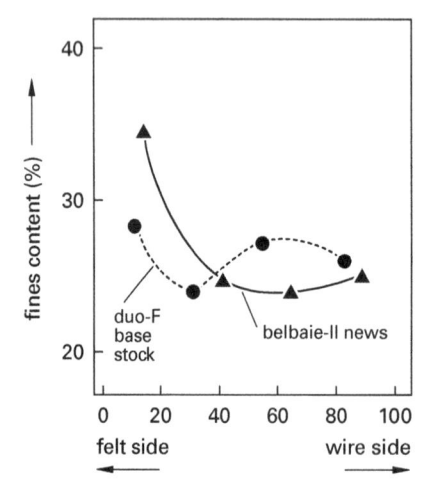

Figure 4.49. Two-sidedness in fines content.

4.9 The Mechanical Properties at Different Moisture Contents

The in-plane stiffness index of paper may be measured by ultrasonic USI and regular tensile testing TSI (Chapter 2).

The development of these different stiffnesses with increasing RH or moisture ratio is different due to the time scale of testing. The TSI is more sensitive than USI. A representative example is given in *Figure 4.50*.

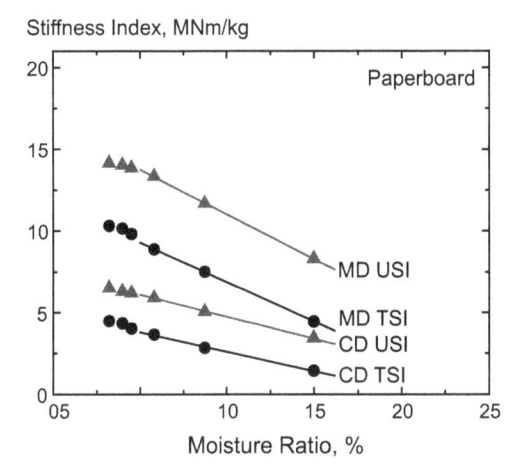

Figure 4.50. The development of these different stiffnesses with increasing RH or moisture ratio is different due to the time scale of testing. The TSI is more sensitive than USI.

The hygroscopicity of the paper means a lot for the mechanical properties of the paper and all paper properties are influenced in different ways by the moisture content. *Figure 4.51* shows the principal change as a percentage for some paper properties at different RH with 50 % RH as reference.

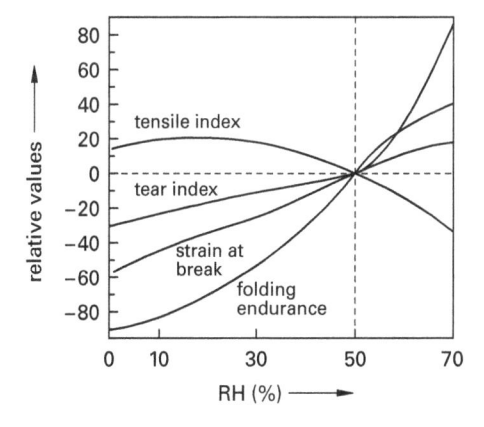

Figure 4.51. Level deviation as a percentage of the value at 50 per cent RH.

As shown earlier in this chapter, the built-in strains in the paper are often released during moisture cycling, which results in a permanent shrinkage of the paper. This release of strains influences the mechanical properties very little, however.

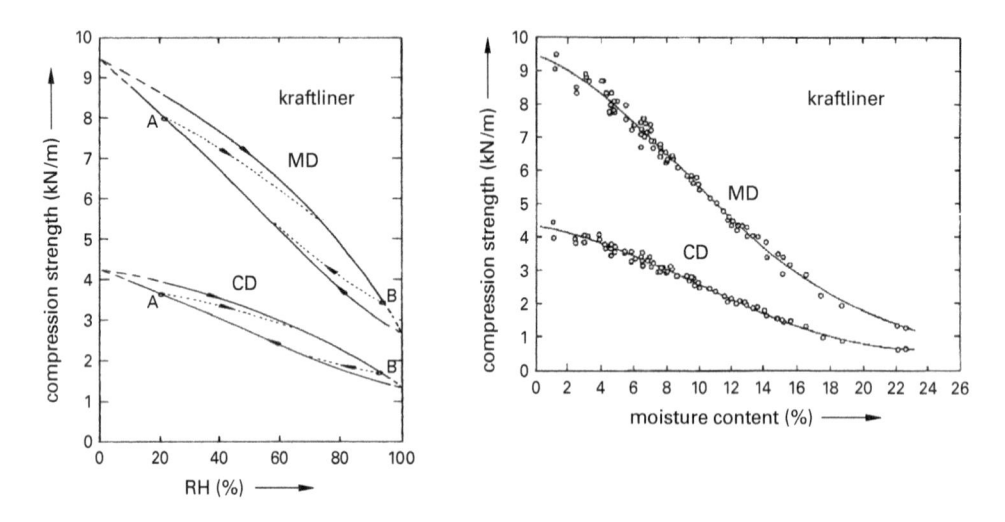

Figure 4.52. Compression index against RH and against moisture content.

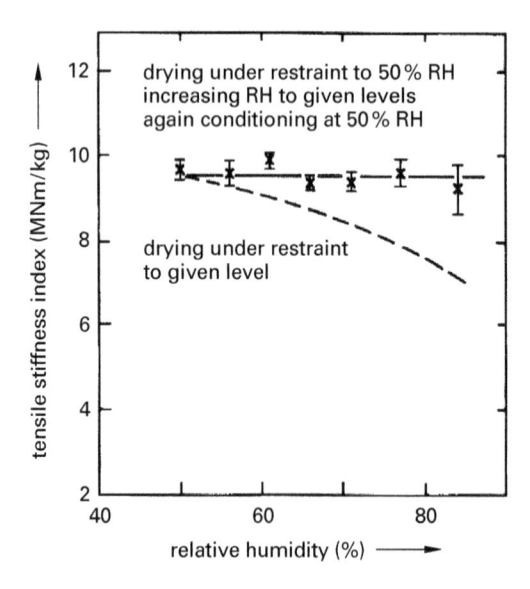

Figure 4.53. Tensile stiffness index against RH.

Figure 4.52 shows an example of this. The compression index of the paper is here plotted against the moisture content of the paper. Independently of how a given moisture content has been reached, whether from dry or from humid conditions, the paper has the same strength at a given moisture content. The same is true for tensile index and tensile stiffness index.

If the corresponding strength is plotted against the RH of the air, the earlier mentioned hysteresis phenomenon is evident since a given RH does not unambiguously determine the moisture content.

The same applies for tensile stiffness index TSI, *Figure 4.53*. The TSI drops with RH, but TSI does not change if the paper is subjected to high RH levels and then conditioned again.

4.10 Mechanosorptive Creep

Creep is the time dependent strain as a result of a constant force (Chapter 2). *Figure 4.54* shows a typical creep result for paper in tension at 80 % RH. If the humidity of the air is lowered to 30 % RH, after a given time, here 4000 and then cycled between these levels, seconds, the creep process accelerates. The phenomenon is called the „mechano-sorptive effect".

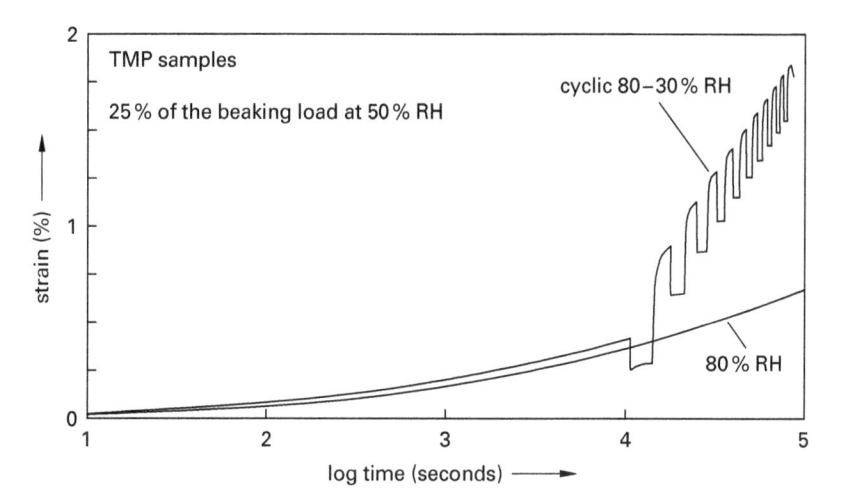

Figure 4.54. Creep in tension at 80 % RH and creep in tension due to a variation in RH between 30 and 80 % RH.

5 Optical Properties of Pulp and Paper

Anthony Bristow
Bristow Consulting AB, Tullinge

5.1 Introduction 146

5.2 Light Absorption and Light Scattering 146

5.3 The Kubelka-Munk Theory 147

5.3.1 Derivation of the Kubelka-Munk Equations 147
5.3.2 Limitations of the Kubelka-Munk Analysis 151
5.3.3 The Absorption and Scattering Coefficients 151
5.3.4 Applications of the Kubelka-Munk Analysis 152

5.4 The Relationship between s, k, R_∞ and R_0 153

5.5 Measurement of R_∞, R_0 etc 154

5.6 Standard Illuminants 155

5.7 Standard Observers 156

5.8 Standardized Measurement Conditions (ISO 2469) 156

5.8.1 Instruments Conforming to ISO 2469 157
5.8.2 Calibration of the Measurement Instruments 158
5.8.3 Calibration within the Paper Industry – the ISO-Hierarchy 158

5.9 Optical Measurements 159

5.10 Colour Perception 161

5.11 Tristimulus Values 161

5.12 Chromaticity Coordinates 162

5.13 Dominant Wavelength and Excitation Purity 163

5.14 CIELAB-Coordinates 165

5.15 Whiteness 165

5.16 Fluorescence 166

5.1 Introduction

Optics is the science of light and human vision, and the optical properties of a paper are the properties connected with its appearance, i.e. its brightness, opacity and colour.

The appearance of a material is not, however, a simple, easily defined, physical property. It is the psychophysical consequence of several different processes. It is the result of:

- *an interaction between the material and the incident light,* which defines the characteristics of the light reflected or radiated from the material,
- *a stimulation of cells in the retina,* when the light reflected or radiated from the material meets the human eye,
- *a perception in the human brain,* as a result of an interpretation of the signals transmitted from the eye to the brain

This means that we must consider this total phenomenon in two different stages:

- *From a papermaking point of view,* we need to consider how light interacts with the material, and how different factors in the raw materials, in the manufacturing process and in the attained structure influence this interaction
- *From a quality control point of view,* we need to consider how the eye and brain react to the stimulus of the light received from the material, so that we can construct a realistic instrument to measure the radiation in a relevant manner and to compute relevant parameters to characterise the appearance of the material.

This chapter includes

- a short presentation of a mathematical theory for the interaction between the different components of light and paper,
- an account of how instruments are designed and calibrated to give measurement values which correspond to the perceptual conception of the appearance of a material, and
- a description of different mathematical ways of treating the measurement data to obtain parameters which give a meaningful representation of properties such as the colour and whiteness of paper.

5.2 Light Absorption and Light Scattering

When a ray of light of a given wavelength strikes a paper sheet, several different things happen. Part of the light is reflected in the specular direction from the surface of the paper (i.e. at an angle to the normal equal to the incidence angle) while another part passes into the material. The light which enters the paper is then partly absorbed and partly scattered in different directions before being reflected back from the surface of the paper or leaving the paper through its underside. The appearance of the material is dependent on the extent to which light of different wavelengths is reflected and absorbed.

- The interaction between the light and the paper can be described with the help of an absorption coefficient, k, and a scattering coefficient, s.

- The *light absorption coefficient, k*, is a measure of that portion of the light which is absorbed. If the absorption is different in different wavelength ranges, the material appears coloured. Unbleached grades have a relatively high absorption coefficient, especially in the blue range, while bleached grades have a very low absorption coefficient. Fillers also have a low absorption coefficient.
- The *light scattering coefficient, s*, is a measure of the free, unbonded surface area of fibres and other particles within the paper. Light is scattered at the interface between two phases with different refractive indices, so that a primary requirement if light is to be scattered in a material such as paper is that the material is porous. The light is reflected at the surfaces of the fibres, both at external surfaces and at internal cracks etc in the fibre.

An example of what a change in the free, unbonded, light-scattering surface does for the optical properties can be seen in the difference between white, newly-fallen snow and a piece of transparent ice. The scattering coefficient in the newly fallen snow is very high because of its porous nature and because of the very large free, light-reflecting surface area in each snowflake. In ice, on the other hand, there are no free, light-reflecting surfaces and the scattering coefficient is close to zero.

5.3 The Kubelka-Munk Theory

In the 1940s, Kubelka and Munk derived equations to describe light reflection from thin layers of paints and similar materials. These equations have since been successfully applied to paper. The Kubelka-Munk equations are valid only when both illumination and measurement take place with diffuse light. Within the field of paper technology, diffuse illumination is therefore used in most optical measurements.

The Kubelka-Munk equations are here derived for a simplified case, where the ray path is drawn perpendicular to the plane of the paper. This derivation gives the same equations as the more complete analysis with diffuse illumination.

5.3.1 Derivation of the Kubelka-Munk Equations

Consider a sheet of paper with grammage W and reflectance factor R placed over a background with reflectance factor R_g. This sheet is illuminated with light having an intensity I, and light having an intensity J is reflected, as indicated in *Figure 5.1*. The reflectance factor R of the sheet is then given by Equation 5.1:

$$R = \frac{J}{I} \tag{5.1}$$

Within this sheet of paper, consider an infinitesimal layer dx, illuminated with intensities i (from above) and j (from below at a distance x from the base). In its passage through this layer, part of the light is absorbed and part of the light is scattered, so that i and j are reduced respectively by $(s + k)\,idx$ and $(s + k)\,jdx$. It is here assumed that all scattered light is scattered at an angle of 180° to its original direction so that, $s \cdot jdx$ and $s \cdot idx$, which respectively represent the light scattered when the light passes in one direction, are added to the intensities i and j when the light travels in the opposite direction.

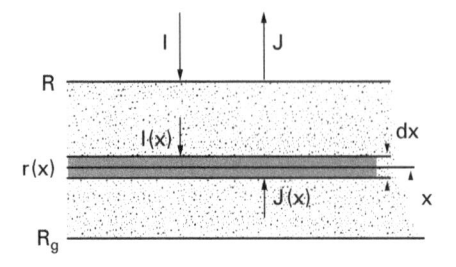

Figure 5.1. Model for derivation of the Kubelka-Munk equations.

The total changes di and dj, calculated in the x-direction, are therefore:

$$-di = -(s+k) \cdot i \, dx + s \cdot j \, dx \qquad (5.2)$$

$$dj = -(s+k) \cdot j \, dx + s \cdot i \, dx \qquad (5.3)$$

Across the infinitesimal layer dx, $r = j/i$, and this gives:

$$dr = d\left(\frac{j}{i}\right) = \frac{i \, dj - j \, di}{i^2} \qquad (5.4)$$

Thus, if the two differential equations, Equation 5.2–5.3, are multiplied by j and i, respectively, and the resulting equations are added, we obtain:

$$-j \, di = -(s+k) \cdot i j \, dx + s \cdot j^2 \, dx$$
$$i \, dj = -(s+k) \cdot i j \, dx + s \cdot i^2 \, dx$$
$$\overline{i \, dj - j \, di = -2(s+k) \cdot i j \, dx + s \cdot i^2 \, dx + s \cdot j^2 \, dx}$$

and division by i^2 gives:

$$dr = \left[-2(s+k)r + s + sr^2\right] dx \qquad (5.5)$$

If, in this equation, we set:

$$a = \frac{(s+k)}{s} = 1 + \frac{k}{s} \qquad (5.6)$$

we obtain:

$$dr = (r^2 - 2ar + 1)s\,dx \tag{5.7}$$

It is necessary to solve this equation by integration between the grammage limits $x = 0$ and $x = w$ and between the reflectance factor limits $r = R_g$ and $r = R$:

$$\int_0^w s\,dx = \int_{R_g}^R \frac{dr}{r^2 - 2ar + 1} \tag{5.8}$$

Integration then gives:

$$sw = \frac{1}{\sqrt{4a^2 - 4}} \left[\ln\left(\frac{2r - 2a - \sqrt{4a^2 - 4}}{2r - 2a + \sqrt{4a^2 - 4}} \right) \right]_{R_g}^R \tag{5.9}$$

This equation can be simplified by considering a special limiting case of Equation 5.5. Assume that the reflectance from the surface R is the same as the reflectance from the background R_g, i.e. $dr = 0$ through the whole sheet. In this situation, i.e. when the paper is placed over a paper pad which is so thick that the reflectance factor is not changed when a further sheet is added to the pad, the reflectance factor is designated $R = R_g = R_\infty$. From Equation 5.5, we then obtain:

$$sr^2 - 2(s+k)\cdot r + s = 0 \tag{5.10}$$

Whence

$$r = R_\infty = 1 + \frac{k}{s} + \sqrt{\left(\frac{k}{s}\right)^2 + 2\left(\frac{k}{s}\right)} \tag{5.11}$$

This is often written in the form:

$$\frac{k}{s} = \frac{(1-R_\infty)^2}{2R_\infty} \tag{5.12}$$

So that

$$a = \frac{1}{2}\left(\frac{1}{R_\infty} + R_\infty\right) \tag{5.13}$$

The reflectance factor R_∞ of a thick pad of paper is called the reflectivity of the paper and is thus a material property, which is dependent only on the ratio between k and s, and not on their absolute values.

As shown in *Figure 5.2*, where the ratio s/k is plotted on a logarithmic scale, large relative changes in this ratio are required to change the reflectivity, R_∞.

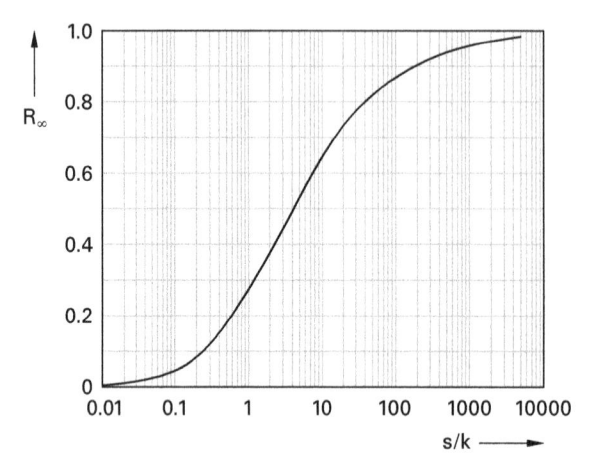

Figure 5.2. The reflectivity R_∞ increases with increasing value of the ratio s/k.

Substitution of Equation 5.13 in Equation 5.9 leads to:

$$sw = \frac{1}{\frac{1}{R_\infty} - R_\infty} \left[\ln\left(\frac{1 - \frac{1}{R_\infty}}{r - R_\infty}\right) \right]_{R_g}^{R} \tag{5.14}$$

After inserting the integration limits, we finally obtain:

$$sw = \frac{1}{\frac{1}{R_\infty} - R_\infty} \ln\left(\frac{(1 - RR_\infty)(R_\infty - R_g)}{(1 - R_g R_\infty)(R_\infty - R)}\right) \tag{5.15}$$

and

$$kw = sw\left(\frac{(1 - R_\infty)^2}{2R_\infty}\right) \tag{5.16}$$

Both s and k have thus been expressed as functions of a number of reflectance values and these equations enable both s and k to be calculated if reflectance measurements can be made against two different backgrounds, i.e. provided the material is not completely opaque, and if the grammage is known. To use the equations given here, one measurement is made over an opaque pad to obtain R_∞ and a second measurement is made over an arbitrary backing R_g to obtain R.

Equation 5.15 is simpler to handle if one measurement is made with an opaque pad of the material to obtain R_∞, and a second is made against a black background with a reflectance factor $R_g = 0$ to obtain the reflectance factor designated R_0:

$$sw = \frac{1}{\frac{1}{R_\infty} - R_\infty} \ln\left(\frac{(1 - R_0 R_\infty) R_\infty}{(R_\infty - R_0)}\right) \tag{5.17}$$

The product of the scattering coefficient and the grammage, *sw*, is called the scattering power of the sheet. The scattering power or covering ability, *sw*, is a very important property for printing paper and is a measure of the ability of the paper to conceal print on the reverse side of the paper. It can also be an important characteristic of a layer of a bleached furnish on a white-top liner or of a coating layer where it is necessary to hide the brown colour of the underlying plies or substrate.

The ratio *100 R_0/R_∞* is called the opacity of the material. In later sections of this chapter, the way in which R, R_0, R_g and R_∞ can be measured is described.

5.3.2 Limitations of the Kubelka-Munk Analysis

It is clear, by consideration of Equation 5.17, that the Kubelka-Munk analysis can yield values for *s* and *k* only if it is possible to obtain values of R_0 and R_∞ which are significantly different from each other. In addition, the following conditions must be observed:

- a single sheet of the material must be *optically thick*, i.e. sufficiently thick to allow multiple scattering to occur, with a reflectance factor over 50 % and a transmittance less than 20 %.
- there must be *negligible surface reflection*, i.e. the material must not be highly glossy and direct surface reflection must be avoided with a gloss trap during measurement
- the *structure must be uniform* throughout the sheet (for laminated structures, special equations are necessary)
- the illumination must be *diffuse*
- the optical phenomena must be limited to absorption and scattering, which means that *no fluorescence* is permitted.

5.3.3 The Absorption and Scattering Coefficients

Although these absorption and scattering coefficients are mathematical parameters derived from reflectance data by application of the Kubelka-Munk theory, the ability of a material to absorb or scatter light of different wavelengths is a physical reality.

Certain molecular groups have the ability to absorb light in certain wavelengths, and the presence of these groups in e.g the lignin means that the paper or pulp appears off-white or coloured, and such groups are called chromophores. It is the task of cooking and bleaching to remove such chromophores and thus give a less yellow pulp, i.e. these operations reduce the k-value. In some cases, dyestuffs are added to give a strongly coloured paper, i.e. to increase the k-value in the complementary wavelength range. In other cases, e.g. in the manufacture of newsprint, small quantities of a bluish dye may be added to counteract the residual yellowness of the pulp and thus give a more neutral shade. In a pulp, the absorption power is essentially an intrinsic property of the sheet and it is not affected by factors such as the degree of consolidation of the material in the sheet.

In contrast, the scattering power is greatly affected by the sheet structure. Surfaces and interfaces scatter light only if the two phases have different refractive indices and if they are not in „optical contact", which means that the distance between two surfaces must be greater than half the wavelength of light. The process of consolidation of the sheet and subsequent calendering operations therefore tend to lower the scattering power.

5.3.4 Applications of the Kubelka-Munk Analysis

If the k and s values of various potential components of a furnish are known, the properties of the furnish can in principle be calculated by simple additivity. For example, the absorption coefficient of a mixed furnish is given by Equation 5.18 where p_A, p_B, p_C etc are the gravimetric proportions of components A, B and C in the stock.

$$k = p_A k_A + p_B k_B + p_C k_C + \ldots \tag{5.18}$$

Similarly, to a first approximation, the scattering coefficient of a mixed furnish is given by:

$$s \approx p_A s_A + p_B s_B + p_C s_C + \ldots \tag{5.19}$$

This equation must be used with caution, since the manner in which the different components bond together may affect the total free scattering surface in the resulting material. This is particularly the case with fillers, where packing of the particles means that the full potential of the filler in this respect is rarely achieved and it is necessary to talk of an „apparent" scattering coefficient. *Figure 5.3* shows this effect for different mixtures of precipitated calcium carbonate (PCC) and ground chalk, where there is a linear additive relationship between the absorption coefficients of the two pulps but where the scattering coefficients are not additive.

The two forms of calcium carbonate have different particle sizes and different size distributions and thus have different scattering coefficients.

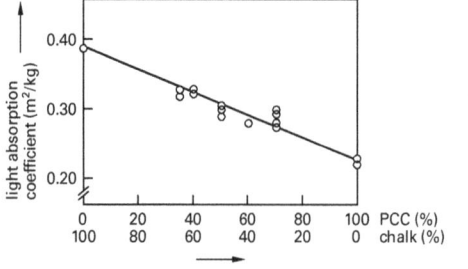

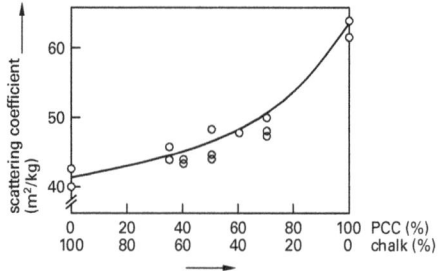

Figure 5.3. (a) The light absorption coefficient and (b) the light scattering coefficient of papers containing different mixtures of precipitated calcium carbonate and ground chalk (after Larsson & Sikker, 1989).

5.4 The Relationship between *s*, *k*, R_∞ and R_0

Figure 5.4 shows, in an optical constitutional diagram, the relationships between some basic optical quantities and how the different quantities are dependent on each other.

The diagram shows R_∞ and the opacity as functions of the light absorption and light scattering coefficients *k* and *s*, and clarifies the principles for the control of papermaking towards improved optical properties. The lines relating to the reflectivity are independent of grammage. The lines relating to opacity have been calculated assuming a grammage of 60 g/m². The diagram is valid independently of wavelength, which means in other words that the relationships hold provided that all measurements are made at the same wavelength.

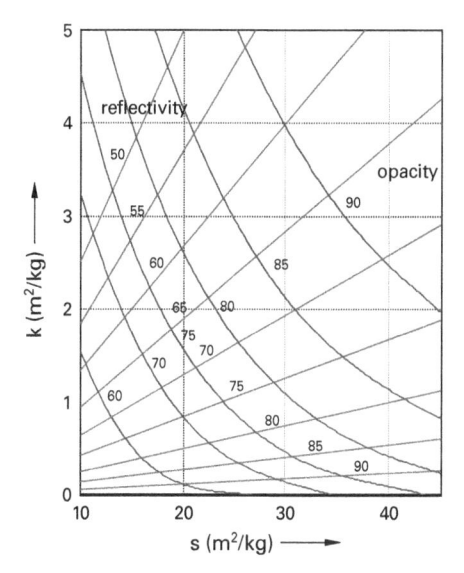

Figure 5.4. The relationships between the optical properties. The opacity data refer to a grammage of 60 g/m².

The optical properties of a material are dependent on the light absorption coefficient and light scattering coefficient of the material. It is thus important for the papermaker to know how the different processes in the papermaking process influence these two material properties. In general, it may be said that pulping and, in particular, bleaching processes seek to reduce the light absorption coefficient, whereas the papermaking process involves efforts to increase the light scattering coefficient.

The first step in the refinement of a pulp is a bleaching process, which lowers the absorption coefficient. In this process, one moves downwards in the diagram and crosses the straight lines for the R_∞-value towards a higher reflectivity.

When paper is produced from the pulp, it must first be mechanically beaten. In the case of mechanical pulps, the refining process leads to a greater number of small particles which give an increase in the scattering coefficient, so that one moves horizontally to the right in the diagram towards a higher R_∞-value and a higher opacity.

In the case of chemical pulps, on the other hand, beating does indeed lead initially to a greater number of small particles, but in the papermaking process these bond together and come into optical contact with each other and the scattering coefficient drops This means that one moves horizontally to the left in the diagram towards a decreasing light scattering coefficient, decreasing R_∞-value and decreasing opacity.

If a filler which has a higher light scattering coefficient than the pulp is added, one moves to the right in the diagram towards both a higher R_∞-value and a higher opacity.

The addition of a dye to the paper always increases the absorption coefficient, so that one thus moves upwards in the diagram towards a lower R_∞-value, but towards a higher opacity.

The fact that the opacity may be raised by increasing k as well as by increasing s is often important. A black paper is more efficient than a white paper if a highly opaque material is required.

The dilemma of the papermaker is that any measure to increase the scattering coefficient normally tends to decrease the strength and vice versa. This can be seen in the effects of beating a chemical pulp, of adding mechanical pulp to a chemical pulp furnish, of adding a filler, of wet pressing etc.

Figure 5.5 shows the curves of *Figure 5.4* divided into two groups that show reflectivity and opacity separately, and approximate positions for a number of different paper grades are indicated in relation to the different parameters.

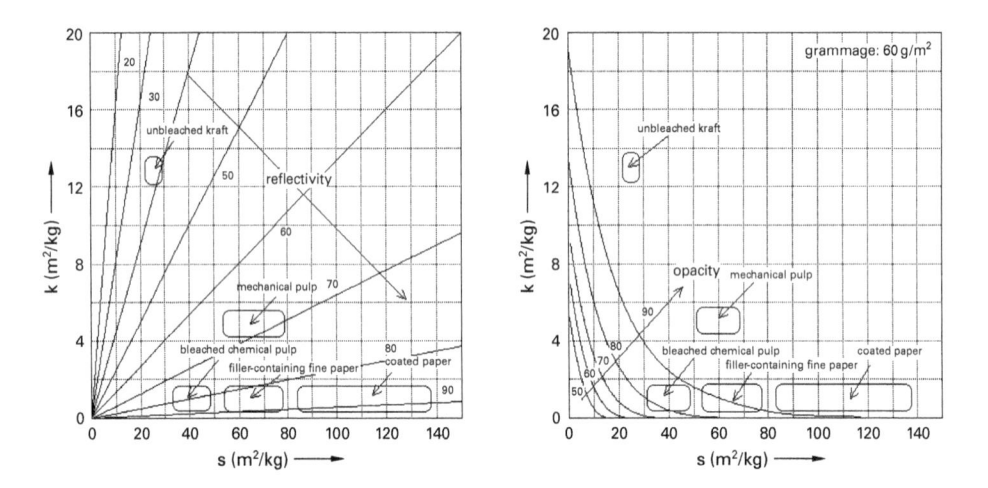

Figure 5.5. The relationship between k and s showing (a) reflectivity contours and (b) opacity contours, with an indication of the approximate positions of different paper grades.

5.5 Measurement of R_∞, R_0 etc

The reflectivity R_∞ and reflectance factors R_0, R_g etc can in principle be measured under arbitrary conditions, but measurements are best carried out under conditions which correspond to the human perception of the appearance of the paper. Before an image is created in the human brain, three components are required:

- radiation from some light source must strike an object
- at least part of this radiation must be reflected from the object
- the reflected radiation must reach the eye and a signal must be sent to the brain.

The signal which reaches the brain and gives rise to a perception of brightness and colour of a material is thus dependent on

- the spectral composition of the incident radiation
- the spectral reflectance characteristics of the material
- the spectral sensitivity of the human eye
- the geometrical conditions of the illumination and viewing situations.

To obtain a meaningful measurement of the optical properties of a material, it is thus not sufficient merely to measure the reflectance of the material in some arbitrary wavelength range. Firstly, it is necessary to select and standardise the type of illumination used. Secondly, it is necessary to filter the radiation using a filter which corresponds to the sensitivity of the eye at different wavelengths. Thirdly it is necessary to decide upon a suitable geometrical presentation.

5.6 Standard Illuminants

To facilitate measurement under standardized conditions, CIE (The International Lighting Commission) has standardized a number of standard illuminants. The most important of these are:

- Standard illuminant A, which corresponds to an incandescent light bulb (Planck-radiator with a temperature of 2854 K),
- Standard illuminant C, which corresponds to daylight on the Northern hemisphere, but which is becoming accepted as representative of indoor daylight,
- Standard illuminant D65, which corresponds to average daylight (Planck radiator with a temperature of 6500 K) which contains considerably more energy in the ultraviolet region than the C illuminant.

The relative spectral power distributions of the standard illuminants A, C and D65 are given in *Figure 5.6*.

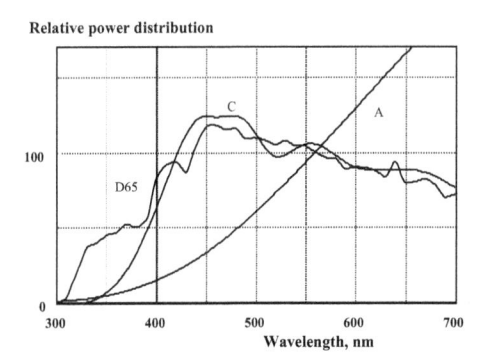

Figure 5.6. The spectral power distributions of CIE standard illuminants A, C and D65.

5.7 Standard Observers

Experiments carried out to determine how humans react to light and how the human brain interprets the stimulus of electromagnetic energy in a specific limited wavelength region in order to perceive colour have led to the concept of the CIE Standard Observer.

In the first place, it has been shown that humans have a *sensitivity to light* which is a maximum at a wavelength of about 550 nm and drops according to a roughly bell-shaped curve to zero at approximately 400 nm and 700 nm. This curve defines what is meant by visible radiation, and is known as the $V(\lambda)$ curve. The region at wavelengths shorter than 400 nm is the ultra-violet (UV) region and the region at wavelengths longer than 700 nm is the infra-red (IR) region. It can thus be said that the eye functions in this respect as a broad-band filter with an effective wavelength of 557 nm.

In the second place it has been shown that the *interpretation of colour* can be described by a combination of three separate sensitivities in the red, green and blue regions of the spectrum. In fact, it has later been shown that the retina of the eye does indeed possess three different types of cones sensitive to different spectral regions. It can thus be said that the eye functions in this respect as a combination of three broad-band filters described by the $\bar{x}(\lambda)$, $\bar{y}(\lambda)$ and $\bar{z}(\lambda)$ functions, where $\bar{y}(\lambda)$ is identical to the $V(\lambda)$ function. These functions are illustrated in *Figure 5.7*.

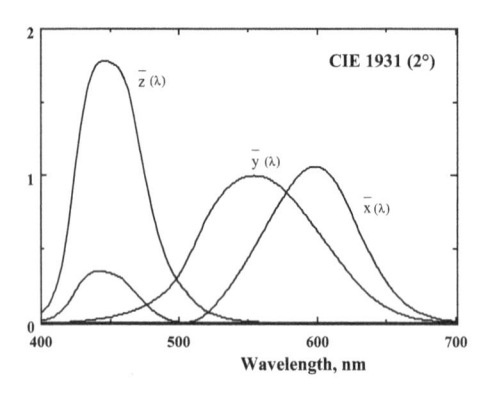

Figure 5.7. The CIE Tristimulus functions representing the colour vision of the human eye

The 1931 Standard Observer was defined in relation to experiments where the half-angle of the viewing cone was only 2°. Later, the 1964 Standard Observer was defined in relation to the results of experiments carried out with a wider viewing cone having a half-angle of 10°. The change in viewing angle makes a slight difference to the human perception of colour.

5.8 Standardized Measurement Conditions (ISO 2469)

For the measurement of the optical properties of pulps and paper, it has been agreed within the International Standardization Organization (ISO) in a document with the number ISO 2469 that measurement shall take place:

- with diffuse illumination and normal observation (d/0°)
- with CIE illuminant C
- according to the CIE 1931 (2°) standard observer
- without the gloss component

In the measurement of whiteness (see 48.15), the CIE illuminant D65 is sometimes used together with the CIE 1964 (10°} standard observer

5.8.1 Instruments Conforming to ISO 2469

The International Standard was originally written in relation to an instrument made by Zeiss and marketed under the name Elrepho. During recent years, new instruments have appeared. The first instruments to replace the Zeiss Elrepho were the Datacolor Elrepho 2000 and the Technidyne microTB1C instruments. A third generation of instruments has now entered the market with the Datacolor Elrepho 3000, the L&W Elrepho, the Technidyne ColorTouch and the Minolta CM-3630.

These instruments all use a sphere to provide a diffuse illumination of the test piece and the test piece is viewed by a suitable detector through an opening in the sphere immediately opposite to the test piece opening, *Figure 5.8*. The sphere is internally lined with a matt white material such as barium sulphate, and the lamps are screened so that direct light from the lamps can fall neither on the detector nor on the test piece. It is thus only diffuse light reflected from the inside of the sphere which strikes the sample, and only the light reflected normal to the surface which reaches the detector.

In the early instruments, glass filters corresponding to the different sensitivity functions of the eye and the desired illuminant were used. The Datacolor E2000 introduced a new technology with an array of diodes which simultaneously detect the reflected radiation at 16 different wavelengths at intervals of 20 nm between 400 and 700 nm. Functions corresponding to the chosen illuminant and to the sensitivity of the eye are then applied in the software where the results are computed. In the third generation instruments, measurements are made at intervals of 10 nm.

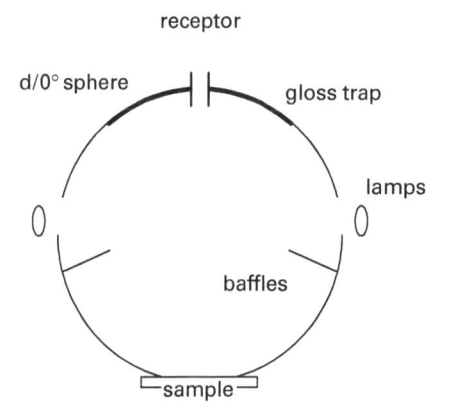

Figure 5.8. Diagram showing the essential features of the instruments used within the pulp and paper industry for optical measurements.

5.8.2 Calibration of the Measurement Instruments

It is an important feature of such instruments that they must be calibrated. A spectral (wavelength) calibration and a photometric (reflectance factor scale) calibration are required.

The manufacturer is responsible for the spectral calibration, i.e. that the measurement takes place at the indicated wavelengths, or that some type of correction is made for any wavelength deviations.

The user is responsible for the photometric calibration, i.e. a decision has to be made as to what is meant by 100 per cent and the instrument has to be adjusted on a daily basis so that it gives a correct reading at 100 per cent. The instrument usually also has to be calibrated at zero per cent. Sometimes, consideration must be given to the fact that the instrument does not necessarily measure linearly across the whole range between 0 and 100 per cent.

The absolute photometric scale is based on *the perfect reflecting diffuser*. This is a concept rather than a physical reality, but several laboratories are able to carry out measurements in relation to this concept and they have also been shown to be in reasonable agreement with each other. (Before this absolute scale could be determined, MgO was used as 100 percent.) As a zero standard, a hollow cavity lined internally with black velvet is usually used.

5.8.3 Calibration within the Paper Industry – the ISO-Hierarchy

For the calibration of instruments for the measurement of paper and pulp, a hierarchy has been established by the ISO organisation. This consists of (a) three Standardizing Laboratories (NIST, PTB, NRCC) and (b) a number of Authorized Laboratories, which supply reference standards of different levels.

The perfect reflecting diffuser (Level 1)
↓
STANDARDIZING LABORATORIES
↓
Reference standards of Level 2 (IR2)
↓
AUTHORIZED LABORATORIES
↓
Reference standards of Level 3 (IR3)
↓
INDUSTRIAL LABORATORIES

Each Authorized Laboratory maintains a reference instrument which is calibrated with a reference standard from one of the Standardizing Laboratories. A special type of tablet is used as IR2 reference standard. To achieve maximum agreement with each other, the different Authorized Laboratories have now reached an agreement to use the same type of tablet for this calibration transfer. 300 IR3 reference standards are sent each month from the Authorized Laboratories to different pulp and paper mill laboratories. These reference standards consist of an optically stable, non-fluorescent, cotton-based paper.

5.9 Optical Measurements

The reflectometer does not measure the total reflectance, i.e. the proportion of the incident light reflected from the sample. Instead, it detects the reflected light only within a narrow cone and it measures the *reflectance factor*, which is defined as the ratio of the reflectance of the sample to the reflectance of the perfect reflectance factor under the same conditions of illumination and detection.

The reflectance factor from an optically infinitely thick pad of sheets is called the reflectivity of the material and is designated R_∞. In practical terms, this means that the pad must be chosen to be sufficiently thick so that a further thickness increase does not influence the result.

The reflectivity can be measured at different wavelengths. In the modern instruments, the light reflected from the sample passes through a monochromator which scatters the light across a number of diodes which make measurements at 400, 420, ..., 700 nm.

From the spectral measurement values, the tristimulus values, X, Y and Z, of the material can be calculated. These three values correspond to the function of the eye in observing colours (see Section 5.13).

The Y-value alone is a quantity which is adapted to the sensitivity of the eye to light and the Y-value is used alone as a measure of the luminance of the material as it would be perceived by an average observer without colour vision. The Y-function covers the whole visible range but it has a maximum effective wavelength of 557 nm. This reflectance factor value is often designated Ry.

A property of particular interest for pulps, but which is also used as a characteristic of paper, is the ISO brightness value. The origin of this property lies in the arbitrary choice of a filter to permit an efficient control of the bleaching process. Since the natural colour of pulp lies in the yellowish range, a measurement in the blue range is more sensitive than the Y-value to how the natural colour of the pulp is removed during the bleaching. The brightness is therefore measured with a filter or calculated with a mathematical function having an effective wavelength of 457 nm, and is often designated $R457$.

The essential differences between the ISO brightness filter and the Y-value filter (based on the $V(\lambda)$ function) are shown in *Figure 5.9*. In this figure, the two filters are shown on relative scales with the peaks at a value of 1.0. To show the advantages and limitations of these two filters, typical curves for a blue and a red paper sample are also shown in the figure. The brightness filter is sensitive in the blue region of the spectrum and thus gives a high brightness value to a blue sample. The Y-value filter, which stretches over and defines the full visual range, has its maximum sensitivity in the mid-spectral range and is insensitive in both the blue and red regions of the spectrum.

The reflectance factor from a single sheet against a black background is designated R_0. An important quantity for printing paper is the ratio between R_0 and R_∞, weighted with respect to the Y-filter and the C illuminant, which is the opacity of the material:

$$opacity = \frac{R_0}{R_\infty} \qquad (5.20)$$

The opacity of a material is a measure of the ability to prevent the reading of a text from being disturbed by the text on an underlying sheet. Note that R_∞ is a material property, whereas R_0 is a single sheet property and is dependent on the grammage.

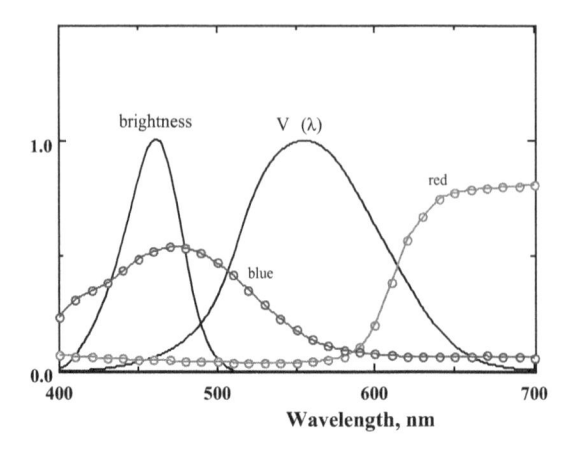

Figure 5.9. The characteristics of the ISO-brightness and Y-value ($V(\lambda)$) filters. Typical reflectance factor curves for a blue and a red sample are also shown to illustrate the advantages and limitations of the two filters

These various properties are listed in *Table 5.1*. The measurement and calculation of colour coordinates and whiteness values are discussed in more detail in later sections.

Table 5.1. Different optical properties.

Property	Symbol	Description
Reflectivity	R_∞	reflectance factor of a pad of the material, so thick that an increase in thickness does not affect the value
Reflectance factor over black	R_0	reflectance factor of a single sheet over a non-reflecting black cavity
Opacity	W	value calculated as 100 times the ratio of R_0 divided by R_∞
ISO brightness	R_{457}	reflectance factor at an effective wavelength of 457 nm
Y-value, luminance	Ry	a measure of the lightness of the material at an effective wavelength of 557 nm
Tristimulus values	X, Y, Z	reflectance factors in the red, green and blue regions which together define the colour of a material
Chromaticity coordinates	x, y	coordinates derived from X, Y, Z which together indicate the chromaticity (colour content) of the material
Dominant wavelength and spectral purity	λ_D, p_E	values derived from the chromaticity coordinates, indicating the monochromatic wavelength which most closely corresponds to the colour of the material and the colour strength
CIELAB values	L^*, a^*, b^*	transformations of the X, Y, Z values to give meaningful values in a three-dimensional colour space
Whiteness	W	transformation of the X, Y, Z values to give a meaningful value for the whiteness of the material
Tint	Tw	tint value associated with the whiteness value to indicate whether there is any colour deviation towards the green or red

5.10 Colour Perception

A colour percept arises in the brain through a complicated process, when electromagnetic radiation meets the retina of the eye. Electromagnetic radiation having wavelengths between 400 and 700 nm, in daily speech called „light", stimulates the light-sensitive cones of the retina. Through a series of physiological processes in the cones and other cells of the retina, analogue potentials arise which are transformed to electrical pulses in the ganglion cells of the retina and these are transmitted via the optic nerve further to the optic centre of the brain. Through a psychological process, a colour percept is created. Colour is thus the result of a physical-physiological-psychological process initiated when light is reflected from the test material and stimulates the colour-sensitive cones and light-sensitive rods of the retina.

We can describe the perceived colour sensation as blue, green, yellow, or red or as a mixture of these hues. We can also characterise the colour as light or dark, as strongly or slightly chromatic etc. Any attempt to order colours thus requires *a three-dimensional system* involving attributes such as hue, chromaticity and lightness.

It is not possible to quantify the physiological and psychological processes which relate a colour percept to its physical stimulus through an explicit mathematical function. The physical stimulus and the percept produced by it can, however, be linked by an empirical psychophysical relationship. The CIE standard observer functions referred to in Sections 5.7 and 5.11 are examples of such relationships, but they are tabulated functions and not mathematical formulae.

5.11 Tristimulus Values

In 1852, Grassman established that any colour percept can be matched by a mixture of three primary colours in suitable proportions, on condition that they have been chosen so that a mixture of two of these primary colours cannot give the third. The system adopted by CIE (Commission Internationale de l'Eclairage, the International Lighting Commission) in 1931 for colorimetry (colour measurement) is based on Grassman's law and this leads to the three spectral colour matching functions $\bar{x}(\lambda)$, $\bar{y}(\lambda)$ and $\bar{z}(\lambda)$ upon which the CIE system for colorimetry is based. These functions are shown in *Figure 5.7*.

For each wavelength, λ, the products of these functions, the reflectance factor $R(\lambda)$ of the sample and energy distribution of the illuminant $S(\lambda)$ at that wavelength give the proportions with which the three functions contribute to the total appearance at the wavelength in question. The whole contribution is given by the summations of these products over the visual range:

$$X = k \sum_{400}^{700} S(\lambda) \, R(\lambda) \, \bar{x}(\lambda) \qquad (5.21)$$

$$Y = k \sum_{400}^{700} S(\lambda) \, R(\lambda) \, \bar{y}(\lambda) \qquad (5.22)$$

$$Z = k \sum_{400}^{700} S(\lambda) \, R(\lambda) \, \bar{z}(\lambda) \qquad (5.23)$$

Where k is a normalization factor:

$$k = \frac{100}{\sum_{400}^{700} S(\lambda) \bar{y}(\lambda)} \tag{5.24}$$

This gives the Y-value = 100 for the perfect reflecting surface. The values X, Y and Z are called the tristimulus values of the colour.

Since the functions included in Equations 5.21–5.24 lack an explicit mathematical form, a summation procedure must be used, based on tabulated values published by CIE for these illuminant and colour matching functions.

5.12 Chromaticity Coordinates

The three tristimulus values give an unambiguous definition of the colour which they represent, but they are difficult to interpret. E.g., the two following sets of figures represent the same hue and the same chromaticness, since the three virtual colours are included in the same proportions.

$$(X, Y, Z) = (10, 20, 30) \tag{5.25}$$

and

$$(X, Y, Z) = (25, 50, 75) \tag{5.26}$$

The second group apply to a much lighter colour than the first group, however. The information is better displayed by a transformation of the data, such that:

$$x = \frac{X}{X+Y+Z} \quad y = \frac{Y}{X+Y+Z} \quad z = \frac{Z}{X+Y+Z} \tag{5.27}$$

Since $x + y + z = 1$, z is defined by the values x and y, and the three tristimulus values can thus be replaced by two values, x and y.

If it is desired to represent colours graphically by their chromaticity coordinates, this is done in the CIE xy-space. an orthogonal coordinate system with x and y as coordinates. In this two-dimensional coordinate system, the point (1,0) indicates the colour X, the point (0,1) indicates the colour Y and the origin (0,0) indicates the colour Z. *Figure 5.10* shows this coordinate system. The horseshoe-shaped curve is the locus of all the pure spectral colours. The straight line joining the extremities of this curve is the purple line, and all real colours are limited to the enclosed area.

It appears here that the representation of colour is reduced to a two-dimensional plane, but this is not the case, the diagram is completed by adding the vertical direction of lightness, the Y-value, as shown in *Figure 5.11*. The dimensions x and y are called the *chromaticity coordinates* of the colour and are together a measure of the colour content of the material while the *Y*-value indicates its luminance (lightness).

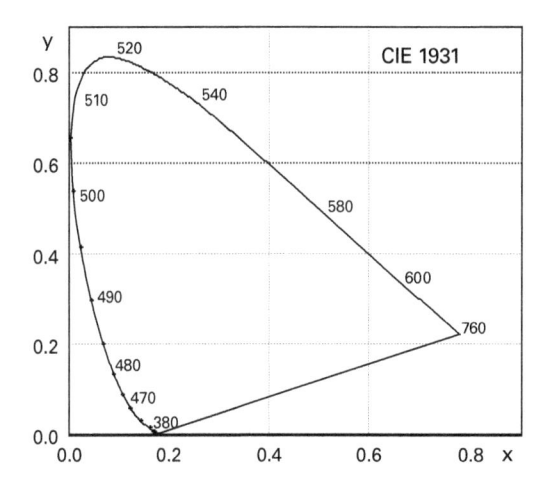

Figure 5.10. The CIE chromaticity diagram.

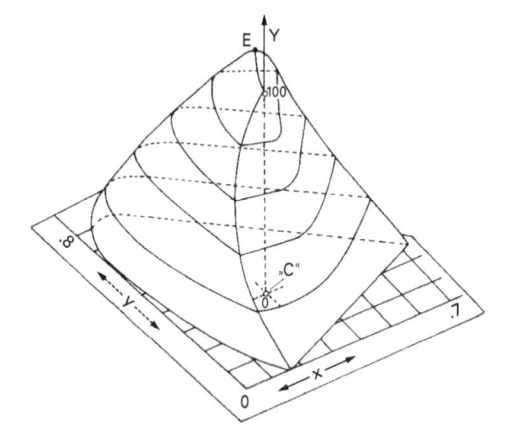

Figure 5.11. The CIE chromaticity diagram in three dimensions, showing the peak at Y = 100.

5.13 Dominant Wavelength and Excitation Purity

Even though the use of chromaticity coordinates facilitates the interpretation of colour values, it is difficult to understand what colour is represented by any pair of values (*x, y*). To give a better understanding of what these numbers mean, a transition can be made from the rectangular coordinate system to a polar coordinate system. With the neutral point defined by the coordinates of the illuminant (x_n, y_n), as origin, a straight line is drawn through the point (*x, y*) representing the colour towards the locus for the spectral colours. All colours which lie on this straight line are said to have the same *dominant wavelength*, which is given by the wavelength where the line intersects the locus for the spectral colours. The ratio of the distance between the coordinates (x_n, y_n) and (x, y) to the distance between (x_n, y_n) and the point of intersection of the line with the spectral locus is called the *excitation purity* of the colour. A chromaticity diagram showing these features is presented in *Figure 5.12*. In this figure the newsprint has a dominant wavelength of 575 nm and an excitation purity of 10 %.

For various reasons, this representation is becoming obsolete. One reason is that the physical quantities „dominant wavelength" and „spectral purity" do not correspond to a sufficiently acceptable extent to the visual characteristics of „hue" and „chromaticness". Colours having the same dominant wavelength do not, for example, have the same visual hue.

Another disadvantage of this diagrammatic representation of colours is that the mathematical distances corresponding to the same visual difference between two colours are different in different regions of the diagram. Although the system gives an unambiguous description which has been valuable in e.g. the standardization of the shade of newsprint, the system is now obsolescent and it is being replaced by the CIELAB coordinate system described in the next section.

The neutral point indicated as (x_n, y_n) is not the equi-energy point ($x = 0.333$, $y = 0.333$). It is the point representing the colour characteristics of the illuminant/observer combination chosen. In *Figure 5.12*, the neutral point represents the C/2° situation. Neutral point data for different illuminant/observer combinations are given in *Table 5.2*.

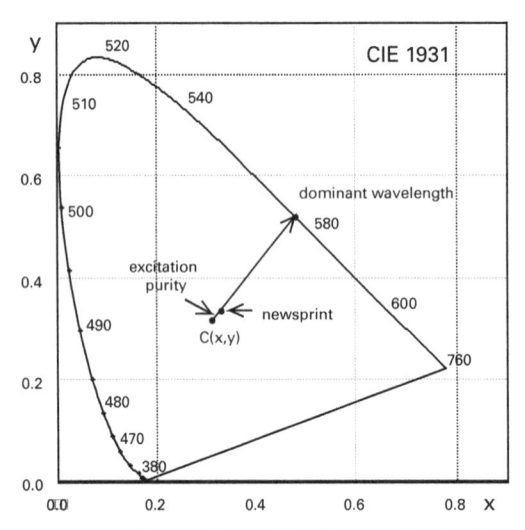

Figure 5.12. The CIE chromaticity diagram, showing for a sample of newsprint how the dominant wavelength and excitation purity are defined.

Table 5.2. Parameters for the neutral points of different illuminant/observer combinations.

Illuminant/observer	A/2°	C/2°	D65/10°
X_n	109,850	98,074	94,813
Y_n	100,0	100,0	100,0
Z_n	35,585	118,232	107,304
x_n	0,4476	0,3101	0,3138
y_n	0,4074	0,3161	0,3310

5.14 CIELAB-Coordinates

In an attempt to establish a system which better mirrors the perceptual experience of colours and colour differences, a new way of mathematically describing different colours has been standardized by CIE in the CIELAB-system. This system is a non-linear transformation of the X, Y, Z-values into a three-dimensional system where L^* corresponds to a lightness value, a^* is the red/green axis and b^* is the yellow/blue axis, according to:

$$L^* = 116\,(Y/Y_n)^{1/3} - 16 \tag{5.28}$$

$$a^* = 500\,[(X/X_n)^{1/3} - (Y/Y_n)^{1/3}] \tag{5.29}$$

$$b^* = 200\,[(Y/Y_n)^{1/3} - (Z/Z_n)^{1/3}] \tag{5.30}$$

Here, X_n, Y_n, Z_n are tristimulus values for the neutral point, which is dependent on the selected illuminant and observer conditions. Tristimulus values of the neutral point for different illuminant/observer combinations are given in *Table 5.2*.

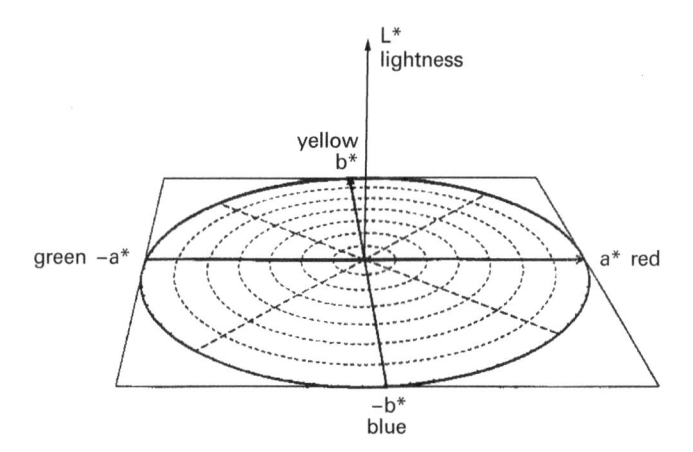

Figure 5.13. A representation of the CIELAB colour space.

This transformation leads to a more or less uniform colour space, as shown in *Figure 5.13*. To a first approximation, the colour difference ΔE between two different colours can be calculated as the distance between two points in this CIELAB-space, according to:

$$\Delta E = \sqrt{\Delta L^{*2} + \Delta a^{*2} + \Delta b^{*2}} \tag{5.31}$$

5.15 Whiteness

Many different attempts have been made to develop a way of mathematically indicating the whiteness of a material. Variations in whiteness are colour differences within a limited range of colours which are regarded as almost white. The brightness value is not sufficient as a measure of whiteness since it takes into consideration only the reflectivity within the blue region.

It is clear that whiteness is in some way a perception of colour and that any attempt to measure the whiteness of a material must be based on a full colour measurement over the whole spectral range, i.e. the tristimulus values. Many different mathematical expressions have been proposed which relate a whiteness parameter to the tristimulus values of the material. Eventually, a committee established within CIE presented an equation for the so-called CIE-whiteness W calculated as:

$$W = Y - 1700(y - y_n) - 800(x - x_n) \tag{5.32}$$

This is complemented by a calculation of a red/green tint value Tw according to:

$$T_w = 650(y - y_n) - 900(x - x_n) \tag{5.33}$$

The conditions for a material to be white in the application of these equations are:

$$40 < W < (5Y - 280) \tag{5.34}$$

and

$$-3 < T_w < +3 \tag{5.35}$$

There is clearly a problem in trying to describe whiteness by a single value, since this conflicts with the understanding that the brain's interpretation of a colour sensation is always three-dimensional, but within the region where the concept of whiteness is meaningful, the equation is becoming accepted.

Methods have now been developed for measuring the whiteness of paper by applying Equation 5.32 to calculate a whiteness value from the measured tristimulus values.

An important feature of this equation is that it recognises that in the subjective human appreciation of whiteness there is a preference for a bluish rather than a yellowish tint. Traditionally in the textile industry, and also in the paper industry, the sensation of whiteness has often been strengthened by the addition of a small amount of a blue dye. More recently, the emphasis has been on the addition of a fluorescent, optical whitening agent, as is discussed in the next section.

5.16 Fluorescence

To produce a higher whiteness than can be attained simply by bleaching or by the addition of a blue dye, a fluorescent whitening agent (FWA) is added to certain products, particularly certain types of fine paper and graphic arts board. This material absorbs energy within the UV-range and re-emits it within the blue visual range at 430–470 nm. This increase in the reflectance factor (radiance factor) in the blue range is observed by the eye as an increase in the whiteness of the material. *Figure 5.14* shows spectral curves for a non-fluorescent paper, for a paper containing a blue dye and for a paper containing a fluorescent whitening agent.

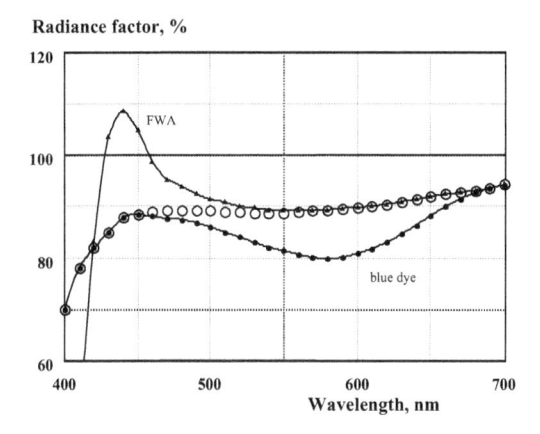

Figure 5.14. Spectral radiance factor curves for a near-white paper, a paper with added blue dye and a paper with added FWA.

Table 5.3 shows calculated data for the curves shown in *Figure 5.14*.

Table 5.3. Effects on optical properties of adding blue dye or FWA.

Property	normal	blue dye	FWA
Y-value (C/2°)	89,4	82,5	90,2
ISO brightness	88,1	87,3	98,4
CIE whiteness (C/2°)	84.1	98.3	118,1

Here it can be seen that a blue dye lowers the Y-value and the ISO-brightness but increases the CIE-whiteness value. The addition of FWA has a negligible effect on the Y-value but leads to an increase in both the ISO-brightness and CIE-whiteness values

The use of a fluorescent additive gives rise to measurement problems, however, since the result of the measurement is to a great extent influenced by the UV-content of the illumination provided by the instrument lamps. To solve this problem, modern instruments include an adjustable UV-cut-off filter, the position of which can be adjusted so that the relative UV-content in the illumination falling on the sample can be made to match that of any desired illuminant.

The Authorized Laboratories referred to in Section 5.8.3 therefore also provide fluorescent reference standards with assigned whiteness values corresponding to the desired illuminant to enable industrial laboratories to adjust the filters in their instruments to the correct positions.

In the initial work when the CIE-whiteness equation was being developed, it was assumed that the D65-illuminant was the most suitable since this has a large UV-content and greatly excites the fluorescent whitening agent, and CIE-whiteness (D65/10°) is indeed now an internationally standardised property.

However, the UV-content in this illuminant exceeds the level which exists in most indoor situations where white paper is used, so that an adjustment of the UV-content to match the C-illuminant is now recommended for indoor conditions, and the concept of indoor whiteness has also been internationally approved, based on the measurement of CIE-whiteness (C/2°). These are the values given in *Table 5.3*. Reference standards with assigned values for both these conditions are now available.

6 On the Mechanisms behind the Action of Dry Strength and Dry Strength Agents

Lars Wågberg
Department of Fibre and Polymer Technology, KTH, Stockholm

6.1 Background   169

6.2 Paper Strength as an Adhesion Problem   170

6.3 Influence of Dry Strength Additives on Paper Strength – The More Traditional View   172

6.3.1 Tensile Properties   172
6.3.2 Compression Properties   178

6.4 References   182

6.1 Background

The increasing use of recycled fibres and other inexpensive furnish components, such as fillers, has led to a decrease in the strength properties of the paper produced from theses raw materials. To overcome this problem it is very common to add different dry strength additives, such as cationic starch [1–15], modified polyacrylamide [16, 17], polyamideamine epichlorohydrine resins (PAE) in combination with carboxymethyl cellulose [18, 19], cationic dialdehyde starches [20] and chitosan [5, 20, 21]. This list can probably be made much more extensive and the references given here can serve more as examples and an introduction to the use of dry strength additives in different paper grades. However, in for example testliner manufacture a large amount of starch is added both in the wet end and in size press applications in order to reach the desired strength properties of the paper. If for example the size press could be avoided it should be possible to significantly increase the productivity of the machines. A general understanding of the mechanism behind the dry strength additives would also be desirable since this would enable an optimised use of these additives.

Wet strength additives are today mostly used for the production of different grades of hygiene papers, filter papers, special grades of board and sack paper [22, 23]. It is also well accepted that the different chemicals are efficient through different mechanisms but there is still a debate regarding the molecular mechanism responsible for the wet strengthening action. The development of new types of wet strength chemicals has been very limited over the last 30 years but the development of new chemicals for preparation of wrinkle free clothing has spurred a development of new chemicals also for the paper industry [24-28]. So far the application of these latter additives has been limited to laboratory trials but very high relative wet strengths can be achieved however so far at the cost of folding endurance of the produced papers.

6.2 Paper Strength as an Adhesion Problem

Regardless of the exact mechanism responsible for the interaction between two fibres it can be concluded that that the interaction between the fibres during consolidation and drying of the paper will be very important for the strength of the paper. During the consolidation and drying process the highly swollen fibre surfaces will be pushed together by the capillary forces formed between the fibres during water removal. This capillary force will deform the external surfaces of the fibres and an intimate contact will be formed between the fibres in the dry sheet resulting in the situation schematically described in *Figure 6.1*.

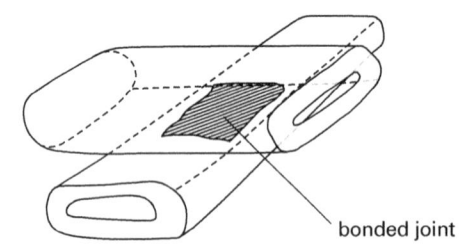

Figure 6.1. Schematic representation of the bonded joint formed between the fibres during consolidation and drying.

The strength of this bonded joint will be determined by

1. The molecular contact area in the contact zone
2. The intermolecular forces
3. Mechanical entanglement between opposing surfaces
4. Possible existence of covalent linkages

In order to determine the relative influence of these factors it is necessary to perform model experiments under well-controlled conditions but this information is unfortunately still lacking.

Traditionally the molecular contact area between the fibres has been determined with light scattering measurements. Since the dimensions that can be determined with light scattering is about half of the wavelength of the light it is easy to realise that the molecular contact area can not exactly be determined with light scattering experiments. This is also one of the reasons why the other factors in connection with *Figure 6.1* not has been determined quantitatively.

A more correct description of the „true" dimensions in the contact zone between the fibres is given in *Figure 6.2* where both the area that can be determined with light scattering has been marked together with „true" contact area. The enlarged picture in *Figure 6.2* also demonstrates that in the area in real contact there might still be areas in real contact and areas where there still exists a small gap between the surfaces.

The specific joint strength in the partly joined area will be determined by several different factors and these different factors are described schematically in *Figure 6.3*.

As was mentioned earlier the mechanical interlocking between the surfaces will contribute to the joint strength in a type of Velcro® organisation of cellulose fibrils on the fibre surface. Another type of interlocking that might occur is the inter-diffusion of polymers across the interface. This inter-diffusion will result in a situation where one part of the polymer resides in the bulk of one phase and with the other part in the bulk of the opposing phase. The total contribution to the joint strength from this inter-diffusion will be dependent on the number of molecules that have crossed the interface and the length of the embedded polymer in each phase. However, if the migration is such that the embedded length is much smaller in one phase than in the other the shorter chain will be the limiting factor for the contribution to joint strength. In practical terms it is likely that short chains of hemicelluloses might migrate across the interface between the fibres and in this way contribute to the specific joint strength.

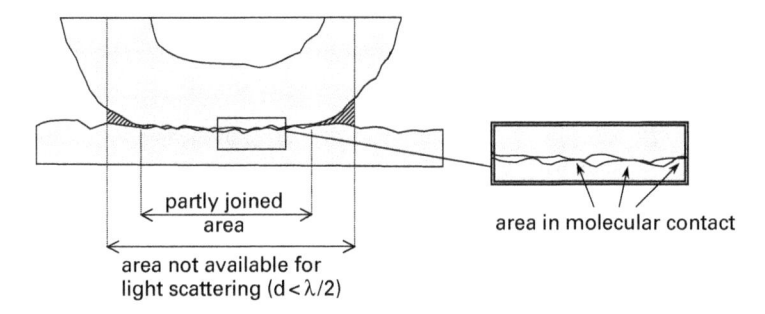

Figure 6.2. Description of which areas that might be detected with light scattering and which areas between the fibres in a fibre/fibre joint that are in real contact. The inset also shows that there still might be areas in poor contact also in the partly joined area.

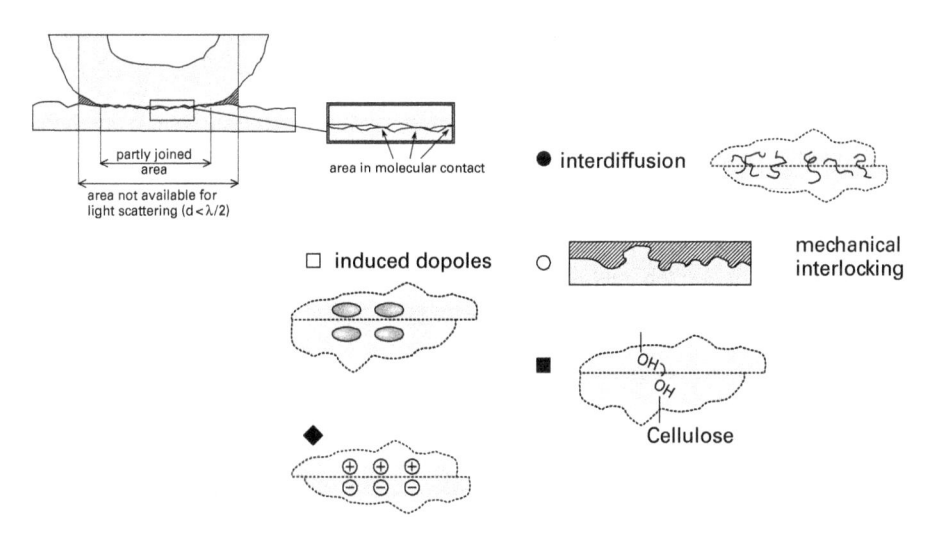

Figure 6.3. Schematic description of the different types of interactions that will determine the specific joint strength in the contact zone between the fibres.

The intermolecular forces that was mentioned in conjunction with *Figure 6.1* has in *Figure 6.3* been divided in three different groups:

1. Hydrogen bonding
2. Van der Waals interactions
3. Ionic bonding

There are no real proofs for which of these that will dominate the contribution from the intermolecular forces but most likely the hydrogen bonding and the van der Waals forces will be the dominating factors. Non-specific van der Waal interaction energy, W_{vdW}, between macroscopic bodies are characterised by the Hamaker constant (A) of the material, are rather long-range and scales as 1/distance between the surfaces according to Equation (6.1), which is valid for crossed cylinders of radius R_1 and R_2.

$$W_{vdW} = \frac{-A\sqrt{(R_1 R_2)}}{6D} \qquad (6.1)$$

Since the force between the crossed cylinders is the distance derivative of the interaction energy the force between the two cylinders will be

$$F_{vdW} = \frac{A\sqrt{(R_1 R_2)}}{6D^2} \qquad (6.2)$$

Since the Hamaker constant for cellulose is in the range of $6 \cdot 10^{-20}$ Nm the van der Waals attraction between two crossed cylinders can be calculated at least for the dry state. In order for a better accuracy the solid cylinders should be exchanged for cylindrical shells.

The hydrogen bonds on the other hand are specific which in turns mean that they are very sensitive to the distance between the hydrogen bonding groups on adjacent surfaces and also the geometric orientation between these groups. This latter fact has during the last years led to doubts about the importance of hydrogen bonding for specific joint strength and hence paper strength since it is very unlikely that to rough fibre surfaces will come into such close proximity that hydrogen bonds might develop between the surfaces. However, recently it has in several publications been shown that the wet transverse elastic modulus of fibres is very low, i.e. in the order of 1–5 Mpa and maybe even lower. This means that the fibre surface resembles the surface of a soft gel and considering the large capillary forces that will pull the fibre together during drying it is not at all unlikely that highly mobile molecules on the fibre-gel surface might orient to form hydrogen bonds with hydrogen bonding groups on the adjacent fibre-gel surface. This area is currently in focus for many research groups and new knowledge about the importance of the different factors composing the intermolecular forces will definitely be available in the near future.

6.3 Influence of Dry Strength Additives on Paper Strength – The More Traditional View

6.3.1 Tensile Properties

Despite the common use of the dry strength additives there is still no mechanism explaining how these additives really work. Several attempts have been made over the years and Davison [29] made a very thorough survey of different mechanisms behind the development of paper strength in general and more specifically how chemicals may affect the strength development.

However, in [29] no conclusions regarding any specific mechanism were given. In another communication [30] Davison suggested that the weak link in paper strength was the fibre-to-fibre bond since the fibre strength was at least twice as large as the dry strength of papers composed of these fibres. This is illustrated in *Figure 6.4*, taken from [30], where data show that despite the fact that a sheet was believed to be „fibre strength limited" it was possible to pull out a large amount of intact fibres from these sheets. The conclusion drawn by Davison [30] was that the fibre-to-fibre bond should be increased in order to improve the paper strength.

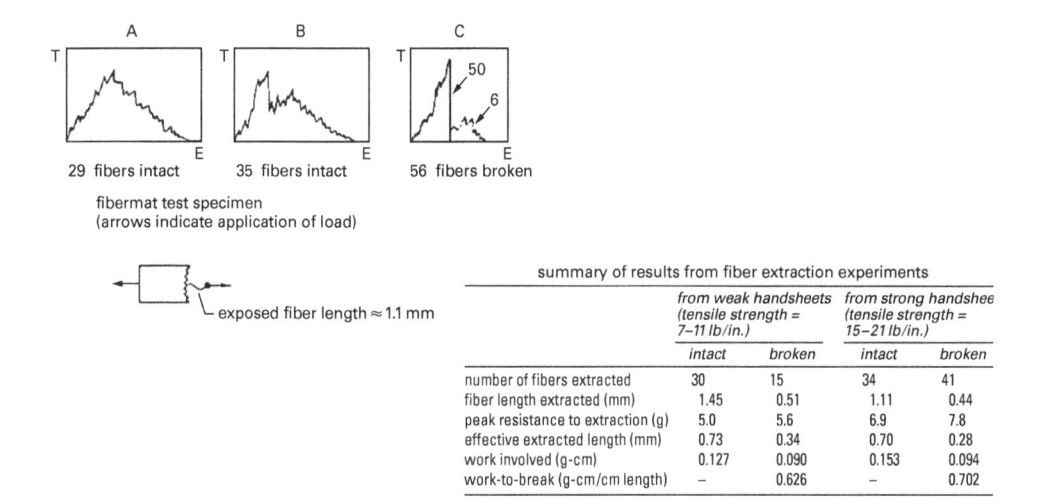

Figure 6.4. Results taken from [30] showing that despite the fact that sheets are strong it is still possible to pull out a large number of intact fibres from the sheets indicating that the joint strength between the fibres is the limiting factor also in these sheets.

In later work [8, 31], where the „shear bond strength" between fibres was determined with the aid of the Page theory [32] which relates the paper strength to the fibre strength and the „shear bond strength" between the fibres according to Equation (6.3).

$$\frac{1}{T} = \frac{9}{(8Z)} + \frac{12C}{(P \cdot L \cdot b \cdot RBA)} \tag{6.3}$$

where
T = Tensile strength of the paper
Z = Zero span of fibres
C = Fibre coarseness
P = Fibre perimeter
b = Shear bond strength
RBA = Relatively Bonded Area

By pressing sheets to different densities with and without cationic starch and inserting relevant values in equation (6.3) it was also found that cationic starch [5] and PAE (polyamideamine epichlorohydrine) polymers increased the paper strength by increasing the „shear bond strength" between the fibres. This procedure demands that the *RBA* can be determined in order to extract the shear bond strength from the slope in a plot of $1/T$ as a function of $1/RBA$. This entity is determined by measuring the light scattering from sheets with zero strength (S_0) (extrapolated values) and from sheets pressed to different densities (S). These values are then used to calculate *RBA* according to Equation (6.4).

$$RBA = \frac{S_0 - S}{S_0} \tag{6.4}$$

Considering the distance over which the van der Waal forces are active according to equations (6.1–6.2) it is easy to realise that the light scattering can not be used to determine a molecular contact area necessary to calculate a correct *RBA* and the Page theory for paper strength is based on the assumption that there is a direct relationship between the real molecular contact area and the area detected by the lightscattering experiments. The use of this theory to calculate a „true" shear bond strength between fibres has been and is being questioned and it is mentioned here not to support or to question its applicability but more since it is the only semi-quantitative theory available today to link paper strength to fibre strength and fibre properties.

Regarding the action of cationic starch on paper strength it has, on the contrary, also been suggested [10] that the cationic starch is efficient through the creation of an increased number of „bonds" between the fibres. This conclusion was drawn from experiments where air-dried and freeze-dried sheets with starch were compared. It was shown that the relative increase in strength was larger for the freeze-dried sheets and this was taken as a „proof" for the formation of new „bonds". This is illustrated in *Figure 6.5* taken from [10].

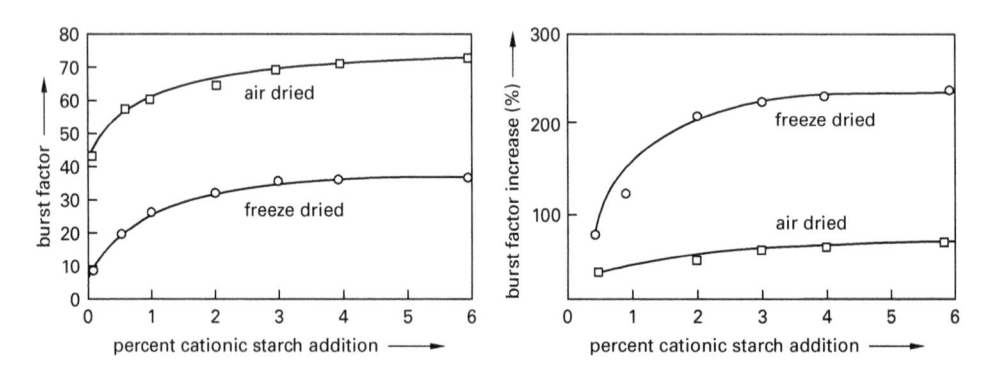

Figure 6.5. By adding cationic starch to fibres and forming sheets that were subsequently either air dried or freeze dried it was found that since the increase in strength, in %, was higher for the freeze dried sheets it could be concluded that the starch could form new joints between the fibres. From [10].

This formation of new „bonds" was directly linked to the cationic nature of the starch since a native starch did not give the same strength response. It was however also stated in [10] that the starch can be efficient through reinforcement of already existing bonds.

The influence of hydrogen bonding on paper strength was investigated in a pioneering work by McKenzie [32]. By acetylating 20 % of the hydroxyl groups of the fibres it was found that the strength of papers formed from water was decreased to 10 % of the strength of the unmodified fibres. If paper was formed from the modified fibres in acetone it was found that sheets as strong as the sheets from unmodified fibres were formed in water, this is shown in *Figure6.6*. McKenzie [33] concluded that hydrogen bonding was important for the formation of strong fibre/fibre joints in water whereas in acetone the mechanism behind the creation of paper strength could be related to "..mechanical interlacing of fibres and cellulose chains to an extent depending on the degree of swelling of the fibres."

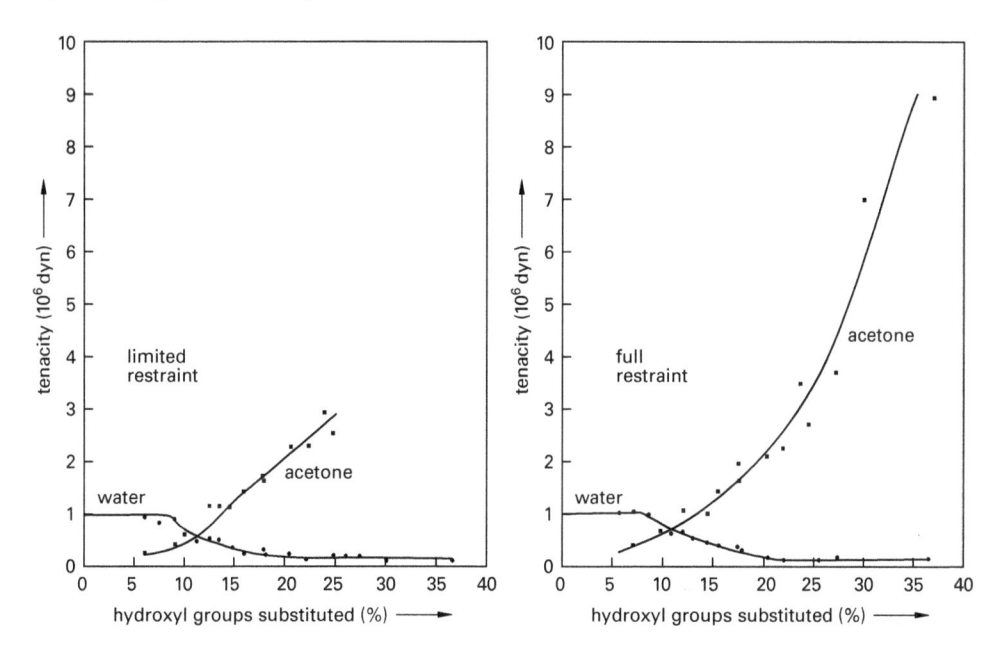

Figure 6.6. By acetylating 20 of the hydroxyl groups of the fibres McKenzie [33] found that the strength of paper made from water decreased to 10 % of the value achieved for unmodified fibres whereas the strength of papers from modified fibres from acetone was even stronger than the original paper made from water.

McKenzie also introduced the concept of looking at the fibre/fibre joint as an adhesion problem and interdiffusion of surface polymers of the fibres into adjacent fibres to increase the strength of the fibre/fibre joint [34]. In a later work by Pelton et al [35] the compatibility of surface polymers, adsorbed to the fibres, on the formation of strong fibre/fibre joints was shown. It was found that when the compatibility of the surface polymers was low, despite full water solubility of the polymers, there was a clear decrease in the joint strength between the fibres. The influence of the outermost layers of the fibres on the formation of strong fibre/fibre contacts has also been shown in recent work [36, 37]. All these results show the large potential of elucidating and optimising the molecular contact area and the bond strength in the molecular contact zone. In *Figure 6.7* the effect of the application of polelectrolyte multilayers to fibres on paper strength is shown [36]. The fibres were prepared with layers of cationic polallyalmine and anionic polyacrylic acid before sheet preparation and as can be seen the application of 4.5 bilayers of polyelectrolytes was sufficient to increase the dry strength of the paper about three times. As can be seen there is a dependence of which polymer that is fixed in the outer layer of the multilayer and this probably depends on the structure of the polyelectrolyte multilayer. There is also an effect on the wet strength of the paper but this will be discussed in Part II of this chapter.

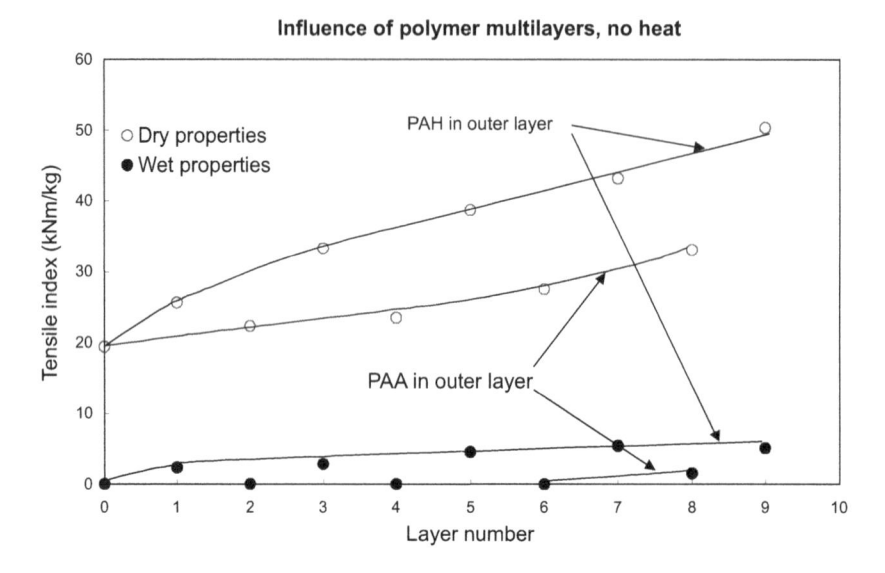

Figure 6.7. Influence of treatments of fibres with polyelectrolyte multilayers on the achieved paper strength. The cationic polymer was a polyallylamine and the anionic polymer was a polyacrylic acid [36].

For the use of starch in clay filled sheets Lindström et al. [1] introduced the stress concentration minimisation (SCM) theory in order to explain how starch will prove the strength. In this theory it is first stated that the introduction of filler produces stress concentrations in the sheet, which subsequently leads to breakage of the sheet. As described in [1] the lifetime of a paper strip under load can be described by the Zhurkov theory for lifetime of materials and in a simplified version the relationship between lifetime, load and the stress concentration factor can be written as

$$t_f = t_0 \cdot \exp\left[\frac{U_0 - j\sigma/\rho}{RT}\right] \qquad (6.5)$$

where
t_f = Life time of the material
t_0 = Empirical constant related to the time of an atomic vibration ≈ 10^{-13} s
U_0 = Activation energy for bond rupture
j = Stress concentration factor
σ = Applied stress
ρ = Density of the sheet

By plotting the ln(life time) as a function of applied stress the stress concentration factor can be determined from slope of the curve in this diagram.

The introduction of starch in the contact point between the filler and the fibres leads to a local yielding when the paper is subjected to a stress increase and a global failure of the sheet can therefore be avoided. This is shown in *Figure 6.8* where the stress concentration factor is presented as a function of addition of cationic starch at different filler concentrations [1].

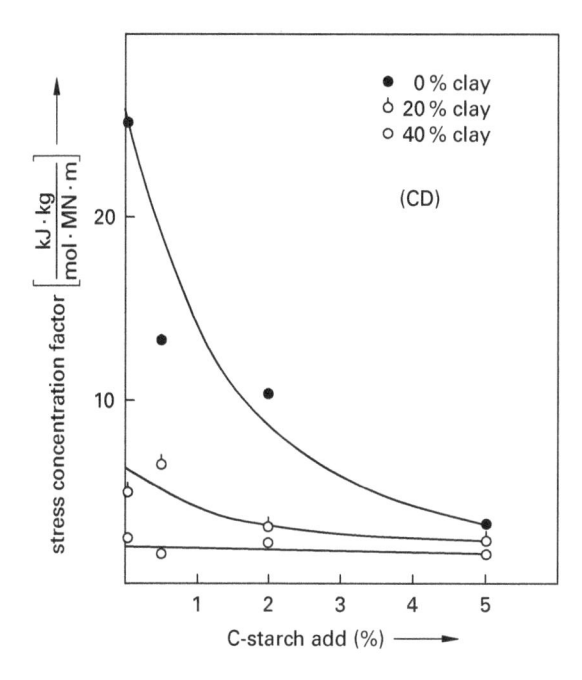

Figure 6.8. The influence of cationic starch and filler concentration on the stress concentration factor in papers containing bleached chemical fibres and filler clay [1].

In an extension of this work [7] it was suggested, as a model, that the fibres were covered with a film of filler particles and that the properties of this film was essential for the strength upon addition of the starch, the experimental results in [1, 7] showed that the incorporation of large amounts of starch in the paper could almost entirely overcome the negative effect of large amounts of filler particles. It was also shown that a wet end addition was more efficient that a size press addition and this was attributed to the need to have the starch between the fibres in order to have full effect of the added starch.

In summary of this first part it may be concluded that for tensile properties and bending stiffness properties it is fairly safe to suggest that the major mechanism behind the dry strength additives is an increase in joint strength between the components in the furnish. It is not clear though if the additives are able to create new joints between the furnish components sine the techniques used so far have been too rough to distinguish what is happening in a size range of 150 Å, which is the range at which the surfaces start to interact with each other in deionised water. In process waters this distance would be even smaller.

6.3.2 Compression Properties

So far the discussion has almost entirely been focused on graphic papers but it is naturally also valid for different types of packaging papers. The literature in this field is very limited and rather few papers are published, which treat the combination of these type of papers and dry strength additives [16–18, 38]. The most important property of these papers is the compression strength but many of the papers written in this area have been focused on tensile properties or burst strength and very little, if any, has been written on the mechanism of dry strength additives in these papers. Furthermore, it is very difficult to find a unified view on the mechanism behind compression failure in additive free raw materials used for packaging products. Fellers [39] has given an extensive review of the possible mechanisms and with reference to general papers on composite materials he suggests that either of the following or combinations of several of the mechanisms can be responsible for the compressive failure in paper

1. Micro buckling of the fibres, matrix still elastic
2. Matrix yielding followed by micro buckling of the fibres
3. Debonding and interlaminar shear followed by microbuckling of the fibres
4. Shear failure (kink, shear band formation)
5. Separation by transverse tension through the thickness (= delamination)

When examining the compression failure zone in paper the situation in *figure 6.9* is usually found. In this figure, presented by C.Fellers STFI, Stockholm, Sweden, it can be seen that there is a kink-band formation where the paper structure has been delaminated but it is not possible to determine whether this caused by a fibre wall damage or a buckling of the segments before the fibre wall delamination.

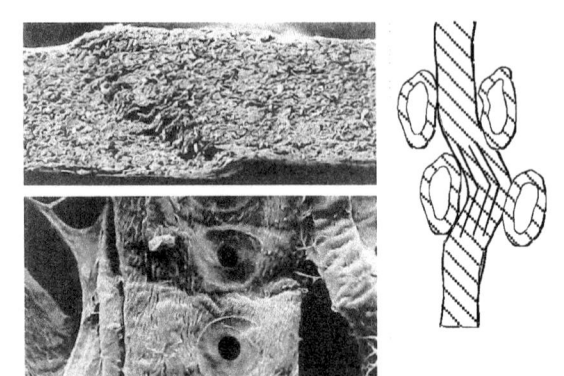

Figure 6.9. Delamination zone of paper that has failed under compressive load. An obvious kink band is formed but the exact mechanism behind the failure is not known.

Paper might be described as a composite material of the components constituting the fibre wall but it has to be remembered that a paper with a density of 750 kg/m^3 still has a void volume of 50 % and a continuum does not exist. Therefore the similarities to composites have to be treated with large caution. This fact has been considered, among others, by Seth [40] who concluded that fibre/fibre „bonding" is very important at low paper densities but as the density increases the influence of the „bonding" in filler free sheets, on the compression strength, decreases and at high densities the compression strength of the fibres becomes the limiting factor. Seth assumed that the „bonding" was increased as the density increased. It should naturally be stated that the „bonding" is as important at high densities as at low densities. This discussion hence suggests that as the papermaking conditions changes there might be a changeover from one compression failure mechanism to another.

With reference to the discussion by Davison [30], mentioned above, this conclusion might be very important for the understanding of the influence of dry strength additives on the compression strength of paper. The possible changeover between different mechanisms can also be seen in *Figure 6.10* where the compression strength of a paper made from a kraft liner pulp is shown as a function of paper density with or without addition of different additives (Wågberg, L. Unpublished results). At low densities there is a large influence of the additives but at higher densities all different papers tend to approach the same SCT (Shortspan Compression Strength) value.

It might be speculated that the joint strength between the fibres has a large influence on paper strength at lower densities whereas the inherent fibre properties will start to dominate at higher densities. It must also be stressed that these questions are currently under intense investigation since recycled linerboard has started to show such poor quality that new additives and new insight into the compression failure mechanism and its dependence on additives is needed.

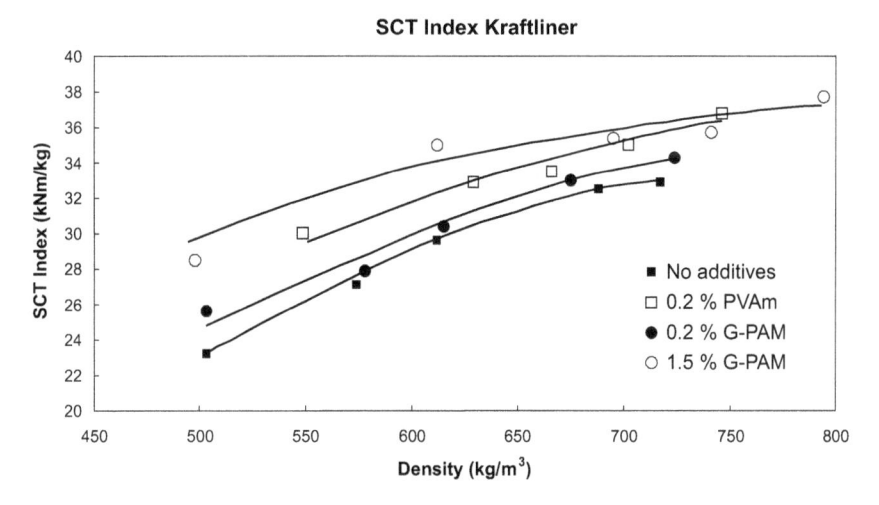

Figure 6.10. Short span Compression Strength of papers made fromk a kraft liner pulp with or without the addition of different additives. PVAm=Polyvinylamine and G-PAM is glyoxal treated polyacrylamide. As the density is increased the relative influence of the additives decreases. (Wågberg, L. Unpublished results).

Sachs and Kuster [41] linked the load deformation curve in compression to the behaviour of cross-sections of the paper as observed in a special device mounted inside a scanning electron microscope. They concluded that „the failure mechanism appears to involve the formation of dislocations in the form of proturberances and fissures in the cell walls, the detaching of fibre wall tissue and separation of microfibrillar bonds (particularly of the S_1/S_2)". From these findings they then concluded that it is this separation of S_1 and S_2 that leads to the delamination of the fibre wall and subsequently to the compressive failure of the linerboard. A typical micrograph from their work is shown in *Figure 6.11* where the fissures in the fibre wall of fibres in the compressive failure zone can clearly be detected.

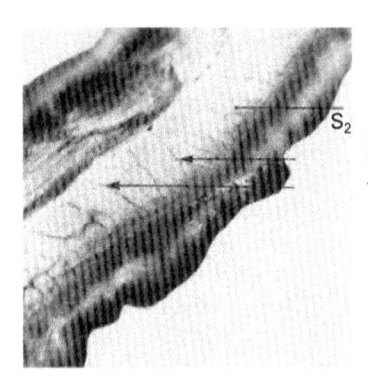

Figure 6.11. Scanning electron micrograph of a fibre that has failed under compressive load. Fissures in the fibre wall were taken as an indication that it was the fibre wall delamination that was the initial process behind the compressive failure of the paper [41].

On the other hand Perkins et al. claimed [42], with support from a theoretical approach and experimental investigations, that the compressive failure is a localised buckling phenomena and that the incremental transverse shear modulus is the most important variable for the compressive strength of linerboard. This theory is however very difficult to apply since some of the parameters in the equations are virtually impossible to determine.

Uesaka [43] has also suggested that the compressive failure is a bending shear buckling phenomena. This was concluded from, among other things, experiments where a relationship between the interlaminar shear modulus and compressive strength was found. It should though be mentioned that Uesaka [43] used a buckling equation, which does not take the transverse shear into consideration and obtained a finite value of the critical load when the slenderness ratio approached zero. However, a more rigorous analysis, including transverse shear, shows that the critical load goes to infinity when the slenderness ratio approaches zero.

As is obvious from this short summary there is no clear single mechanism explaining the reason to the compression failure in packaging papers. On the contrary it is the opinion of the author that it has been demonstrated that there might be several explanations depending on the papermaking conditions, additives and papermaking raw materials.

Dry strength additives might affect several properties of the fibres that might have of large influnec on the paper strength. First of all the additives can increase the number of active joints in the paper. This will decrease average length of the free segments between fibre joints and naturally this will increase the buckling resistance of the paper. Secondly the additives might form a thin very stiff layer on the fibre surfaces that might increase the buckling resistance of the fibre segment between two fibre/fibre joints. A third possibility is that the additives might block crevices and inhomogeneities on the fibre wall and in this way remove possible initiation points cracks that might propagate throughout the fibre wall under compressive load. Finally the additives might cross link the entire fibre wall enabling the avoidance of a delamination of the fibre wall when the paper is subjected to compressive loading.

All these mechanisms are schematically shown in *Figure 6.12* and if it would be possible to tailor chemicals that could be specialised to handle a single mechanism of those mentioned in *Figure 6.12* it would also be possible to determine the relative influence of the different mechanisms causing the compressive failure in a paper under compression load.

a) improvement in number of active bonds, shorter segments between joints

b) stiff polymer layer on the fibre surface

c) filling of crevices on the fibre surface, elimination of starting points for cracks

d) crosslinking of the fibre wall, prevention of kink band formation

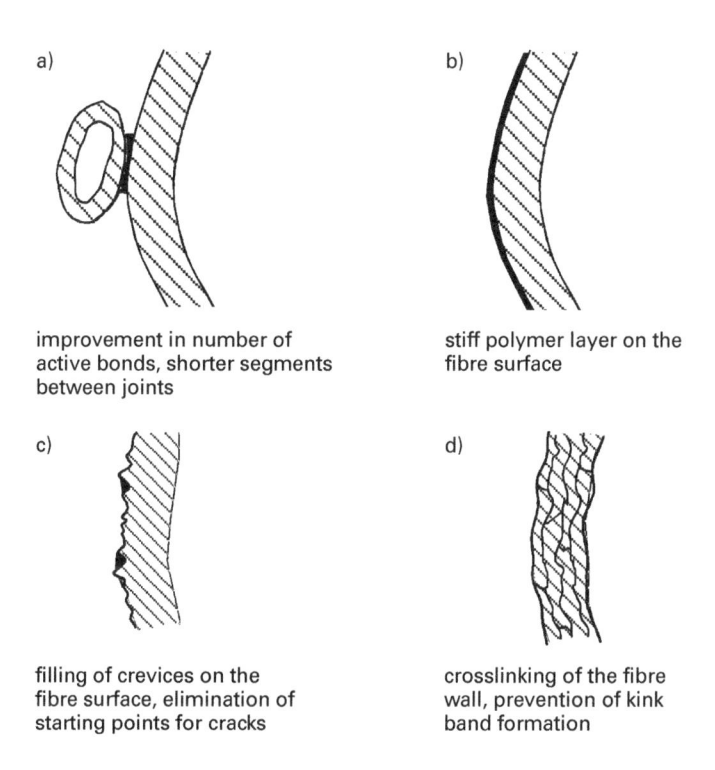

Figure 6.12. Different possible influences of additives on the improvement in compression strength of the paper following addition of dry strength additives. a) Improvement in the number of active joints and hence a decrease in the efficient segment length betweemn two joints. b) Formation of a stiff polymer layer on the fibre surface that might improve the buckling load of a single segment. c) Blocking of crevices and inhomogeneities that might be initiation points for cracks. d) Crosslinking of the fibre wall to prevent fibre wall delamination.

The reason for summarising the possible mechanisms for compression failure of packaging papers is to draw the attention to the possibilities for improvements with an appropriate chemistry. Astonishingly little work has been conducted to investigate if it would be possible to improve the compression strength of the paper through fibre wall strengthening, improved bending stiffness of the fibre segments and increased joint strength between the fibres.

6.4 References

[1] Lindström, T., Kolseth, P., and Näslund, P.: In V. Punton (Ed.) *Papermaking Raw Materials*. Proceedings of the Confernce held at Oxford 1985, Mech. Eng. Publ., London, 1985, pp.589.
[2] Lindström, T., and Florén, T.: *Sven. Papperstidn.* 87(1984)R97.
[3] Björklund, M.: Lic Thesis, KTH, 1994.
[4] Retulainen, E., Nieminen, K., and Nurminen, I.: *Appita* 46(1993)1,33.
[5] Ghosh, A.K.: Appita 47(1994)3,227.
[6] Formento, J.C., Maximino, M.G., Mina, L.R., Sray, M.I., and Martinez, M.J.: *Appita* 47(1994)4,305.
[7] Rigdahl, M., Lindström, T., and Kolseth, P., in H. Giesekus and M.F. Hibberd (Eds): *Progress and Trends in Rheology*. Proceedings of the Second Conf. of European Rheologists, Prague, June 17–20, 1986, pp.240.
[8] Howard, R.C., and Jowsey, C.J.: *J.Pulp Paper Sci.* 15(1989)6, J225.
[9] Gaspar, L.A.: Proc. *Tappi Annual Meeting, Atlanta, GA, 1982*. Tappi Press, 1983, pp.89–91.
[10] Moeller, H.W.: *Tappi* 49(1966)5,211.
[11] Lindström, T., Hallgren, H., and Wågberg, L.: *Proceedings from the EUCEPA Conference held in Florence 1986*, pp.13:1.
[12] Reynolds, W.F. (Ed.) *Dry strength additives*. Tappi Press, Atlanta, 1980.
[13] Laleg, M., Pikulik, I.I., Ono, H., Barbe, M.G., and Seth, R.S.: *Proceedings from the 77th Annual Meeting of CPPA/TS, Montreal, 1989*, B159.
[14] Laleg, M., Ono, H., Barbe, M.G., Pikulik, I.I., and Seth, R.S.: *Paper Techn. Ind.* 32(1991)5,24.
[15] Marton, J.: *Tappi* 63(1980)4,87.
[16] Smith, D.C.: *Proceedings from Tappi Papermakers Conference 1992*, Tappi Press Atlanta, 1992, pp.393.
[17] Kimura, Y., and Hamada, M.: *Proceedings from Japan Tappi Conference 1993*, Tappi Press, Atlanta, 1993, pp.83.
[18] Stratton, R.A.: *Nordic Pulp Paper Res.J.* 4(1989)2,104.
[19] Stratton, R.A., and Colson, N.L.: Materials Research Society, *Mat. Res. Soc. Symp. Proc.* Vol. 197(1990)173.
[20] Laleg, M., and Pikulik, I.I.: *Nordic Pulp Paper Res. J.* 8(1993)1,41.
[21] Allan, G.G., Fox, J.R., Crosby, G.D., and Sarkanen, K.V.: Fibre-Water interactions in papermaking. *Transactions of the symposium held at Oxford:Sept.1977*, Technical Division of the Paper and Board Industry, London, U.K., pp.765.
[22] Dunlop-Jones, N., in J.Roberts (Ed.): *Paper Chemistry*. Blackie Academic & Professional, ISBN 0 7514 0236 2, 1996, pp. 98.
[23] Bates, R., Beijer, P. and Podd, B., in L. Neimo (Ed.): *Papermaking Chemistry*. Fapet Oy, ISBN 952 52160407,1999, pp. 289.
[24] Welch, C.M., and Kottes Andrews, B.A.: *Textile Chemist and Colorist* 21,2(1989)13.
[25] Welch, C.M.: *Textile Res. J.*58,8(1988)480.
[26] Caulfield, D.F.:*Tappi J.* 77,3 (1994) 205.

[27] Luner, P., Zhou, Y.J., Caluwe,P., and Tekin, B., in C.F. Baker (Ed.): *Products of Papermaking. Transactions of the 10th Fundamental Research Symposium held at Oxford, Sept.1993*, Pira International, 1993, ISBN 1 858020549,pp.1045.
[28] Xu, Y., Yang, C.Q., and Chen, C-M.: *Tappi J.* 82,8(1999)150.
[29] Davison, R.W., in W.F. Reynolds (Ed.): *Dry strength additives*. Tappi Press, Atlanta, 1980, pp.1–31.
[30] Davison, R. W.: *Tappi* 55(1972)4, 567.
[31] Wågberg, L., and Björklund, M.: *Nordic Pulp Pap. Res. J* 8(1993)1, 53.
[32] Page, D.H.: *J.Pulp Paper Sci.* 115(1989)6,J229.
[33] McKenzie, A.W., and Higgins, H.G.: *Australian J. Appl. Sci.* 6,2(1955)208.
[34] McKenzie, A.W.: *Appita* 37(1984)580.
[35] Pelton, R., Zhang,J., Wågberg, L., and Rundlöf, M.: *Nordic Pulp Paper Res. J.*15,5 (2000) 440–445.
[36] Wågberg, L., Forsberg, S., and Juntti, P.: *J. Pulp Paper Sci.* 28,2(2002)222.
[37] Laine, J,. Lindström, T., and Glad-Nordmark, G.: *Nordic Pulp Paper Res. J.* 15,5(2000)520.
[38] Glittenberg, D.: *Paper Tech.* Vol(1992) 34.
[39] Fellers, C., in R. Mark (Ed.): *Handbook of Physical and Mechanical testing of Paper and Paperboard, Vol.1*. Marcel Dekker, Inc., New York and Basel, 1984, pp.349–383.
[40] Seth, R. S., Soszynski, R.M., and Page, D.H.: *Tappi* 62(1976)10, 97.
[41] Sachs, I.B., and Kuster, T.A.: *Tappi* 63(1980)10,69.
[42] Perkins, R.W.Jr, and McEvoy, R.P.Jr.: *Tappi* 64(1981)2,99.
[43] Uesaka, T., and Perkins, R.W.Jr.: *Sven.Papperstidn.* (1983)R191.

7 On the Mechanisms Behind the Action of Wet Strength and Wet Strength Agents

Bo Andreasson
SCA Graphic Research, Sundsvall
Lars Wågberg
Department of Fibre and Polymer Technology, KTH, Stockholm

7.1 Introduction 185

7.2 Wet-Strength Mechanisms 186

7.3 The Chemistry of Commercial Wet-Strength Resins 189

7.3.1 Urea-Formaldehyde 189
7.3.2 Melamine-Formaldehyde 190
7.3.3 Alkaline-Curing Resins 192
7.3.4 Glyoxalated Polyacrylamide Resin (G-PAM) 196
7.3.5 Starch 200

7.4 Functional Groups in Fibres 203

7.5 Future 204

7.5.1 Future Use of Existing Chemicals 204
7.5.2 Future Chemistry 205

7.6 References 206

7.7 Patents 207

7.1 Introduction

During World War II, the need for wet-strength papers initiated the development of wet-strength resins. In Europe, the first wet-strength paper was produced in Germany in 1939 by adding polyethyleneimine (PEI) to the pulp. The following years, intensive development work led to wet-strength resins based on formaldehyde. The formaldehyde resins are more effective and cheaper than PEI is. In 1960s alkaline curing wet-strength resins were developed and replaced the formaldehyde resins. Today, alkaline curing wet-strength resins are still dominant.

In 1988, about 65 million tons of paper, card and board were produced in Europe. About 5–6 % of this was paper accounted for wet-strength grade [1]. The most important types are wet-strength sanitary papers like kitchen towels, paper handkerchiefs, serviettes, wipes, etc. Consumption of tissue paper products has been growing at around 4 % per year since the beginning of the 1980s [1]. Other important grades of wet-strength paper produced in Europe are: sacks, packaging paper, wallpaper, map paper, bank note paper, etc.

This chapter reviews the chemistries of wet-strength additives starting with formaldehyde resins. Many comprehensive reviews have been published in the field of wet-strength chemistry and the mechanisms giving wet-strength in paper [2–7] and this is an attempt to summarise these. In Section 7.2, wet-strength mechanisms, in general, will be described. In Section 7.3, the chemistry of commercial wet-strength resins is described. Section 7.4 describes functional groups in fibres, which are very important in this context. Finally, in Section 7.5, the future in this area is discussed

7.2 Wet-Strength Mechanisms

It seems generally accepted that the fibres that make up a paper are held together by intermolecular forces. A widely spread opinion [4] is that the main parts of these forces are hydrogen bonds between hydroxyl groups on adjacent fibres. However, hydrogen bond forces are very short-ranged (< 5 Å) and will probably be responsible for holding the fibres together only when the fibres are very close to contact. When we come to consider long-range interactions between macroscopic particles and surfaces we shall find that the two most important forces are the van der Waals- and electrostatic-forces [8]. Between molecules, van der Waals forces are fairly weak, much weaker than coulombic or hydrogen bonding interactions.

However, between macroscopic bodies the van der Waals forces become large due to additivity. Contrary to hydrogen bonding forces, van der Waals forces between surfaces are always present and have a range of at least 100 Å. A small fraction of covalent bondings between fibres are probably also present [4]. When a paper is saturated with water, the fibres swell and all the bonds are broken except the covalent bonds. However, the number of covalent bonds is small resulting in a drop in strength leaving only a few percentages of the original dry strength. This process is very quick; an untreated paper saturated with water loses all its strength within a few seconds.

Since many products need wet-strength, chemical wet-strength resins have been developed. Wet-strength is usually expressed relative its dry-strength as a percentage. It has been suggested that papers with relative wet-strength more than 10–15 % should be considered wet-strength papers [9]. Wet-strength can be divided in temporary and permanent. Papers saturated with water losing its wet-strength gradually within a couple of hours are said to have temporary wet-strength whereas papers saturated with water keeping its strength for much longer times are said to have permanent wet-strength. In *Figure 7.1*, temporary and permanent wet-strengths are shown. The strength of paper with permanent wet-strength is almost constant upon soak time whereas paper with temporary wet-strength loses its strength much faster.

The purpose of the wet-strength chemical is to [4, 9, 10]:

1. Protect existing bonds by preventing fibre swelling.
2. *Form* bonds that are insensitive to water i.e. covalent bonds.

In the literature [4, 9–11], at least two methods are invoked to explain the development of wet-strength in paper:

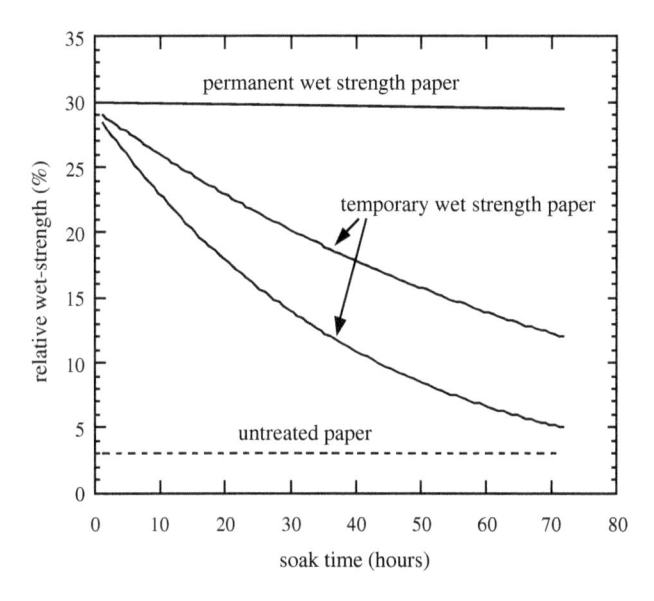

Figure 7.1. Relative wet-strength vs. soak time.

1. *Protection mechanism*: the wet-strength chemical diffuses into the fibres and cross-links with itself to form an insoluble network around and through the fibre contacts. When the treated paper is rewetted, this network inhibits fibre swelling in the fibre-fibre contact areas and fibre separation. The existing bonds hold the fibres together and a part of the original dry-strength is preserved. This is schematically shown in *Figure 7.2*.

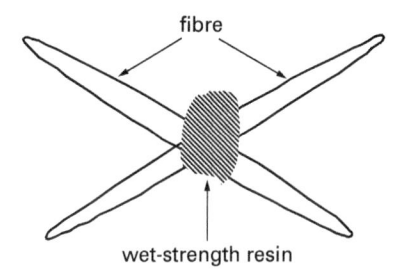

Figure 7.2. Protection mechanism.

2. *Reinforcement mechanism*: The wet-strength chemicals react with the hydroxyl groups or carboxyl groups on the fibres and form linkages of covalent bonds between fibres (fibre-wet-strength chemical-fibre). The covalent bonds are insensitive to water. The linkages reinforce the fibre-fibre contacts when the paper is rewetted as shown in *Figure 7.3*.

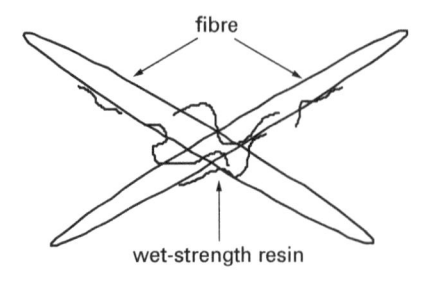

Figure 7.3. Reinforcement mechanism.

During the late 1930s and 1940s the first wet-strength resins were developed. From that time up to now, there has been a rapid growth in the use of wet-strength resins and wet-strength papers. Commercial wet-strength resins share four attributes:

1. *Water-soluble or water dispersible*, thus allowing even dispersion and effective distribution on the fibres. However, there is a class of chemicals that are not water-soluble but have been used to give wet-strength to paper [4]. These are latexes such as the polyvinyls, acrylics and styrene-butadiene's. Important differences between these and the water soluble products are the fact that they give rise to other properties in the paper, higher dosage is required to give wet-strength and the mechanism by which they operate are different. This is partly due to their different form (latex compared to dissolved polymers) and partly due to their low reactivity to cellulose. Probably they form a continuous phase of polymers in and around the fibre-fibre contacts.
2. *Cationic* thus facilitates adsorption onto anionic fibres. It seems generally accepted that the initial attraction of the resin to fibres is primarily electrostatic. The cationic resin molecules are attracted to the negatively charged surface of fibres and fines. Once the resin is attracted to the pulp, its retention appears to be due to ion exchange with the counterions of the negatively charged pulp [6]. A few anionic wet-strength resins are also mentioned in the literature. The retention of these requires the use of a cationic promoter.
3. *Polymeric*. With high molecular weight polymers being more completely adsorbed to the surface of the fibres and forming stronger bonds at the same adsorbed amount.
4. *Reactive* with formation of cross-links with itself and/or with cellulose thus promoting networks.

Of the numerous patents describing compositions for wet-strength chemicals, relatively few have been exploited and developed into products that have gained commercial acceptance. Initially Urea-formaldehyde- and melamine-formaldehyde-resins set the standard for performance and cost. Later, resins were developed which cured under neutral or alkaline conditions. The following sections review each of the major chemical groups of wet-strength resins in chronological order of their appearance in the patent literature.

7.3 The Chemistry of Commercial Wet-Strength Resins

7.3.1 Urea-Formaldehyde

Early experiments showed that the impregnation of paper with formaldehyde gave rise to wet-strength, but the offensive odour and the brittleness of the paper resulting from its low pH were drawbacks [4]. Later on, the reaction product of urea and formaldehyde, dimethylolurea, was found to produce wet-strength in paper. The structure of dimethylolurea is shown in *Figure 7.4*.

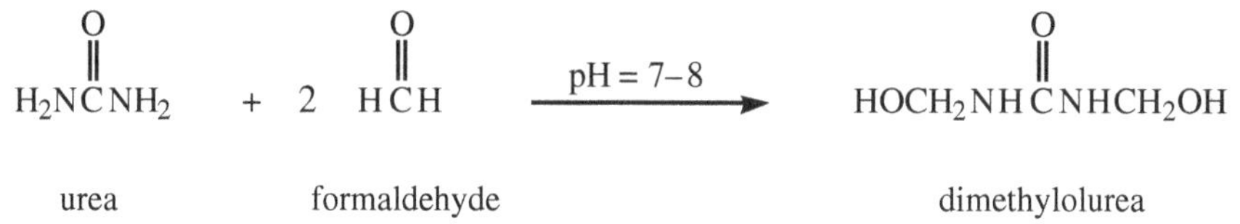

$$H_2NCONH_2 + 2\ HCHO \xrightarrow{pH = 7-8} HOCH_2NHCONHCH_2OH$$

urea   formaldehyde   dimethylolurea

Figure 7.4. Formation of dimethylolurea.

Dimethylolurea self-cross link to a polymer when pH is lowered to 4–5 [5]. When this polymer is added to the pulp it will further cross-link to an insoluble, three-dimensional network [4]. However, this product is not very substantive to fibres. This was solved by reacting dimethylolurea with water soluble polyfunctional amines such as ethylenediamine, diethylenetriamine and triethylenetetramine. This resin is cationic and by that, substantive to pulp.

Since urea-formaldehyde is the earliest wet-strength resin developed for commercial use, it has been used extensively in the study of wet-strength mechanism in paper. In a review article by Chan [5], many studies over the years are presented. The main conclusion is that the insoluble, three dimensional network can protect existing fibre-fibre bonds and restrict hydration and swelling when the paper is rewet, i.e. the UF resin does not form covalent bonds with groups on the fibres. Obviously, the UF-resin is supposed to produce wet-strength by the protection mechanism mentioned earlier (see *Figure 7.2*). The true nature of the physical form of UF resins in solution is not totally clear. Helmer [12] showed that the higher molecular mass the more efficient was the UF resins. In later work [13], Helmer also found that the higher molecular mass detected could be an artefact of a change from a true solution to a colloidal form more than being due to a cross-linking to a high molecular mass single polymer.

UF resins are thermosetting and the pH in the system has a dramatic effect on the rate of cure. For example, paper made at a pH of 5.5 will eventually reach the same wet-strength level as paper made at a pH of 4.5 but will require much greater ageing time. UF resins may be used in the pH range 3.8–4.5 [4] with an optimum pH of 4.0. At this pH the resin achieves most of its wet-strength property after about two weeks of ageing. Addition as close to the head box as possible is recommended when UF resins are used. Typical resin addition levels vary between 0.5 % and 2.5 %, depending on the type of paper produced desired. The advantages and drawbacks of UF resins are the following:

Advantages:

- UF resins have no organochlorine in the polymer system and will not contribute any absorbable organic halides (AOX) to the paper making system.
- Low cost
- Easy repulping (compared to PAE)

- Permanent wet-strength

Drawbacks:

- The resin, produced in a conventional manner, contains approximately 6 % of free formaldehyde [5] that is complicated for health and safety reasons. However, using modifiers and additives [5] the amount of free formaldehyde can be reduced to about 1 %. Acid resistant handling system (i.e. machine parts) is required to deal with UF resins.
- UF resins are relatively low in cost. They are still used in making paper towels, tissue paper, bag paper and wet-strength linerboard.

7.3.2 Melamine-Formaldehyde

Next development was melamine-formaldehyde (MF) resins that were first patented for use in paper in 1944 [4]. Melamine reacts with formaldehyde at pH of 7–8 to form a series of methylol melamines. In *Figure 7.5*, reaction between melamine and formaldehyde of 1:1 by molar ratio is shown.

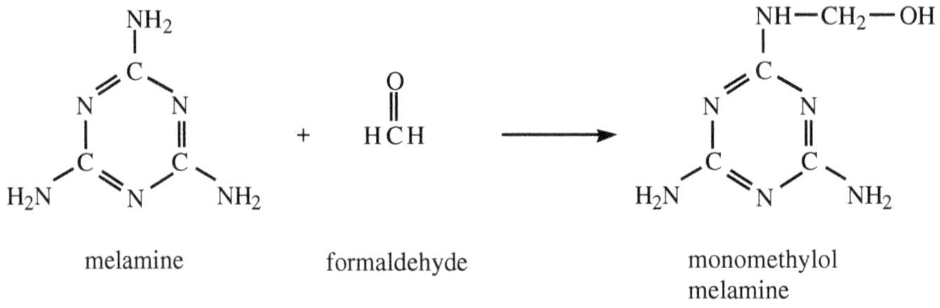

Figure 7.5. Reaction between melamine and formaldehyde.

In aqueous solution at acidic conditions the methylol melamine's are cationic and very substantive to negatively charged fibres. It has been found that the use of 3:1 by molar ratio of formaldehyde to melamine gives the best MF resins for wet-strength [5]. Like UF resin, MF resin has been found to not form any specific bonds, like covalent bonds, with groups on fibre [5]. It seems likely that the main reaction in wet-strength development is the polycondensation of the resin itself. To a great extent the resin functioned to protect the existing bonds by reducing the swelling in the bonded areas [4, 5, 9]. The polycondensation mentioned above is the result of ether and methylene linkages resulting in melamine-formaldehyde colloids of about 20 methylol melamine monomers. The formations of ether and methylene linkages are shown in *Figure 7.6a and Figure 7.6b*.

The formations of the colloids takes place at pH lower than 5.5. It is recommended to add the resin to a dilute stock near the head box. The rate of cure of the resin is increased by low pH and high temperature. MF resin can be used in the pH range 4.0–5.5 with an optimum at 4.5. MF resins have been used in combination with polymers [5]. For example, addition of polyvinyl alcohol to MF resin gives a product with better wet-strength performance [5]. Also a greater water absorption rate for the treated paper is observed in this case.

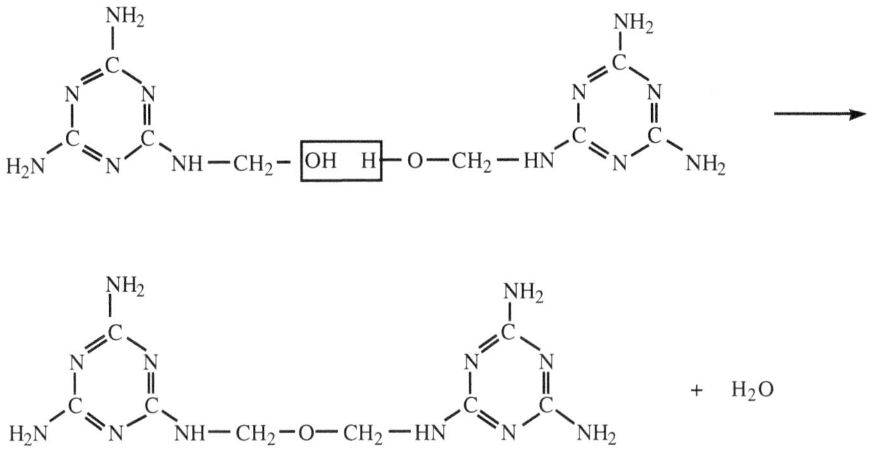

Figure 7.6a. Ether linkage between two monomethylol molecules.

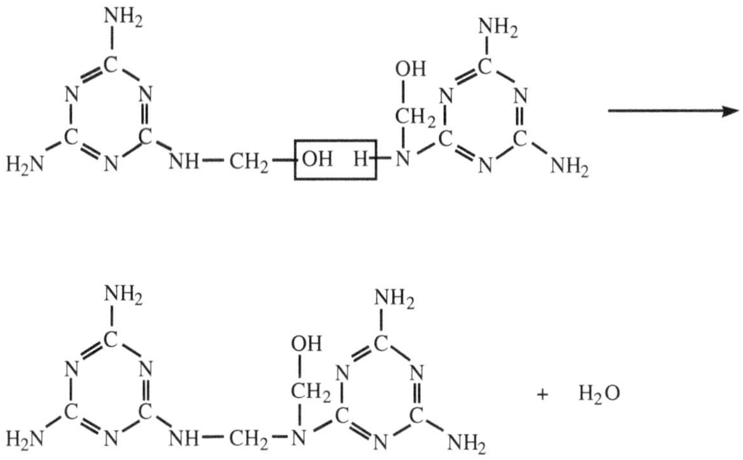

Figure 7.6b. Methylene linkage between two monomethylol molecules.

Advantages:

- Also MF resins are cheaper than PAE but a bit more expensive than UF resins.
- Permanent wet-strength
- The broke handling is easy

Drawbacks:

- Commercial MF resins have a pH value of about 2 that makes it very corrosive.
- An acid-resistant handling system is required.
- Like the UF resins, MF resins contain about 2–5 % free formaldehyde, which is complicated for health and safety reasons.

MF resins are still used, especially in products where high permanent wet-strength is needed for example in currency paper, map paper and photographic paper.

The drawbacks are serious since it is very difficult to solve the problem with free formaldehyde in both UF and MF resins. As mentioned earlier, experiments have shown that it is possible to reduce the level of free formaldehyde to about 1 % [5]. The UF resin from Eka Chemicals contains less than 0.5 % free formaldehyde [14] and the future will tell us if further reduction is possible. Another drawback with using UF or MF resins in tissue manufacture is the need for a low curing pH. This will largely disturb the coating on the yankee cylinder and it is definitely not possible to run these chemicals at their optimum working pH.

7.3.3 Alkaline-Curing Resins

As mentioned earlier, UF and MF resins require acid conditions to cure effectively in paper. In the 1950s, alkaline-curing epichlorohydrin products were investigated [6]. The first alkaline-curing wet-strength resins to become commercially practical were poly (aminoamide)-epichlorohydrin (PAE) resins. These rapidly started to replace UF and MF resins in many applications. They offered improved absorbency and reduced machine corrosion and other benefits. Later, polyalkylenepolyamine-epichlorohydrin (PAPAE) and amine polymer-epichlorohydrin (APE) were introduced commercially. These are, like PAE, cationic and thermosetting at near neutral and alkaline pH conditions. The terms PAE, PAPAE and APE describe the backbone polymers of the resins. Therefore PAPAE and APE resins are sometimes categorised together as polyamine-epichlorohydrin resins [6]. The poly (aminoamide) backbone of the PAE resin is a result from the reaction between adipic acid and diethylenetriamine, *Figure 7.7*.

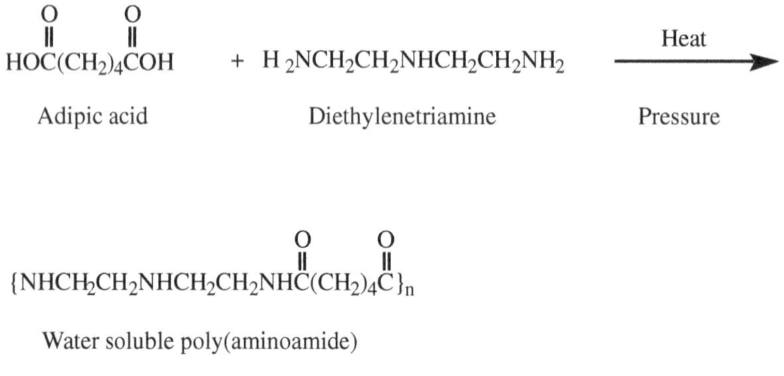

$$HOC(CH_2)_4COH \ (Adipic\ acid) + H_2NCH_2CH_2NHCH_2CH_2NH_2 \ (Diethylenetriamine) \xrightarrow{Heat,\ Pressure}$$

$$\{NHCH_2CH_2NHCH_2CH_2NHC(CH_2)_4C\}_n$$

Water soluble poly(aminoamide)

Figure 7.7. Formation of poly (aminoamide) from adipic acid and diethylenetriamine.

The resulting poly (aminoamide) is then reacted with epichlorohydrin. In the PAPAE resin, a polyalkylenepolyamine is reacted directly with epichlorohydrin and in the APE resin an amine polymer is reacted with epichlorohydrin. The structure of the final wet-strength resin will depend on whether the reaction partner of epichlorohydrin is a primary, secondary or tertiary amine [6]. Secondary amines react with epichlorohydrin to form tertiary aminochlorohydrins, which form cyclic structures of the 3-hydroxy-azetidinium salt type, *Figure 7.8*. Since these structures are fairly constrained they are also fairly reactive.

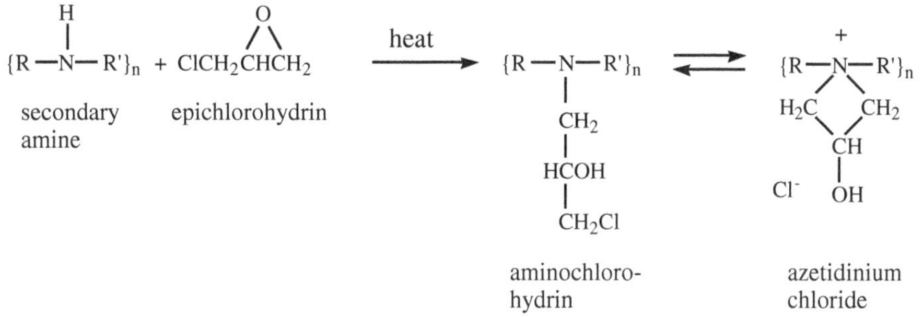

{R—N—R'}ₙ + ClCH₂CHCH₂(O) —heat→ {R—N—R'}ₙ with side chain CH₂—HCOH—CH₂Cl ⇌ {R—N⁺—R'}ₙ with azetidinium ring (H₂C–CH(OH)–CH₂), Cl⁻

secondary amine | epichlorohydrin | aminochlorohydrin | azetidinium chloride

Figure 7.8. Reaction of epichlorohydrin and secondary amines.

The most important PAE resins are derived from secondary amines and the 3-hydroxy-azetidinium ring is their principal reactive group. Tertiary amines react with epichlorohydrin to form a glycidyl (2,3-epoxypropyl) ammonium salt. Only a few wet-strength resins based on the reaction product of primary amines and epichlorohydrin exist and they will not be described in this report. The functional groups can occur independently of the category of backbone polymer. The most important resins with 3-hydroxy-azetidinium as reactive groups are of PAE and PA-PAE type and the most important resins with glycidyl (2,3-epoxypropyl) ammonium as reactive groups are of PAE and APE type. On weight basis, PAPAE resins are less effective than PAE resins, while many APE resins, especially of epoxide type, are more effective. When PAE is mentioned in the literature [3, 4, 6, 9–11] it is often PAE with 3-hydroxy-azetidinium as reactive group. PAE is the most important alkaline curing wet-strength resin. PAPAE and APE resins that were later introduced to the paper industry have not reached the same volume of use. In this report PAE with 3-hydroxy-azetidinium as reactive group will be in focus and from now it will be denoted only PAE. As mentioned earlier, the groups formed when the poly (aminoamide) is reacted with epichlorohydrin is aminochlorohydrin that self-alkylate to form 3-hydroxy-azetidinium rings, *Figure 7.8*.

The chemistry of this PAE resin has been extensively studied and a number of comprehensive reviews have been published [4, 6, 9, 10]. The 3-hydroxy-azetidinium ring confers both reactivity and permanent (pH-independent) cationic charge [4]. As indicated in *Figure 7.8* the reaction between aminochlorohydrin and 3-hydroxy-azetidinium is reversible and at equilibrium approximately 80–85 % is 3-hydroxy-azetidinium and 15–20 % is aminochlorohydrin [15]. The 3-hydroxy-azetidinium ring may react in different manner:

1. Reaction with other PAE macromolecules (homo-cross linking)
2. Reaction with cellulose fibres (co-cross linking)
3. Reaction with water

The different reactions of the 3-hydroxy-azetidinium ring are shown in *Figure 7.9a–Figure 7.9c*.

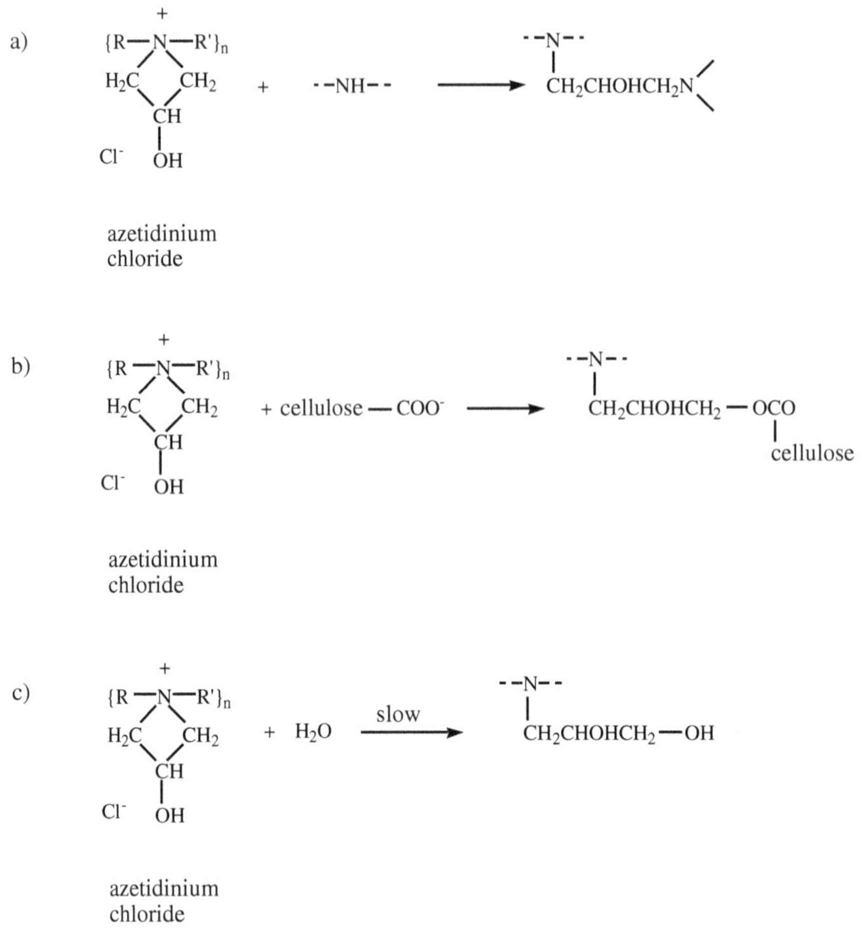

a) [azetidinium chloride structure] + --NH-- → --N--CH₂CHOHCH₂N<

azetidinium chloride

b) [azetidinium chloride structure] + cellulose—COO⁻ → --N--CH₂CHOHCH₂—OCO—cellulose

azetidinium chloride

c) [azetidinium chloride structure] + H₂O —slow→ --N--CH₂CHOHCH₂—OH

azetidinium chloride

Figure 7.9a-c. Different reaction mechanisms of azetidinium chloride from PAE. Reaction with secondary amines, cellulose carboxyl groups and water.

Model compound studies using sucrose or methylglucoside indicate that PAE resins do not react with cellulose hydroxyl groups [10]. In contrast, direct and indirect evidence supports the reaction of 3-hydroxy-azetidinium groups with the carboxylate groups of pulp [10]. For instance, direct spectroscopic evidence for the reaction of 3-hydroxy-azetidinium groups with pulp carboxylate to form ester groups has been reported [16]. In the final resin, about 35 % of the active groups are involved in homo-cross linking, 50 % are free as 3-hydroxy-azetidinium groups and 15 % are free as aminochlorohydrin [15]. A picture of crosslinked PAE is shown in *Figure 7.10*.

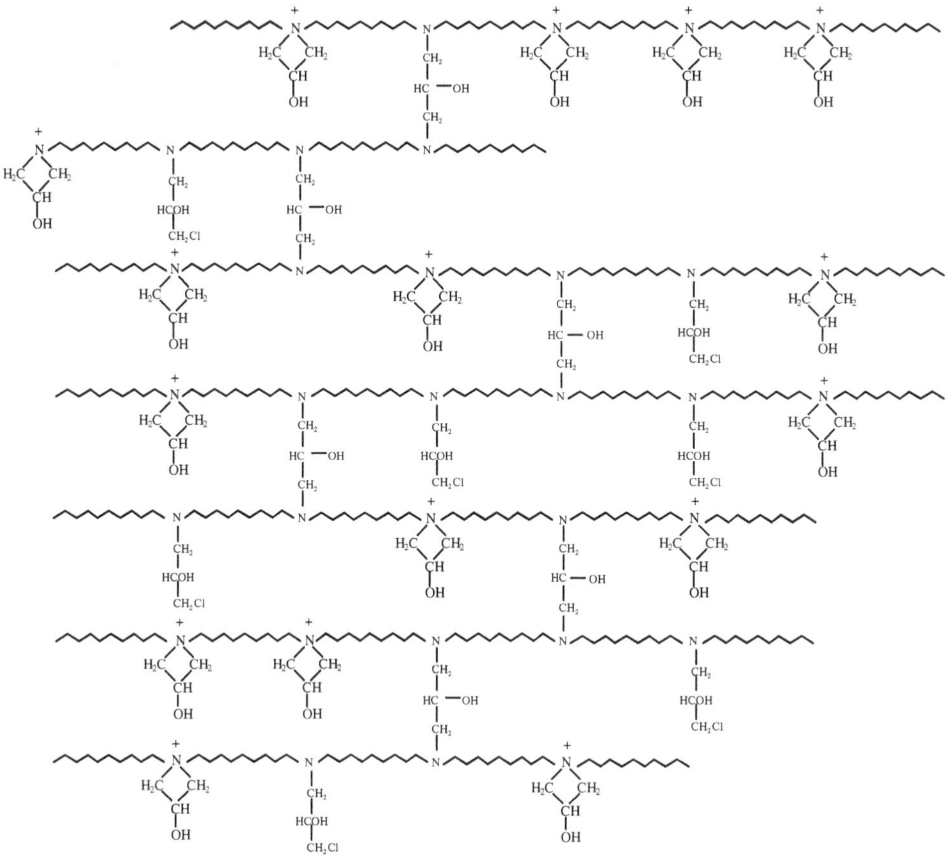

Figure 7.10. The structure of crosslinked PAE.

Since the resin reacts with carboxylic groups but not with the hydroxylic groups on fibres, the carboxyl content of the pulp affects performance of the resin, and typically a higher wet-strength can be achieved with unbleached kraft than bleached kraft. Bleached sulphite pulps, which have a lower carboxyl content are the most difficult to treat [4].

The presence of soluble lignosulphonates reduces the efficiency of PAE resins by reaction with the resin at the expense of the fibres [17]. Poorly washed neutral sulphite pulp contains high levels of sodium lignosulphonates (SLS) which reduces the efficiency of PAE. However, by addition of alum or calcium chloride to the neutral sulphite pulp, SLS concentration may be reduced [17]. The degree of pulp refining can effect the performance of PAE resins. A more highly refined stock will develop wet-strength because the higher surface area of the fines, in particular, allows it to retain more resin [6, 18]. This effect should also be pronounced in mechanical pulp where the fraction of fines is high.

The optimal pH for alkaline-curing resins in paper manufacture is 6 to 8. The resin itself is very stable down to about pH = 5 when self-cross linking of the resin will be severe. Also, at low pH the fraction of anionic carboxylate groups on the pulp is reduced, resulting in lower surface charge on the fibres. The upper limit is about pH = 9 [4]. The residence time is also an important factor. Up to a point, increasing residence time improves the retention of the resin molecules. However, PAE tend to lose efficiency at very long exposure times. The reason for this behaviour is probably due to polymer migration away from the surface and into the pores of the fibre. PAE may also interact with inorganic and organic substances that can affect its interaction with the cellulose fibres, for a review see reference 6. PAE can be used in combination with other polymers. For example, PAE can be used in combination with cationic and neutral polymers like starch, dialdehyde starch, polyacrylamide-glyoxal resin, etc. [6]. Even better performance can be reached in combination with anionic polymers. Beneficial effects require the use of controlled ratios of anionic to cationic polymer so that the formed complexes have a cationic net charge. Synergistic effects have been reported for complexes formed between PAE and carboxymethyl cellulose (CMC), polyacrylates, anionic starch, etc. [6].

As mentioned earlier, PAE resins are widely used as wet-strength chemicals. It is frequently used in tissue and towel, liquid packaging board, food packaging, currency, maps, etc. PAPAE resins find somewhat more limited use but they are used in unbleached pulps, mostly in packaging. Dosages between 0.5 %–1.5 % are recommended depending on the type of paper produced desired.

Advantages:
- Neutral or slightly alkaline conditions which reduces machine corrosion (compared to UF and MF).
- Permanent wet-strength
- High relative wet-strength

Drawbacks:
- Difficult to repulp
- Organic chlorine containing
- Less absorption capacity of PAE treated papers (compared to untreated paper and paper treated with G-PAM)

Today the companies are trying to solve the three problems mentioned above. Paper containing PAE is very difficult to repulp. For example, for unbleached paper, a high pH (11–12 using sodium hydroxide) and high temperature (77 °C) in addition to mechanical treatment, is usually necessary. Today, Hercules and Eka Chemicals have PAE resin almost free from organic chlorine. In a patent from Hercules [19], Espy presents much improved absorbency in tissue papers containing PAE. Instead of using only PAE or PAE/CMC, a combination of PAE, CMC and a tertiary-amino polyamide epichlorohydrin resin was used. The mechanisms giving the faster absorption are not outlined or at least not mentioned in the patent.

7.3.4 Glyoxalated Polyacrylamide Resin (G-PAM)

G-PAM was introduced on the market in the 1960s and has grown to be an important additive for wet-strength in paper. Since G-PAM gives temporary wet-strength, the principal use is in tissue grades. G-PAM can be prepared in different ways, as an example, preparation of G-PAM in two steps [4, 9] is presented in *Figure 7.11 and Figure 7.12*. In the first step, shown in *Figure 7.11*, a cationic polyacrylamide (PAM) is formed by the reaction of acrylamide and quaternary ammonium cationic monomers.

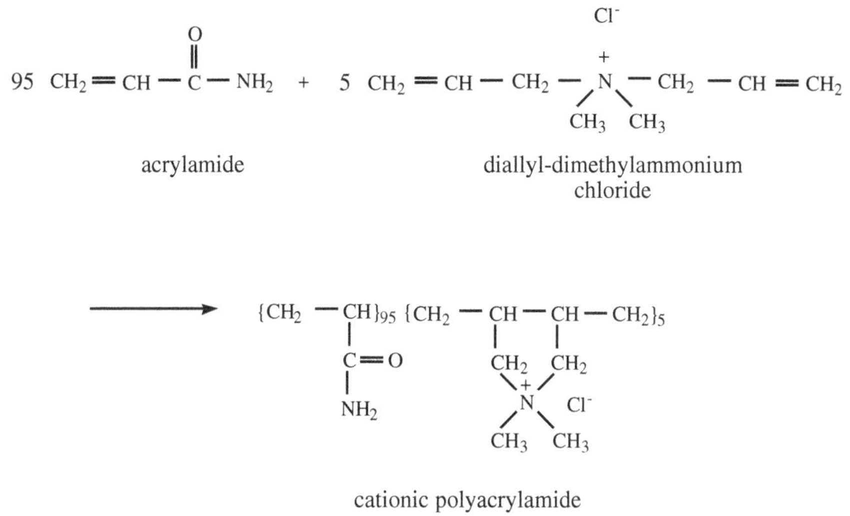

$$95 \ CH_2\!=\!CH-\overset{\overset{\displaystyle O}{\|}}{C}-NH_2 \ + \ 5 \ CH_2\!=\!CH-CH_2-\overset{\overset{\displaystyle Cl^-}{+}}{\underset{\underset{\displaystyle CH_3 \ \ CH_3}{}}{N}}-CH_2-CH\!=\!CH_2$$

acrylamide   diallyl-dimethylammonium chloride

$$\longrightarrow \ \{CH_2-CH\}_{95}\{CH_2-\underset{\underset{\displaystyle NH_2}{\underset{\displaystyle C=O}{|}}}{CH}\} \ \{CH_2-\underset{\underset{\displaystyle CH_3 \ \ CH_3}{}}{\underset{\underset{\displaystyle N}{+}}{\underset{\displaystyle CH_2 \ CH_2}{|\ \ \ |}}}CH-CH-CH_2\}_5 \ Cl^-$$

cationic polyacrylamide

Figure 7.11. An example of the first step in the preparation of G-PAM. Reaction of acrylamide and diallyl-dimethylammonium chloride.

In the second step, shown in *Figure 7.12*, G-PAM is formed by the reaction of the cationic polyacrylamide (PAM), formed in the first step, and glyoxal.

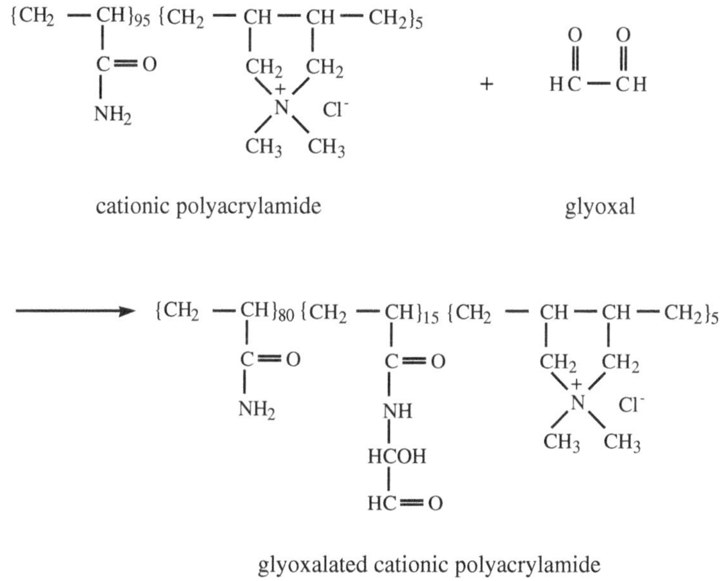

$$\{CH_2-CH\}_{95}\{CH_2-CH-CH-CH_2\}_5 \ \ \ \ + \ \ \ \ \overset{\overset{\displaystyle O}{\|}}{HC}-\overset{\overset{\displaystyle O}{\|}}{CH}$$

cationic polyacrylamide   glyoxal

$$\longrightarrow \ \{CH_2-CH\}_{80}\{CH_2-CH\}_{15}\{CH_2-CH-CH-CH_2\}_5$$

glyoxalated cationic polyacrylamide

Figure 7.12. The second step in the preparation of G-PAM. Reaction of cationic polyacrylamide and glyoxal, resulting in glyoxalated cationic polyacrylamide.

As can be seen in *Figure 7.12*, the final G-PAM resin contains three active groups, namely: unreacted amines, amides reacted with glyoxal and quaternary ammonium cations. The unreacted amines are free to form hydrogen bonds with hydroxyl groups on cellulose resulting in increased dry strength. The quaternary ammonium cations are important for interaction with negatively charged fibres. The amides reacted with glyoxal are the groups that will form homo- and co-cross links. Obviously, in the preparation of G-PAM the amount of glyoxal used can control the reactivity of the final resin. Evidence [4, 7, 9] strongly suggests that G-PAM imparts wet-strength to paper primarily through covalent bond (hemiacetal and acetal) formations between free aldehydes on its reactive group and hydroxyl groups on fibres, see *Figure 7.13*.

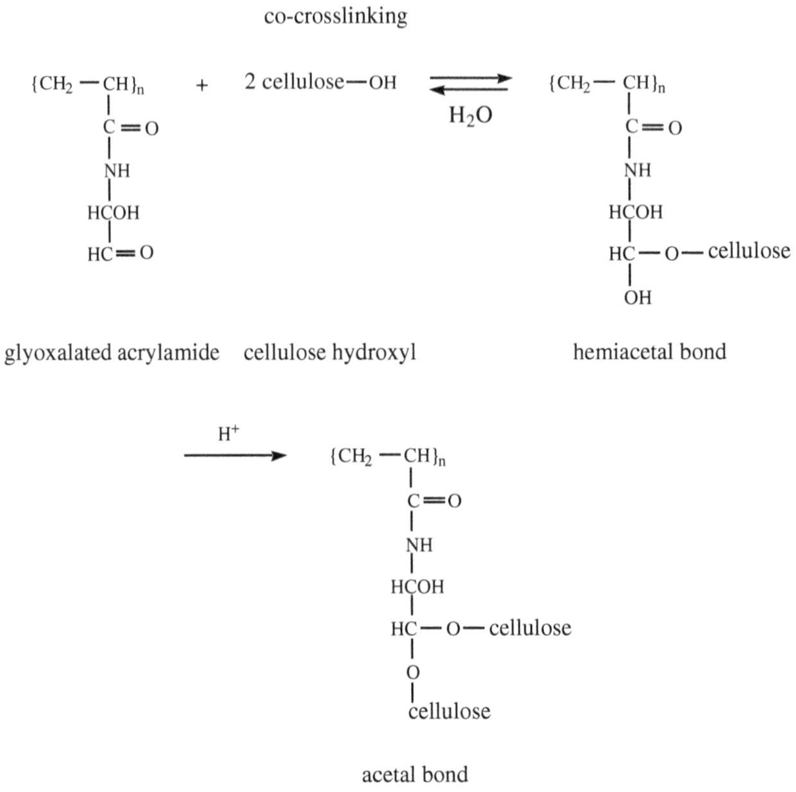

co-crosslinking

{CH$_2$—CH}$_n$ +   2 cellulose—OH   ⇌ {CH$_2$—CH}$_n$
  | H$_2$O |
  C=O   C=O
  | |
  NH   NH
  | |
  HCOH   HCOH
  | |
  HC=O   HC—O—cellulose
                                                     |
                                                     OH

glyoxalated acrylamide   cellulose hydroxyl   hemiacetal bond

                    H$^+$
          ─────────────→   {CH$_2$—CH}$_n$
                              |
                              C=O
                              |
                              NH
                              |
                              HCOH
                              |
                              HC—O—cellulose
                              |
                              O
                              |
                              cellulose

                          acetal bond

Figure 7.13. Co-cross linking reactions between the active group on G-PAM and hydroxyl groups on cellulose. Hemiacetal and acetal bonds may be formed depending on pH.

Also homo-cross links are formed between aldehydes and free amines as indicated in *Figure 7.14*.

Probably also fibre-resin-fibre bonds are formed. Obviously, G-PAM gives wet-strength with the reinforcement mechanism. As indicated in *Figures 7.13* and *Figure 7.14*, the reaction of G-PAM with cellulose is helped by acidic conditions. The reaction is very rapid at pH around 5 but rapid enough at neutral conditions. Curing of the paper gives little or no additional wet-strength. Also indicated in *Figure 7.13* is the reversibility of the reaction of G-PAM with cellulose in the presence of water. As a consequence, paper treated with G-PAM gradually loses a portion of its wet-strength on prolonged soaking in water; G-PAM hence gives temporary wet-strength [7]. It has been shown that if a laboratory hand sheet is soaked for four hours it still retains over 85 % of its wet tensile strength after drying [4]. However, the action of alkali is immediate and complete. The wet-strength loss with alkali is not reversible, probably due to a destruction of the aldehyde functionality [2, 7]. Sulphite or bisulphite ions can react to form anionic bisulphite adduct which react with cationic charge of the resin, through internal „salt" formation, resulting in poor retention, see *Figure 7.15*.

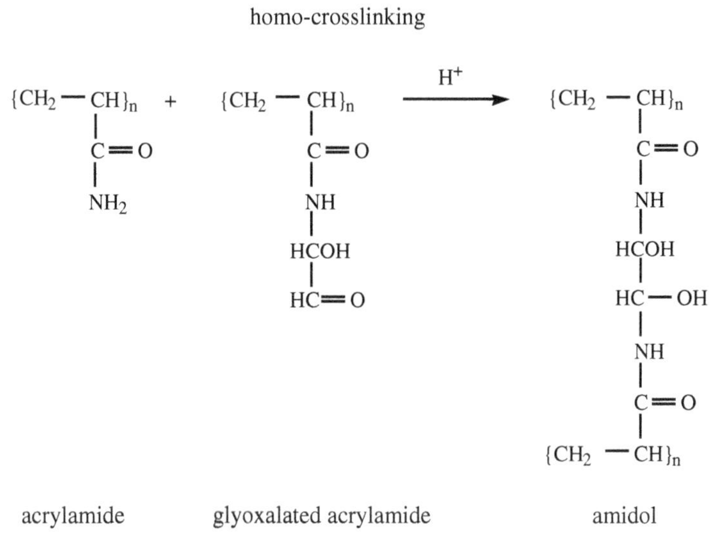

homo-crosslinking

$$\{CH_2-CH\}_n + \{CH_2-CH\}_n \xrightarrow{H^+} \{CH_2-CH\}_n$$

acrylamide   glyoxalated acrylamide   amidol

Figure 7.14. Homo-cross linking reactions between different groups on G-PAM forming an amidol.

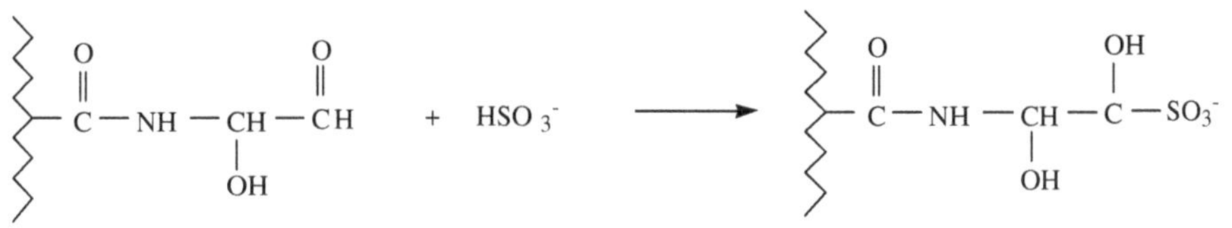

Figure 7.15. Reaction of bisulphite with the active group of G-PAM.

This can be avoided by addition of a high charge density cationic polymer. Also bicarbonate, HCO_3^-, may have negative effect on the performance of G-PAM. HCO_3^-, decompose with heat to CO2 and OH⁻ G-PAM may then react with OH⁻ instead of hydroxyl groups on cellulose. This negative effect can be minimised or even eliminated by simply reducing pH. pH ≤ 6.7 eliminates efficiency losses due to bicarbonate [7]. This is another reason to work at pH levels slightly acidic. G-PAM, with its relatively low cationic charge, is susceptible to interference from anionic contaminants. On storage, the resin continues to cross-link and can gel, therefore, in order to achieve the desired stability, the resin should be stored much diluted.

Since paper treated with G-PAM has temporary wet-strength, it is repulped more readily than paper made with PAE. Broke repulping normally requires no special treatment, but can be accelerated by heat and alkali. As mentioned in the previous section, PAE have negative effect on the absorbency of paper. This is not the case with G-PAM. The time to absorb a water drop is almost the same in a paper treated with G-PAM as an untreated paper [7]. The tendency to retain hydrophobic materials by G-PAM are reduced compared to PAE is one of the reasons for this behaviour [7]. Papers treated with G-PAM get increased dry strength, which can have a negative effect on sheet softness, which is important in tissue. However, dry strength can also be an advantage, for example increased utilisation of weaker fibres [7]. Also sheets with low basis weight can be produced if the dry strength is high.

Manufacturers give the following application advises for best performance: Thick stock to minimise resin adsorption to fines; Rapid mixing; Addition after all refining; Stock pH< 8; Contact time less than five minutes.

G-PAM can be used at pH between 4.5 and 7.5 with optimum between 6.0 and 7.0 [4]. Farley [7] recommend as low pH as possible, but not below pH = 4.0. At high concentrations of anionic material a large portion of the resin is consumed in neutralising the anionic materials. A good solution of this problem is to pre-treat the furnish with a relatively inexpensive highly charged cationic polymer, for example a quaternary ammonium polymer.

Advantages:
- Easy repulping
- High absorption capacity of the treated paper (applicable for tissue grades)
- Dry strength, which allow the use of weaker fibres

Drawbacks:
- Temporary wet-strength
- Decomposition of paper at alkaline contact

As mentioned earlier, dry strength is sometimes listed as a drawback. Too much dry-strength can give stiff papers, which is a drawback in tissue grades. Temporary wet-strength is not a serious drawback in tissue products, wiping up spills or drying ones hand normally takes a few seconds. Loss of strength in a couple of seconds which is the case when using alkaline household cleaners is probably more serious.

7.3.5 Starch

Starch is a natural polymer. It is present in roots, tubers and seeds. The main sources for starches are corn, wheat, rice and potato. Normally a plant synthesises two types of starch molecules, amylose and amylopectin. Amylose is a linear polysaccharide whereas amylopectin is a branched polysaccharide. Amylose consists of 100–10000 glucose monomers whereas amylopectin consists of approximately 100000 glucose units, so the degree of polymerisation is much higher for amylopectin. Both amylose and amylopectin are built up by glucose monomers. Potato starch consists of 79 % amylopectin and 21 % amylose and maize starch consists of 72 % amylopectin and 28 % amylose. Waxy maize starch contains only amylopectin [20]. In order to use starch as wet-strength agent, the natural starch must be modified [4, 21]. Starch has been used as wet-strength additive since the 1960s [4]. By oxidation of the adjacent hydroxyl groups with periodic acid, aldehyde groups will be formed *Figure 7.16*.

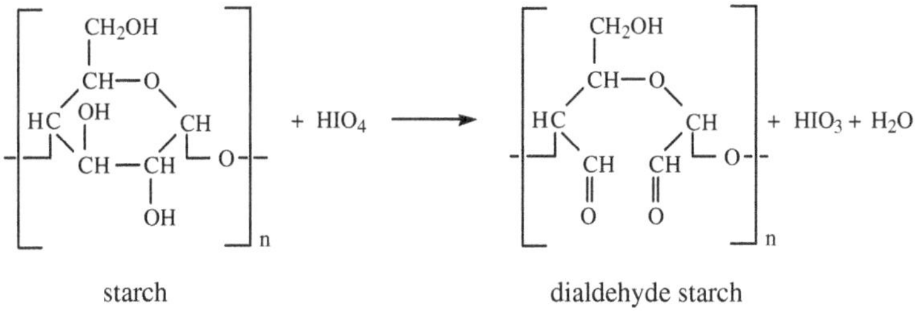

[Chemical reaction scheme showing oxidation of starch to dialdehyde starch with HIO₄, producing HIO₃ + H₂O]

Figure 7.16. Oxidation of a glucose unit in starch with periodic acid resulting in formation of aldehyde groups.

This oxidation can be very selective and effective and in most cases 80–90 % conversion to dialdehyde starch is found [4]. The reaction with the hydroxyl groups on cellulose is very similar to that of G-PAM mentioned in the previous part. The similarities are many, also the fact that DAS give temporary wet-strength. In order to make DAS substantive to pulp, cationic aldehyde starch can be formed by reacting DAS with betaine hydrazide hydrochloride in the presence of water and heat [22]. This reaction is shown in *Figure 7.17*.

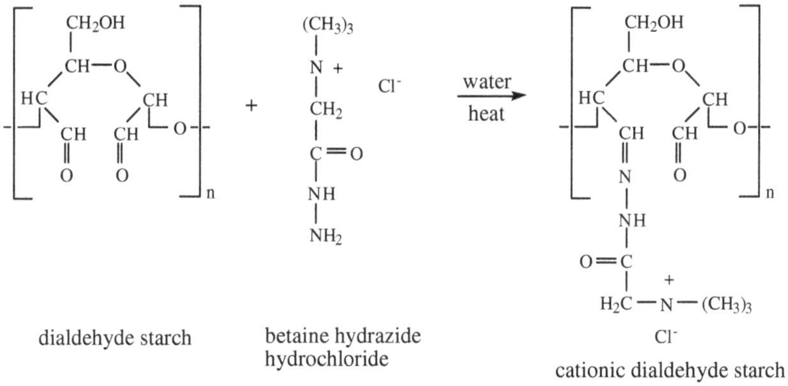

[Chemical reaction scheme showing dialdehyde starch + betaine hydrazide hydrochloride → cationic dialdehyde starch, with water/heat]

Figure 7.17. Cationic dialdehyde starch formed by reaction of dialdehyde starch and betaine hydrazide hydrochloride.

As indicated in *Figure 7.18* both hemiacetal- and acetal-bonds may be formed.

Hemiacetal bonds, formed at high pH, are easily broken by water and give temporary wet-strength. Acetal bonds, formed at low pH, are more stable to normal tap water but hydrolysed at acidic conditions and give wet-strength that is more permanent [4]. Of course, starch macromolecules can also bind to other starch molecules by hemiacetal bonds. Unreacted groups of starch may form hydrogen bonds with free hydroxyl groups on cellulose, this contributes to increased dry strength [21]. One of the problems with DAS is the self cross-linking of the polymers upon storage. This has been solved by Solarek et al. [21]. The idea they introduced was to add the aldehyde groups to the starch backbone in a blocked and unreactive form. Then, at the time the starch was cooked the blocked aldehyde groups could be unblocked and activated and available for cross-linking reactions (homo- and/or co-crosslinkings). This resin is called blocked reactive group starch (BRG-starch). In the example shown in *Figure 7.19*, dialdehyde starch has been modified to cationic starch by addition of a quaternary ammonium group. The cationic starch is then reacted with N-(2,2-dimethoxy-ethyl)-N-methyl-chloroacetamide (DMCA).

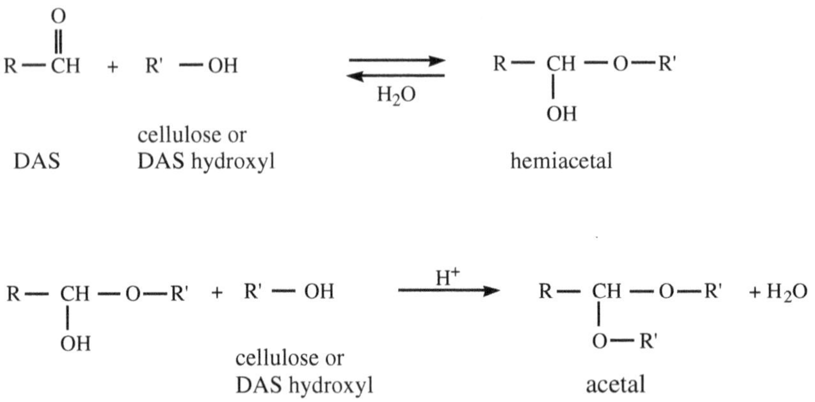

$$R-\overset{O}{\overset{\|}{CH}} + R'-OH \underset{H_2O}{\rightleftarrows} R-\underset{\underset{OH}{|}}{CH}-O-R'$$

DAS   cellulose or DAS hydroxyl   hemiacetal

$$R-\underset{\underset{OH}{|}}{CH}-O-R' + R'-OH \xrightarrow{H^+} R-\underset{\underset{O-R'}{|}}{CH}-O-R' + H_2O$$

cellulose or DAS hydroxyl   acetal

Figure 7.18. Hemiacetal and acetal formation of dialdehyde starch.

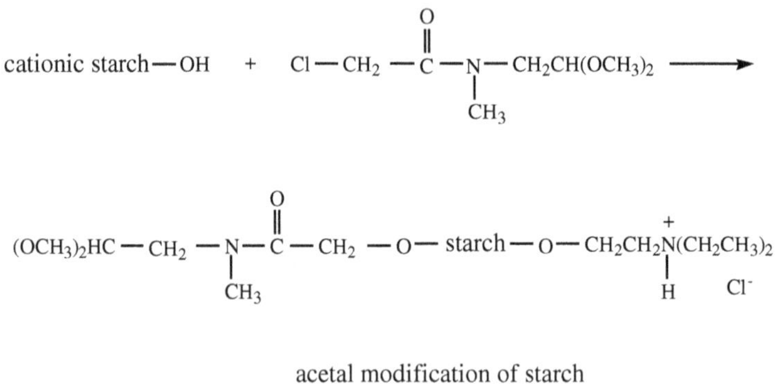

$$\text{cationic starch}-OH + Cl-CH_2-\overset{O}{\overset{\|}{C}}-\underset{\underset{CH_3}{|}}{N}-CH_2CH(OCH_3)_2 \longrightarrow$$

$$(OCH_3)_2HC-CH_2-\underset{\underset{CH_3}{|}}{N}-\overset{O}{\overset{\|}{C}}-CH_2-O-\text{starch}-O-CH_2CH_2\overset{+}{\underset{\underset{H}{|}}{N}}(CH_2CH_3)_2 \quad Cl^-$$

acetal modification of starch

Figure 7.19. Blocked reactive group (BRG) starch.

The starch monomers consist of a cationic group and the acetal substituent group. This is a dry product delivered as a powder [21]. To activate the reactive aldehyde group a starch solution must be adjusted to pH = 2.5 and cooked at 95–100 °C for 20–30 minutes *Figure 7.20*.

It is supposed that the aldehyde groups in starch react with cellulose and other starch molecules in the same manner as DAS and G-PAM. As already mentioned, DAS and BRG starches give temporary wet-strength and it seems likely that a faster decay as for G-PAM is observed [4, 21]. Relatively high wet-strengths have been reported [21, 23] but not as high as for PAE, G-PAM and MF.

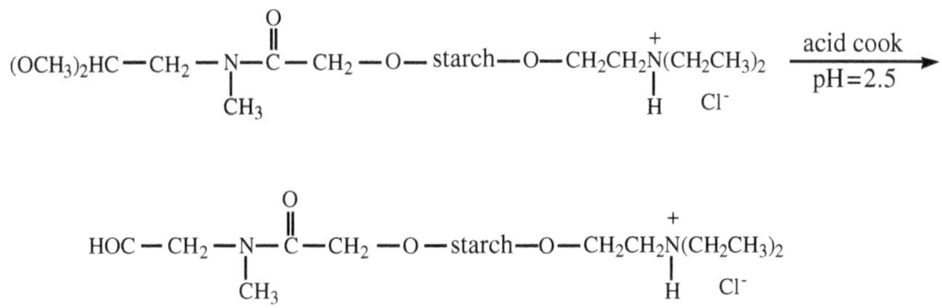

$$(OCH_3)_2HC-CH_2-\underset{CH_3}{N}-\overset{O}{\underset{\|}{C}}-CH_2-O-starch-O-CH_2CH_2\overset{+}{\underset{H}{N}}(CH_2CH_3)_2 \quad Cl^- \xrightarrow[pH=2.5]{acid\ cook}$$

$$HOC-CH_2-\underset{CH_3}{N}-\overset{O}{\underset{\|}{C}}-CH_2-O-starch-O-CH_2CH_2\overset{+}{\underset{H}{N}}(CH_2CH_3)_2 \quad Cl^-$$

Figure 7.20. Activation of the reactive aldehyde group by cooking at pH = 2.5.

Advantages:
- Easy repulping
- Biologically degradable
- Dry strength, which allow the use of weaker fibres
- From a renewable resource

Drawbacks:
- Less effective than PAE, G-PAM and MF
- Temporary wet-strength
- Decomposition of paper at alkaline contact

More research work must be spent to make starch more effective. Temporary wet-strength is not a very dramatic drawback in tissue products. Since the chemistry behind the mechanisms is the same as for G-PAM, paper treated with starch is decomposed by alkaline which is a serious drawback. Starch can also be used in combination with other wet-strength chemicals and exhibit synergistic improvements. The most studied combinations are probably with PAE, for a review, see reference 6. Exciting results have also been presented by Laleg and co-workers [23]. Cationic starch in combination with cationic zirconium-oxychloride, ZrOCl2, gave synergistic effects on relative wet-strength.

7.4 Functional Groups in Fibres

Roughly speaking wood consist of cellulose, hemicellulose and lignin in the proportions 40 %, 30 % and 30 % respectively [24]. Since the functional groups in these macromolecules are involved in hydrogen bonds and covalent bonds they are of great importance. The functional groups are also responsible for the negative charge in the fibres that is important for the attraction of cationic ligands.

In cellulose, the functional groups are hydroxyl groups (–OH and –CH2–OH). pKa values for these groups are high (pKa ≈ 15) and at normal pH the amount of ionised groups can be neglected. Even at extremely high pH values, only a small fraction of ions are present (0.1 % at pH = 12). However the hydroxyl groups may be involved in hydrogen bonds responsible for holding fibres together. Like in cellulose, hemicelluloses like glucomannan and galactan have the functional groups –OH and –CH2–OH and for the same reason, they will not be ionised at normal pH.

The functional groups in the hemicellulose xylan (about 10 % in spruce) [25] are carboxyl groups (–COOH). Carboxyl groups have pKa = 4–5 and are highly ionised at normal pH (99.7 % –COO– at pH = 7).

The functional groups in lignin are methoxyl (–OCH$_3$), phenol groups like C$_6$H$_5$–OH and C$_6$H$_5$–CH$_2$–OH and some terminal aldehyde groups. Among these, the phenol groups can contribute to the charge (pKa ≈ 10). At pH = 7, 0.1 % of these groups are ionised but at higher pH a considerable fraction may be in a dissociated form. It should also be mentioned that many of the phenolic groups are occupied through linkages to neighbouring groups [24]. After wood pulping and the following bleaching (ECF or TCF), the chemical composition may be dramatically changed. In the sulphate process followed by bleaching, most of the lignin is removed. Also, a considerable amount of hemicellulose is removed resulting in a large drop of charged carboxyl groups. Obviously the total charge and the surface charge of the fibres are very dependent upon pH which must be taken into account in interaction studies between fibres and charged ligands.

The total charge of bleached sulphate pulp has been determined at pH = 7, to about 30 μekv/g [26]. In the sulphite process, lignin is attacked by sulphite or bisulphite and form sulfonic acids (C$_6$H$_5$–HSO$_3$). pKa for sulfonic acid is about –6.5 [27], so sulfonic acid is always ionised at normal conditions. In the following bleaching step most of the lignin residues are removed leaving only a small fraction of sulfonic acid. The total charge of bleached sulphite pulp has been estimated at pH = 7 to about 35 μekv/g [26]. Probably a small part of this total charge originates from sulfonic acids.

In chemithermomechanical (CTMP) pulping, wood chips is pre-treated with alkaline solution of sodium sulphite prior to the mechanical defibration. The lignin is attacked by sulphite resulting in a large amount of charged sulfonic acid groups. Since most of the xylan is preserved in the CTMP process, the pulp is also rich of charged carboxyl groups. After bleaching with H$_2$O$_2$, the total charge was estimated at pH = 7 to about 250 μekv/g [28]. Also the important surface charge (for ligand interaction) has been estimated by polyelectrolyte titration to about 32 μekv/g (at pH = 8) [28]. Using conductometric titration, the amount of sulfonic acid groups and carboxyl acid groups have been estimated in bleached (H$_2$O$_2$) CTMP. [26]:

- Sulfonic acid groups   40–70 μmol/g
- Carboxyl acid groups   150–250 μmol/g

As have been described earlier in this chapter, the functional groups can also form hydrogen bonds and covalent bonds with charged or uncharged ligands.

7.5 Future

7.5.1 Future Use of Existing Chemicals

In the time of environmental awareness, some of the wet-strength chemicals mentioned earlier in this report have some „bad" properties. The chemical industry has reacted responsibly. UF resins have been produced with lower „free" formaldehyde. PAE resins with reduced quantity of organic chlorine-compounds are available. Among the existing wet-strength resins, PAE and G-PAM seem to be most widely spread. PAE is very established and paper with high relative wet-strength can be produced with this chemical. As mentioned earlier, PAE give rise to paper that is stiffer than untreated paper, which is a drawback in tissue grades. PAE treated paper does also have less absorption capacity and longer absorption times. Researchers seem to work a lot in trying to minimise the bad paper properties caused from the use of PAE.

For instance Espy [19], presented (in a patent assigned to Hercules Inc.) a method to improve the absorption time considerably. Using a combination of PAE/CMC together with a poly (amino amide) polymer without azetidinium groups, much better absorption times were achieved. Being highly charged, the azetidinium free polymer competes with PAE for the binding sites on the negatively charged fibres. This probably led to a less bonded structure between the fibres, which result in shorter absorption time. It is probable that the use of the azetidinium free polymer also result in lower levels of relative wet-strength.

The absorption capacity is an important factor in tissue grades. By definition, the use of wet-strength resins, whose purpose is to prevent fibre swelling, reduce the absorption capacity. A general believe is that the use of CTMP- or HT-CTMP-pulp, which have high absorption capacity, could solve this problem. However, lab studies definitely show that the use of HT-CTMP and wet-strength resins alone is not the solution to this specific problem.

In a very interesting patent by Phan [29], assigned The Procter & Gamble Company, a combination of many chemicals was used. 75 % of kraft pulp was treated with PAE (for wet-strength), CMC (for dry-strength) and a mixture of dihydrogenated tallow dimethyl ammonium methyl sulfate and polyethylene glycol, (DTDMAMS/PEG-400) (for better absorbency and softness). 25 % CTMP was treated with non-ionic surfactant and DTDMAMS/PEG-400. The two fractions were then mixed. No results were presented, but keeping in mind the excellent products from the Company, this patent is very interesting and important. The use of the TAD technique that seem to be standard in The Procter & Gamble Company does of course also have an influence on their results, this technique is described in numbers of patents and will not be further discussed.

7.5.2 Future Chemistry

Many patents in the field of new wet-strength resins do only include modifications of PAE or other existing wet-strength resins. This accentuates how difficult this field is and how difficult it will be to replace PAE and G-PAM. To find wet-strength resins that are biodegradable is an interesting part of the future chemistry. As already mentioned, starch is an alternative.

ATO-DLO (Instituut voor Agrotechnologisch Onderzoek) in The Netherlands, have improved the process for oxidation of starch to dialdehyde starch. Dialdehyde starch is an excellent starting point for future modification of the starch molecule. An interesting part is the possibility to repulp paper treated with starch by amylase and/or amylopectinase. This enzymatic break down is very interesting and should of course give advantages when dealing with the repulping problem. Anyway, much more research work must be spent to make starch more effective as a wet-strength resin.

Another alternative is chitosan which is an amino-polysaccharide and, as can be seen in *Figure 7.21*, the structure of chitosan is very similar to that of starch.

Chitosan is cationic without chemical modifications, however, the charge is dependent upon pH. Chitosan is available from shellfish wastes and has been evaluated as a paper strength additive [30]. It has been suggested [31] that chitosan interact with cellulose fibres that is more or less a requirement for a potential wet-strength resin. Anyway, chitosan is expensive and must be much cheaper to be a realistic alternative in the future.

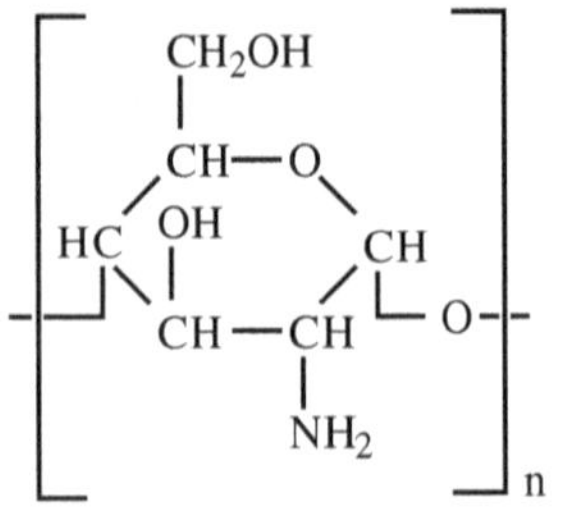

chitosan

Figure 7.21. The chemical structure of chitosan.

In the patent literature, an abundance of new wet-strength resins are continuously presented. It is always interesting to read about new ideas and new chemistry but it is a very time consuming work to map all new ideas and even more time consuming to make experiments.

7.6 References

[1] Stange, A. M. W. (1994): in *Wet-Strength Resins and Their Application.* Tappi Press, Atlanta. p. 101–108.
[2] Farley, C. E.: Glyoxalated Polyacrylamide wet strength resin. In *1988 Tappi Wet and Dry Strength Seminar Notes.* 1988: Tappi Press, Atlanta.
[3] Espy, H. H., and Rave, T. W.: The mechanism of wet-strength development by alkaline-curing amino polymer-epichlorohydrin resins. *Tappi Journal* 71 (1988) 133–137.
[4] Dunlop-Jones, N. (1991): in *Paper Chemistry.* Blackie & Sons Ltd., Glasgow and London. p. 76–96.
[5] Chan, L. L., and Lau, P. W. K. (1994): in *Wet-Strength Resins and Their Application.* Tappi Press, Atlanta, Georgia. p. 1–11.
[6] Espy, H. H. (1994): in *Wet-strength Resins and Their Application.* Tappi Press, Atlanta. p. 13–44.
[7] Farley, C. E. (1994): in *Wet-Strength Resins and Their Application.* Tappi Press, Atlanta. p. 45–61.
[8] Israelachvili, J. N. (1985): in *Intermolecular and surface forces. With applications to colloidal and biological systems.* Academic Press inc. (London) Ltd., London.
[9] Neal, C. W.: A review of the Chemistry of wet strength development. In *Tappi wet & dry strength short course.* 1988. Chicago: Tappi Press, Atlanta.
[10] Espy, H. H.: The mechanism of wet-strength development in paper: a review. *Tappi Journal* 78 (1995) 90–99.
[11] Devore, D. I., and Fischer, S. A.: *Wet-strength mechanism of polyaminoamide-epichlorohydrin resins. Tappi Journal* 76 (1993) 121–128.
[12] Helmer, U.: *Investigation of cellulose modifying chemicals.* In Department of polymer technology. 1985, The royal institute of technology.: Stockholm.
[13] Helmer, U.: *Private Communication.*

[14] Wilde, M. (1996): *Private Communication*.
[15] Gorzynski, M. (1996): *Private Communication*.
[16] Wågberg, L., and Björklund, M.: On the mechanism behind wet strength development in papers containing wet strength resins. *Nordic Pulp & Paper Research Journal* 8 (1993) 53–58.
[17] Merrett, K. J.: Factors affecting the performance of polyamide type wet strength resins. *Appita* 37 (1983) 233–236.
[18] Espy, H. H.: The effects of pulp refining on wet-strength resin. *Tappi Journal* 70 (1987) 129–133.
[19] Espy, H. H.: Absorbance and permanent wet-strength in tissue and toweling paper. *United States Patent* United States 5,316,623 (1994) Hercules Incorporated.
[20] Svegmark, K.: *Stärkelse. Struktur och reologi.* (1987) Svenska livsmedelsinstitutet (SIK). 547 (Swedish).
[21] Solarek, D., Tessler, M. M., Jobe, P., and Peek, L.: *Cationic Starch Aldehydes.* in *Paper Chemistry Symposium 1988*. 1988. Stockholm: STFI.
[22] Mehltretter, C. L., Yeates, T. E., Hamerstrand, G. E., Hofreiter, B. T., and Rist, C. E.: *Preparation of cationic dialdehyde starches for wet strength paper. Tappi* 45 (1962) 750–752.
[23] Laleg, M., and Pikulik, I. I.: *Unconventional strength additives. Nordic Pulp and Paper Research Journal* (1993) 41–47.
[24] Sjöström, E. (1993): in *Wood chemistry. Fundamentals and applications.* Academic Press, Inc., San Diego.
[25] Annergren, G.., and Wilhelmsson, K.: *Sågtimmer/fiberråvaror-användning i SCAs svenska fabriker.* (1996) SCA Graphic Research. K1135
[26] Wågberg, L., and Annergren, G.: *Physiochemical characterization of papermaking fibres.* (1997) SCA Research AB. F2325.
[27] Solomons, T. W. G. (1992): in *Organic Chemistry.* John Wiley & Sons, Inc., Toronto.
[28] Wågberg, L., and Eriksson, M.: Effekt av våtstyrkemedel på arkegenskaper i papper av CTMP och blekt sulfatmassa. (1990) *SCA Research AB. K961* (Swedish)
[29] Phan, D. V., and Trokhan, P. D.: Soft absorbent tissue paper with high permanent wet strength. *European Patent* France EP 0 610 337 B1 (1996) *The Procter & Gamble Company*.
[30] Allan, G. G., Crosby, G. D., and Sarkanen, K. V.: Evaluation of chitosan as a strength additive for alpha-cellulose and unbleached sulphite papers. In *International paper physics conference, 1975, Atlanta*.
[31] Domszy, J. G., Moore, G. K., and Roberts, G. A. F. (1985): in *Cellulose and its derivatives.* Halstead Press England, p. 463–473.

7.7 Patents

[32] Allen, A. J.: Azetidinium polymers for improving wet strength of paper. *United States Patent* United States 5,510,004 (1996) Hercules Incorporated.

[33] Clungeon, N. S., Devore, D. I., Fischer, S. A., and Giordan, J. C.: Wet strength resin composition and methods of making the same. *International Patent* United States WO 95/27008 (1995) *Henkel Corporation.*

[34] Darlington, W. B., and Lanier, W. G.: Repulpable wet strength paper. *United States Patent* United States 5,427,652 (1995) *The Mead Corporation.*

[35] Dickerson, J. A., Goldy, H. J., Smith, D. C., and Staib, R. R.: Improving the strength of paper made from pulp containig surface active carboxyl compounds. *European Patent* Germany EP 0 723 047 A2 (1996) *Hercules Incorporated.*

[36] Espy, H. H., and Putnam, S. T. Cationic water soluble polymeric reaction product of poly(diallylamine)-epihalohydrin and nitrogen compound. *United States Patent* United States 3,966,694 (1976) *Hercules Incorporated.*

[37] Espy, H. H.: Absorbance and permanent wet-strength in tissue and toweling paper. *United States Patent* United States 5,316,623 (1994) *Hercules Incorporated.*

[38] Fu, Y. L., Huang, S. Y., and Dexter, R. W.: High strength wet webs for the production of paper and process for producing paper making fiber webs having wet web strength. *European Patent* Germany 0 289 823 (1988) *American Cyanamid Company.*

[39] Jansma, R. H., Begala, J., and Furman, G. S.: *Strength resins for paper. United Stated Patent* United States 5,490,904 (1996) *Nalco Chemical Company.*

[40] Phan, D. V.: Soft absorbent tissue paper with high temporary wet strength.

[41] *European Patent* France EP 0 610 340 B1 (1996) *The Procter & Gamble Company.*

[42] Phan, D. V., and Trokhan, P. D.: Soft absorbent tissue paper with high permanent wet strength. *European Patent* France EP 0 610 337 B1 (1996) *The Procter & Gamble Company.*

[43] Ramakant, T. S.: Sulfonated cellulose and method of preparation. *UK Patent* United Kingdom 2 289 695 (1995) *Kimberly-Clark Corporation.*

8 The Surface of Paper

Anthony Bristow
Bristow Consulting AB

8.1 Introduction 210

8.1.1 Two-sidedness 210

8.2 Surface Roughness 211

8.2.1 Different Roughness Measures 211
8.2.2 Surface Contact Area 212
8.2.3 Surface Volume 212
8.2.4 Air-leak Methods: Bendtsen, PPS, Sheffield 213

8.3 Surface Compressibility 214

8.4 Surface Porosity 215

8.5 Surface Permeance 216

8.6 Surface Absorptivity 216

8.7 Surface Wettability 217

8.7.1 Surface Contact Angle 217
8.7.2 Wetting Delay 218

8.8 Surface Printability 218

8.8.1 Ink Transfer 218
8.8.2 Print Density 220
8.8.3 Print Through 221
8.8.4 Set-off 222
8.8.5 Mottle 223

8.9 Surface Strength, Linting 223

8.9.1 Pick Strength 224
8.9.2 Fibre Lifting 224

8.10 Surface Friction 225

8.11 Surface Gloss 225

8.11.1 Reflection from a Plane Surface 226
8.11.2 Reflectance from a Rough Surface 228
8.11.3 Measurement of Gloss 229
8.11.4 Gloss Variations 230

8.12 The Creation of the Surface 230

8.1 Introduction

A paper is a three-dimensional network consisting primarily of fibres but, although it can often be considered to be an infinite network in two dimensions, the network is always finite in the thickness direction. In the thickness direction, there is a clearly defined boundary between the network and the air, and the nature and structure of the network at this boundary define the surface of the paper.

The surface is naturally relatively flat because the fibres lie preferentially in the plane of the sheet, but it is not intrinsically sufficiently flat to fulfil its purpose and to meet end-use requirements. Various steps must therefore be taken in the papermaking process to ensure that the surface is indeed as flat and smooth as is desired.

The surface of the paper is an extremely important region, particularly in printing papers. The function of the surface is to accept and carry a printed message and it is sometimes said in this context that the main task of the fibre network is merely to act as a support for the surface.

In a mathematical sense, a surface is a plane with zero thickness and thus zero volume, but in a paper technology context the surface of the paper refers to a region with a certain thickness and a certain volume possessing well-defined and important properties. If we consider a paper web to consist of fibres lying in the plane of the sheet, then the surface is a region which is one and in some cases perhaps two fibres thick.

In this chapter, we consider surface properties such as surface roughness, surface porosity, surface compressibility, surface strength, surface wettability and surface gloss, and how these properties are defined and measured. In the final section of the chapter, we consider briefly some of the important papermaking processes required to ensure that the surface meets the end-use specifications. These processes are discussed in more detail in other chapters of this textbook.

8.1.1 Two-sidedness

A sheet of paper has two sides and therefore two surfaces. For some purposes, e.g. in packaging, papers are made which are intentionally different in furnish or in finish on their two sides, but for other purposes, e.g. in printing, the two sides must be as similar as possible. It is in the latter context that the term „two-sidedness" is used, in a negative sense, indicating a problem to be solved or preferably a problem to be avoided.

In the traditional fourdrinier machines, there is more than one cause of two-sidedness. In the first place, the fact that the paper sheet is formed on a wire means that the side of the paper in contact with the wire is embossed with the pattern of the woven wire. In the second place, the unidirectional drainage of the water from the sheet through the wire means that there is a flow through the sheet which tends to give a non-uniform distribution of fines, filler particles etc through the sheet, with either an enrichment or a depletion on the wire side. In this case there is a corresponding depletion or enrichment on the top side. The top side of the sheet often reflects the texture of the felt in the press section and may be referred to as the felt side.

The desire to avoid this two-sidedness was one of the factors which led to the development of twin-wire formers with a similar wire on each side of the wet paper web and simultaneous dewatering in both directions.

In some cases, a two-sidedness may be desired as for instance in wrapping papers made on a machine with a yankee cylinder, where one surface is fairly rough while the other is highly glazed.

In other cases, differences between the two sides can be introduced by differences in furnish. A liner destined to be used on the outside of a corrugated board has an inner surface which must glue efficiently to the corrugated medium and an outer surface which must resist weather and mechanical handling and must also bear a printed message concerning the contents of the box. This outer surface will normally have a higher quality kraft pulp furnish, but it may also be bleached in so-called white-top board, and may even be coated.

Many paper grades are now coated and in these products there is usually a desire that the two sides shall be the same, so that the coating must be applied to the two sides in a manner which will avoid the development of two-sidedness.

8.2 Surface Roughness

The surface of a paper is never perfectly smooth, but it is necessary to be able to specify and control the smoothness. This means that it has been necessary to consider how to define the smoothness or roughness of the surface and how to make reproducible measurements of this property. The name given to this property depends on the measurement technique adopted. If the value of the property increases as the surface becomes smoother, the method is referred to as a smoothness measurement; if the value of the property increases as the surface becomes rougher, the method is referred to a roughness measurement

8.2.1 Different Roughness Measures

The roughness is instinctively defined in relation to deviations from a plane surface. The measurement of the roughness is however usually related particularly to the behaviour of the paper in a printing process. It is therefore natural to define the roughness in terms of the deviation from a plane surface pressed against the surface under a pressure corresponding to the nip pressure in the printing press.

a) contact fraction (%)

b) surface volume (cm^3/m^2)

c) surface pit distribution

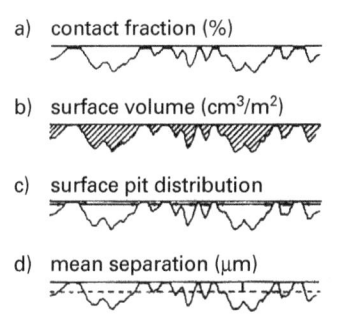

d) mean separation (μm)

Figure 8.1. Different ways of characterising the surface roughness.

Figure 8.1 illustrates this situation, and shows that several different measures of the roughness are possible. Parameters which have been suggested are

- the surface contact area, expressed as a percentage,
- the surface volume, expressed in cm^3/m^2
- the surface pit distribution
- the mean surface displacement, expressed in µm

Here it is interesting to note that measures (b) and (d) are equivalent, since 1 cm^3/m^2 is equal to 1 µm.

8.2.2 Surface Contact Area

Although it is not widely used, the method developed by Chapman is worth mention. In this method, a glass prism is placed on top of the paper and the surface is illuminated in a manner which leads to total internal reflection in the regions where the glass is in contact with air and not with the paper. The result is a pattern of contact and non-contact regions. This pattern may be visually assessed or photographed and analysed to give a mean contact area as a proportion of the total area and also detailed information about the contact pattern.

Similar information can be obtained by printing under a low pressure in a laboratory press with a very small amount of ink. Under these conditions, the ink is transferred only to the raised regions of the paper surface, and the print obtained shows the nature of the surface in a manner which can be quantified by image analysis.

8.2.3 Surface Volume

The surface volume is a useful concept, but it is usually difficult to measure because any liquid used to assess this volume is usually absorbed into the paper. The surface volume may therefore best be assessed as the extrapolation of an absorption curve to zero time. The Bristow absorption tester – commonly known as the Bristow wheel – is one method which provides such a measure. The principle of the method is illustrated in *Figure 8.2*.

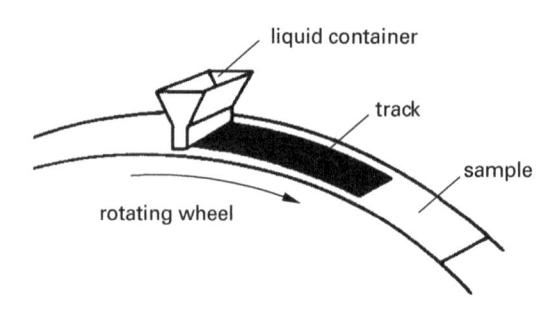

Figure 8.2. The principle of the Bristow absorption method.

In this method, the absorption is assessed as the length of a stain left by a given volume of liquid when it is absorbed into the paper through a slit in a liquid container which passes over the surface at a given speed. The absorption time is varied by altering the speed of rotation of the wheel on which the test piece is fixed.

A plot of the amount of a non-swelling liquid absorbed as a function of the square root of the absorption time is usually a straight line, as shown in *Figure 8.3*. The intercept of this line on the ordinate at zero time is the amount of liquid transferred to the surface in zero time before the absorption can begin, i.e. the surface volume, in cm^3/m^2. The significance of the horizontal section of the curve for water in this diagram is discussed in section 8.6.2.

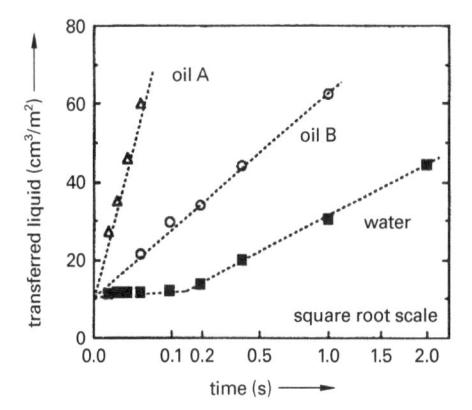

Figure 8.3. A plot of transferred liquid versus time, on a square root scale, obtained using the Bristow absorption tester.

8.2.4 Air-leak Methods: Bendtsen, PPS, Sheffield

The most commonly used method of assessing the roughness of a paper surface is an indirect method, involving an air-leak concept. The basic principle is that a cup is inverted over the surface and air is introduced into the cup under pressure. If the surface is rough, the seal between the edge of the cup and the surface is imperfect, and the air tends to leak out between the surface and the edge of the cup. The rate at which the air flows out is thus a measure of the roughness. In the Bendtsen method, the roughness of the paper is expressed simply as a flow rate in ml/min.

Parker refined the measurement technique in an instrument known as the Parker-Print-Surf (PPS) instrument. He developed a more sophisticated measuring head, shown in section in *Figure 8.4*, he used a clamping pressure in the measurement nip which was similar to that in the nip in a printing press, and he developed an equation for converting the air-flow rate into a measure of roughness expressed in meaningful units.

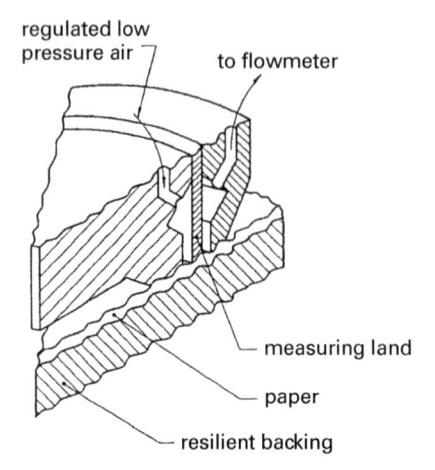

Figure 8.4. A section showing the main features of the measuring nip in the Print-surf instrument.

His equation was based on the concept that the flow through a gap is proportional to the cube root of the thickness of the gap. According to the model adopted by Parker, the spaces through which the air flows between the paper and the measuring land can be considered to be a uniform gap, so that the mean deviation G of the surface from a plane is given by the expression:

$$G = [(12 \, \eta \, b \, Q)/(l \, \Delta p)]^{1/3} \tag{8.1}$$

where
η is the viscosity of air
b is the width of the measuring land
Q is the volume of air flowing in unit time
l is the length of the measuring land
Δp is the pressure difference across the measuring land, corrected for the air compressibility

The Print-Surf instrument thus gives a roughness measure expressed in µm. The Sheffield instrument used in North America is similar in its concept to the Bendtsen instrument, but uses different dimensions, different pressures etc..

8.3 Surface Compressibility

The roughness of paper is of particular concern when the paper passes through a printing nip and ink on the printing forme is expected to transfer from the forme to the paper, but the fact that paper is a compressible material is also important. In the 1970s, the concept of „surface compressibility" was introduced when it was realised not only that the compression of the paper during its passage through a printing nip may be important but that the surface region may compress differently from the bulk as a whole.

The surface compressibility can be assessed by measuring the change in surface roughness when the paper is compressed under different pressures, as shown in *Figure 8.5*.

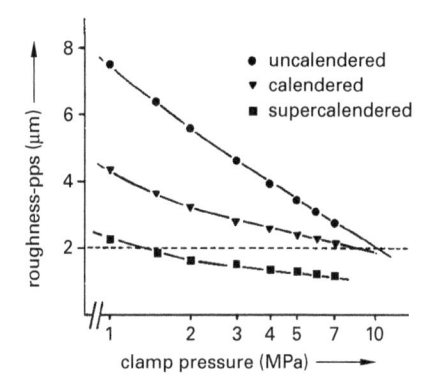

Figure 8.5. The decrease in surface roughness with increasing pressure is a measure of the surface compressibility.

It is routinely assessed by measuring the roughness under two different clamping pressures without moving the paper, and it is then expressed as

$$K = 100\,(G_1 - G_2)/G_1 \tag{8.2}$$

where G_1 and G_2 are the roughness values at clamping pressures of 1 MPa and 2 MPa respectively

8.4 Surface Porosity

The porosity of a paper is properly expressed as the fractional void volume. If a sheet has a density ρ_P, and if the density of the cellulosic solid phase is ρ_C, then for a given mass m of material, the volume of the paper is equal to m/ρ_P and the volume of the cellulosic solid phase is equal to m/ρ_C. The difference between these two volumes is the volume of the void space and the porosity is then given by expressing the void volume as a fraction of the total volume:

$$\text{porosity} = (m/\rho_P - m/\rho_C)/m/\rho_P = 1 - \rho_P/\rho_C \tag{8.3}$$

It is clear from a philosophical point of view that the porosity of the surface region of the paper may differ from that of the bulk sheet, but it is difficult to make measurements to demonstrate this. Attempts have been made to study thin cross-sections of a paper sheet by image analysis and to determine the relative amounts of material and of voids in successive layers through the cross-section, but there is always some doubt as to how much distortion may have been introduced by the sectioning in the microtome.

A method for measuring the distribution of surface pores has been proposed which is essentially the inverse of the Chapman method (section 8.1.2), where the non-contact areas are analysed, but it is not possible in this method to distinguish between surface pores leading into the interior of the paper and surface pits, which are merely indentations in the surface.

8.5 Surface Permeance

The term „porosity" is often incorrectly used to refer not to the amount of air in the material, but to the rate of air flow through the material. This property should however more correctly be referred to as the air permeance of the material. The word „permeance" is used rather than „permeability" to emphasise that this is a sheet property, and not a material property normalised with respect to thickness.

The air permeance through a sheet is measured in a device such as the Gurley or Bendtsen apparatus where a single sheet of the material is clamped between a metal edge and a rubber gasket, a pressure difference is applied across the material, and the flow of air in unit time is determined. Under steady-state conditions, the flow through the bulk of the sheet is the same as the flow through the surface, and this means that it is not possible to make any distinction between the bulk permeance and the surface permeance.

Although the permeance is dependent on the thickness and structure of the sheet, it may provide some information about the density and tightness of sealing of the surface region.

8.6 Surface Absorptivity

In assessing the effect on the surface of a surface treatment such as a calendering of the paper, it may not be sufficient merely to measure the change in roughness or the change in gloss. Some means of assessing the extent to which the surface pores have been closed up, especially in the case of a coated surface, may be required.

The K&N ink test is such a method. The K&N ink is a proprietary ink manufactured for the purpose of assessing ink absorption and it consists of a pigment extender and a purple dye in an oil. A quantity of ink is applied to the surface of the paper and the excess is removed after a given time (2 minutes). The extent to which the ink has been absorbed is assessed optically by determining the intensity of the stain. Measurements are made using a spectrophotometer conforming to ISO 2469 and incorporating C/2° and the CIE-Y-function as illuminant/observer conditions and the *K&N*-value is calculated as:

$$K\&N = 100\ (R_\infty - R_p)/R_p \tag{8.4}$$

where R_∞ is the reflectivity of a pad of the unstained paper and R_p is the reflectance factor of the stained area of a single sheet of paper, obtained by placing the stained sheet over an opaque pad of the unstained paper.

This test has its limitations. The method was developed as a qualitative method, but it has been found to be sufficiently useful to justify its acceptance as a quantitative procedure. The value obtained is to some extent dependent on the batch of ink used and it is clear from equation (8.4) that the value obtained is also dependent on the reflectivity R_∞ of the substrate. The method is, nevertheless, sensitive to changes in the surface structure, and a decrease in the *K&N*-value after, for example, a calendering operation is a clear indication that the surface pores have been closed up so that the absorption through the surface has been reduced. Further valuable information can often be obtained by carrying out the test over a range of different contact times.

8.7 Surface Wettability

The wettability of the surface by liquids is important in many situations. Some products are designed to absorb water and others are required to repel water. An essential factor in any gluing operation, for example, is that a liquid phase – whether it be a solution, an emulsion or a pure liquid – must wet the substrate. Similarly, any printing operation requires that the paper be wetted by the ink, and this may be oil-based or water-based. The lithographic offset process requires a correct balance between an oil-based ink and a water-based solution which is used to provide a demarcation of the non-image areas. The lithographic process is essentially an exercise in surface chemistry, even though the process often operates under conditions where the surface chemical aspects are not a critical problem.

8.7.1 Surface Contact Angle

A basic measure of the wettability of the surface is the contact angle between a liquid and the surface. If a drop of liquid is placed on a surface, an equilibrium is established, as indicated in *Figure 8.6*.

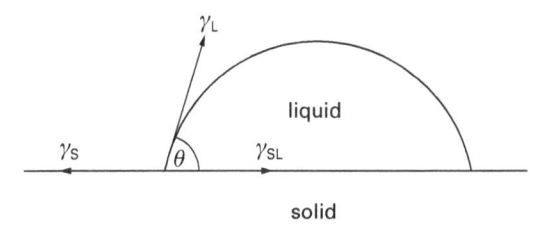

Figure 8.6. The equilibrium of forces determining the contact angle of a liquid on a surface.

The contact angle is then defined by the Young equation:

$$\gamma_{sv} = \gamma_{lv} \cos\theta + \gamma_{sl} \tag{8.5}$$

where γ_{sv} is the surface free energy of the solid/vapour interface, γ_{lv} is the surface tension of the liquid/vapour interface and γ_{sl} is the surface free energy of the solid/liquid interface. This means that the contact angle is dependent on the relationship between the different surface tensions according to:

$$\theta = \arccos\left[(\gamma_{sv} - \gamma_{sl})/\gamma_{bl}\right] \tag{8.6}$$

It is not easy to measure the angle itself, but instruments have been devised which project an image of the drop onto a screen so that the height and base of the drop can easily be measured. Assuming that the projected image is a sector of a circle, the contact angle θ can then be calculated as:

$$\theta = 2\arctan(2h/b) \tag{8.7}$$

where b is the base and h is the height of the image

Modern instruments are computerised so that this is done automatically. Such instruments should not, however, be used uncritically. One problem is that the paper must be held sufficiently flat to ensure that the base is properly visible on the screen. Another problem is that on a porous substrate such as paper, the liquid may be progressively absorbed into the material, so that it is necessary to define the measurement conditions if reproducible results are to be obtained.

A more advanced treatment of the surface wettability takes into consideration the fact that the surface tension can be divided into polar and non-polar components.

8.7.2 Wetting Delay

The thermodynamic analysis of wettability in terms of contact angles and surface tensions is based on equilibrium conditions, but printing, gluing or similar operations take place under dynamic conditions and consideration must in practice be given to the dynamic wettability. There is some indication that there may sometimes be a significant wetting delay. In Figure 8.3, the curve representing the uptake of water has a horizontal region, and this had been interpreted as indicating that there is a wetting delay, with a wetting time characteristic of the surface, before the absorption can begin. Such a wetting delay is not observed in the case of the oils shown in the same diagram.

8.8 Surface Printability

One of the important properties which a printer requires when he purchases a paper is good printability. Normally a printer talks simply of a paper's printability but in this chapter we talk of „surface printability" to emphasise that it is essentially the nature of the surface which determines whether or not the paper can be said to have a good printability. The printability of the surface can be summarized under three headings as the ability of the paper to contribute to

- a good transfer of ink from the printing forme to the paper
- a satisfactory interaction between the ink or its components and the paper
- a good appearance of the ink on the paper, i.e. good print quality

8.8.1 Ink Transfer

The importance of the surface is emphasised in the theory of ink transfer. The transfer of ink to a paper in a printing nip is dependent on a number of press-related and ink-related factors, but it is primarily dependent on the chemical and physical characteristics of the paper surface. Fetsko and Walker developed a theory for the transfer of ink to a paper surface where they proposed that a certain amount of ink b is immobilised by the paper at the moment of contact in the printing nip. If the amount of ink on the printing forme is x (whether g/m^2 or µm), then the free quantity between the forme and the paper when the paper leaves the nip is equal to $(x-b)$. This means that the total amount y transferred to the paper is given by:

$$y = b + f(x-b) \tag{8.8}$$

where f is a splitting factor which depends on e.g. the nip geometry. This is shown diagrammatically in *Figure 8.7*.

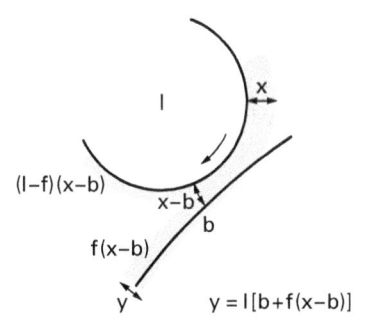

Figure 8.7. Model of ink transfer according to Fetsko and Walker.

This transfer can however only take place when there is contact between the ink-covered forme and the paper. If the fraction of the paper surface in contact with the forme is l then the amount transferred is, after some rearrangement, equal to:

$$y = l\left[fx + b(1-f)\right] \tag{8.9}$$

A further adjustment is required to take into account the fact that this relationship cannot apply for small amounts of ink when $x < b$. To take this into account, l and b are replaced respectively by the exponential functions:

$$l = \left(1 - e^{-kx}\right) \tag{8.10}$$

and

$$b = b\left(1 - e^{-\frac{x}{b}}\right) \tag{8.11}$$

so that the full Fetsko-Walker equation is:

$$y = \left(1 - e^{-kx}\right)\left[fx + b(1-f)\left(1 - e^{-\frac{x}{b}}\right)\right] \tag{8.12}$$

Many attempts have been made to improve on this equation, but it remains valid as a general description of the process, and it emphasises that the transfer of ink is related to a surface-roughness parameter k and a pore-structure or absorptivity factor b which governs the initial immobilisation of the ink. In this treatment, the wettability of the paper is not considered to be a problem.

8.8.2 Print Density

In studies of ink transfer to paper, it is usual to evaluate the result by considering the optical density of the ink on the paper. This is done by measuring the reflectance factor of the printed area R_p and the reflectance factor $R\infty$ of the paper and calculating the density D as

$$D = \log_{10}\left(R_\infty/R_p\right) \tag{8.13}$$

An important parameter is then the ink requirement (ink demand), defined as the amount of ink needed on the paper to achieve a given print density. The density criterion depends on the process and on the type of paper being printed.

Figure 8.8 shows a typical graph of print density against amount of ink on the paper. The figure shows how the ink requirement to achieve, in this case, a density of 1,2 with this particular ink on this particular paper is determined.

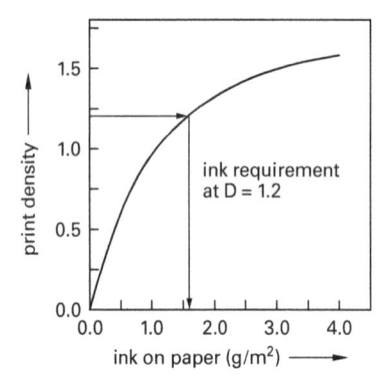

Figure 8.8. Print density plotted against the amount of ink on the paper.

When two papers are being compared with respect to ink requirement, it is important to note that there may be other criteria to consider. In all half-tone printing, pictures and other images are produced by breaking down the picture into a pattern of dots. Good print quality demands that all the dots are transferred completely and as discrete dots to the paper. With increasing ink quantity, however, the ink tends to spread out and flow together so that essential information is lost. This sets an upper limit to the quantity of ink which may be used. An example is given in *Figure 8.9*, which shows print density curves for four different types of paper. On each curve, a limit is shown above which it is not possible to print without an unacceptable filling in of dots.

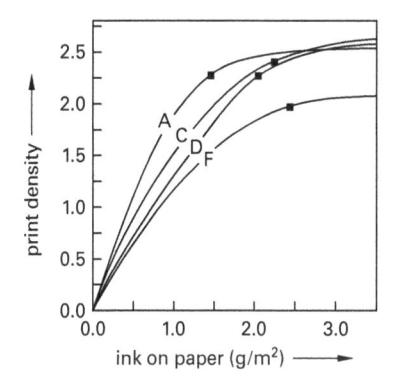

[Figure: print density vs ink on paper curves labeled A, C, D, F]

Figure 8.9. Limits on the print density curve.

8.8.3 Print Through

An important type of interaction between ink and paper is that where the ink, or components of the ink, penetrate into the paper. This leads to a so-called print-through effect, where the image on the front face is more or less visible on the reverse side.

Print through, G, is quantified by determining the reflectance factor R_q on the reverse side of the print, and is calculated by analogy with the calculation of print density as:

$$G = \log_{10}\left(R_\infty/R_q\right) \qquad (8.14)$$

This phenomenon can be divided into different components, as indicated in *Figure 8.10*.

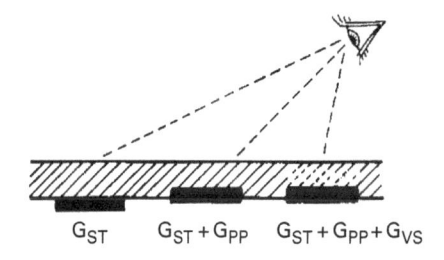

[Figure: diagram showing print-through components labeled G_{ST}, $G_{ST}+G_{PP}$, $G_{ST}+G_{PP}+G_{VS}$]

Figure 8.10. The different components of the print-through phenomenon.

- a show-through component, G_{st}, corresponding to the opacity of the paper, i.e. an unavoidable component unless the paper is totally opaque
- a pigment-penetration component, G_{pp}, due to the penetration of the ink into the paper, i.e. a component which increases in magnitude the more the ink penetrates into the paper
- a vehicle-separation component, G_{vs}, due to the fact that the ink releases oil which penetrates further into the paper and reduces the opacity of the paper behind the print.

This means that the opacity can be mathematically represented as the sum of three terms

$$G = G_{st} + G_{pp} + G_{vs} \tag{8.15}$$

and that the components can each be expressed by a term resembling equation (8.14) as follows:

$$G = \log_{10}(R_\infty/R_a) + \log_{10}(R_a/R_b) + \log_{10}(R_b/R_q) \tag{8.16}$$

Techniques have been developed for determining meaningful values of R_a and R_b so that these components can be isolated.

Figure 8.11 shows data corresponding to that shown in Figure 8.8. Here the print through on the reverse side of the paper corresponding to the print density of 1,2 on the front surface is indicated. The calculated show-through component G_{st} is also indicated. The difference between the two curves is thus the print through due to the penetration of pigment and oil into the paper.

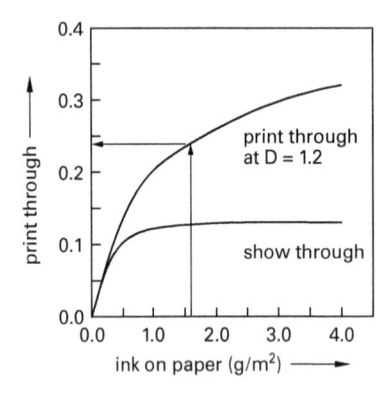

Figure 8.11. Print through plotted against the amount of ink on the paper. The magnitude of the show through due solely to the opacity of the sheet is also shown.

8.8.4 Set-off

The interaction between the ink and the paper also affects the drying of the ink. Some inks are dried by the application of external energy in the form of heat or UV-radiation. Others dry slowly by atmospheric oxidation. In the latter case, it is important that the ink passes rapidly through a so-called initial setting stage. During the time before this setting is complete, the ink can easily transfer from the paper to any other surface into which it comes into contact.

Set-off is the name given to this transfer phenomenon when it takes place under direct contact without any rubbing or shearing action. It is measured by pressing the print against a recipient substrate a short time after printing. It is calculated by analogy with print density according to the equation:

$$SO = \log_{10}(R_{\infty, SO}/R_{p, SO}) \tag{8.17}$$

where *SO* is the set-off and $R_{\infty,SO}$ and $R_{p,SO}$ are respectively the reflectance factors determined on the set-off substrate and on the stain on the set-off substrate. The values depend on the substrate chosen. The substrates commonly used are either a smooth cast-coated paper or a second sheet of the paper being tested.

In some cases, it is important to determine the set-off as a function of time after printing in order to assess which ink/paper combination shows the fastest setting. In other cases, it may be important to determine the magnitude of the set-off at a given time as a function of the amount of ink on the original print. *Figure 8.12* shows such set-off data corresponding to the print density data shown in *Figure 8.8*. The degree of set-off corresponding to a print density of 1,2 on the front face of the paper is also shown.

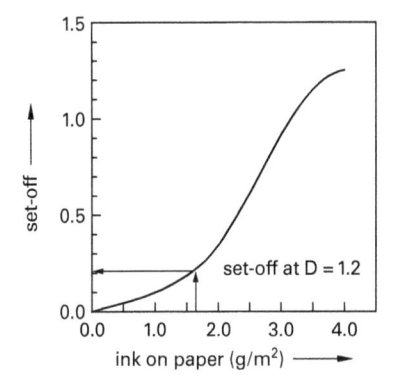

Figure 8.12. Set-off plotted against the amount of ink on the paper.

8.8.5 Mottle

The calculated print density is a mean value corresponding to the area measured. The spectrophotometer described in Chapter 5 has a circular measurement area with a diameter of 30 mm so that, although the Fetsko-Walker equation considers the possibility of uneven transfer in a physical sense, the measurement of print density provides a mean value and ignores the fact that an uneven transfer or a subsequent uneven interaction can lead to an uneven appearance. In general an uneven appearance is referred to as mottle, and the tendency for a paper to give a mottled print is an important quality criterion. The mottle is assessed visually or by an image analysis technique, which makes it possible to consider the mottle in different spatial wavelengths.

8.9 Surface Strength, Linting

In the traditional letterpress and lithographic offset printing operations, the surface of the paper is exposed to considerable outward tensile forces as the paper leaves the printing nip, due to the high-viscosity and high-tack printing inks in use. This means that particles (lint) may be torn from the paper and these then collect on the printing form or rubber blanket and eventually impede ink transfer and lead to a deterioration in print quality. The printer must then stop and clean the press and lose valuable productivity. This is a problem particularly in the multicolour lithographic offset process where the application of water in the non-image areas in an early printing station can lead to a weakening of the surface in areas to which ink is transferred in a later printing station.

Special measures must be adopted by the papermaker in his choice of furnish, additives and surface treatment to ensure that the surface strength of the paper meets the requirements of the printing process.

8.9.1 Pick Strength

The need to be able to measure a paper's propensity to leave quantities of lint in the press has led to the development of a special test procedure involving the use of the IGT Printability tester.

The principle of this test is that a disc, 1 cm wide and carrying a thin oil film on its circumference, runs over a strip of paper at an accelerating speed. The surface of the paper strip is then examined and the position where there are the first signs that particles are being lifted from the surface is recorded. The speed associated with this position is then determined. The force exerted by the oil on the surface of the paper as it leaves the nip is proportional to the product of the velocity and the viscosity. This product – or, in the case of a single standard oil, the speed – is a measure of the strength of the surface, and is known as the IGT-pick strength. Different oils with different known viscosities are available and these enable the strength to be determined over different ranges. It is important that this test is carried out under standard conditions of temperature and relative humidity, not only because the surface strength of the paper is moisture-dependent but also because the viscosity of the oil is temperature-dependent.

In practice, any increase in printing speed leads to a greater tendency for fibres or other particles to be pulled out from the paper surface. The tendency within the printing industry towards ever-increasing press speeds thus means that there is a continual demand for papers with higher pick strength.

8.9.2 Fibre Lifting

In recent years, interest in the phenomenon of a partial separation of fibres known as fibre lifting (fibre rising) has grown, especially in connection with heat-set offset printing. In this process, the paper is exposed to both water and heat and this process weakens the anchoring of the fibre in the surface so that the tendency for one end of the fibre to break away from the surface may increase.

Extensive work has been done to clarify the nature of the problem, and a method has been developed where a paper is treated with a defined amount of water and then heated and dried. Thereafter it is bent over a curved rod and the surface is viewed with a video-camera. This enables the number of fibres which rise from the surface as the paper is bent over the curved rod to be determined.

8.10 Surface Friction

When two surfaces rub against each other, forces of friction always oppose the movement. These forces may be critical for the performance of a paper in any given situation. In some cases, friction is a desirable and necessary property. In other cases, friction may be a disadvantage. When paper sacks containing some product are stacked in a warehouse, it is important that there is sufficient friction to prevent the sacks from sliding. When paper sheets are being fed through a printing press, it is important that they pass through singly and do not block into pads containing several sheets.

Surface friction is measured in a device which measures the stress when a sled to which one sheet of paper is fixed is drawn at constant speed over a second sheet of the same paper. Tests have shown that it may be important in this test to distinguish between the initial stationary friction and the dynamic friction. The force needed to start the sled moving may be greater than the force required to keep it moving at a constant speed over the surface. It has also been found that the force required tends to decrease if the sled is drawn repeatedly over the same position on the paper. To increase the precision of the test, it is common practice to record the frictional force for the third traverse of the sled across the paper sheet.

It is known that paper friction is dependent on the roughness, that it can be affected by the presence of lubricant substances on the surface and that it may be affected by the moisture content of the paper and thus indirectly by the atmospheric relative humidity.

8.11 Surface Gloss

The concept of gloss falls into the same category as colour (cf. Chapter 5) in that it is a psychophysical attribute rather than a simple physical property. The human brain considers a material to be glossy if there is a glare from the surface, a lustre in the structure of the surface or highlights in the surface showing a direct reflection of a light. Gloss is, however, a multidimensional phenomenon which cannot easily be measured.

Steps are taken to impart a high gloss to many grades of paper, but gloss is not always desirable, particularly in books for leisurely reading, as is clear perhaps in the following quotation:

> „The demand of the reading public for a paper of less gloss, which shall cause less strain on the eyes, has already been met by a number of magazine and textbook publishers, and it is apparently only a question of time before the use of highly calendered paper will be a rarity for ordinary reading matter."

Although these words sound fairly modern, it is instructive to note that they were written by L. R. Ingersoll as long ago as 1914 in a journal called Electrical World. In part, they represent a true forecast. There are, for instance, product standards requiring low-gloss paper in school textbooks.

Here, we are more concerned to consider how the psychophysical phenomenon of gloss can be understood and converted into an instrumental measurement. The gloss of a paper is assessed by measuring the amount of light which is reflected from the surface of the paper in a narrow cone at an angle (the specular angle) equal to the incidence angle.

8.11.1 Reflection from a Plane Surface

The reflectance R is defined as the ratio of the reflected light J to the incident light I:

$$R = \frac{J}{I} \tag{8.17}$$

The amount of light reflected from a smooth flat surface of a material with a given refractive index is dependent on the polarisation of the light, and is given by the Fresnel equations. When a ray light strikes an optically flat non-metallic surface with an angle of incidence θ, some light is reflected from the surface and some is refracted into the material. The amount of light reflected at the specular angle is dependent on the refractive index. The refractive index defines the angle of refraction ϕ within the material according to Snell's law:

$$n = \frac{\sin\theta}{\sin\varphi} \tag{8.18}$$

The reflection at the surface is dependent on the angle of refraction and on the plane of polarisation of the light. For light oscillating parallel to the plane of the incident ray, the reflectance R_p is given by:

$$R_p = \frac{\sin^2(\theta - \varphi)}{\sin^2(\theta + \varphi)} \tag{8.19}$$

and for light oscillating perpendicular to the plane of the incident ray, the reflectance R_s is given by:

$$R_s = \frac{\tan^2(\theta - \varphi)}{\tan^2(\theta + \varphi)} \tag{8.20}$$

These equations can be transformed to show the reflectance as a function of the angle of incidence θ for a given refractive index n. For light polarised parallel to the incidence plane, the reflectance R_p is given by:

$$R_p = \frac{\left[\cos\theta - \left(n^2 - \sin^2\theta\right)^{1/2}\right]^2}{\left[\cos\theta + \left(n^2 - \sin^2\theta\right)^{1/2}\right]^2} \tag{8.21}$$

and for light polarised perpendicular to the incidence plane, the reflectance R_s is given by:

$$R_s = \frac{\left[n^2\cos\theta - \left(n^2-\sin^2\theta\right)^{1/2}\right]^2}{\left[n^2\cos\theta + \left(n^2-\sin^2\theta\right)^{1/2}\right]^2} \tag{8.22}$$

The total reflectance in the case of a non-polarised light source is then given by:

$$R = \frac{R_p + R_s}{2} \tag{8.23}$$

i.e. by the expression:

$$R = \frac{1}{2}\left[\frac{\left[\cos\theta - \left(n^2-\sin^2\theta\right)^{1/2}\right]^2}{\left[\cos\theta + \left(n^2-\sin^2\theta\right)^{1/2}\right]^2} + \frac{\left[n^2\cos\theta - \left(n^2-\sin^2\theta\right)^{1/2}\right]^2}{\left[n^2\cos\theta + \left(n^2-\sin^2\theta\right)^{1/2}\right]^2}\right] \tag{8.24}$$

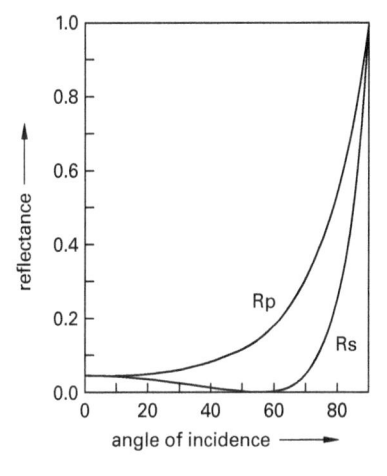

Figure 8.13. The reflectance as a function of angle of incidence.

Equations (8.21) and (8.22) are shown graphically in *Figure 8.13* for a material having a refractive index of 1.5. The curve for light oscillating parallel to the incident ray shows a progressively increasing reflectance, whereas the curve for light oscillating perpendicular to the plane of incidence drops to zero at the so-called Brewster angle θ_B which is defined by the expression

$$\theta_B = \frac{\pi}{2} - \varphi_B \tag{8.25}$$

Note that the curves do not pass through the origin but that they give a reflectance of about 4 % for normal incidence ($\theta = 0$).

These relationships are important for the definition of any gloss scale and in all the instruments used, the reflectometer value is converted into a gloss value by defining a scale such that the reflectance from the surface of black glass of a given refractive index at a given wavelength is defined as a gloss value of 100.

8.11.2 Reflectance from a Rough Surface

The Fresnel equations lead to interesting relationships describing the reflectance of an optically flat surface for different angles of incidence, but they are not of primary concern when the gloss of papers is being measured. The Rayleigh criterion for a surface to be considered optically smooth is that the height h of any asperities in the surface should not exceed the distance given by:

$$h < \frac{\lambda}{8\cos\theta} \qquad (8.26)$$

This means that for measurement at an angle of 75°, the limit is 250 nm, whereas values of the order of several μm are commonly given by the PPS method for determining surface roughness.

It has been suggested that the effect of the surface roughness on the gloss can be assessed by the equation:

$$\ln(R_R / R_F) = -16\pi^2 \cos^2\theta (s/\lambda)^2 \qquad (8.27)$$

where R_R and R_F are respectively the reflectances of the rough and the corresponding flat surface, θ is the angle of incidence, λ is the wavelength of incident light, and s is the mean roughness of the surface in μm.

Gloss can also be studied with a gonioreflectometer, i.e. with an instrument which allows a free choice of both incidence and measurement angle.

If a surface is not flat, then light is reflected not only in the specular direction but in all directions, and the light reflected from the surface is mixed with light reflected after multiple scattering from within the material. It is, however, extremely rare for the light to be reflected uniformly in all directions. If the intensity of the light reflected from a surface is measured in different directions when an incident light ray strikes a surface at an oblique angle, it is found that this intensity is direction-dependent. The light reflected from the point of incidence of the light can then be described with vectors, the length and direction of which show the directional dependence of the intensity. The curve which connects the end points of such vectors is called the indicatrix of the vectors.

For a completely matt surface, the indicatrix is a circle, whereas for a perfect mirror it consists of a point, in accordance with the law of specular reflection. *Figure 8.14* shows the appearance of the indicatrices for surfaces with different gloss levels.

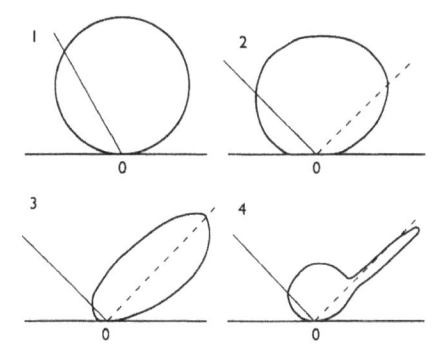

Figure 8.14. The reflectance indicatrix for different types of surfaces. (1) a Lambertian surface, (2) a matt paper, (3) a drawing paper (4) a calendered coated art paper.

- *Indicatrix 1*, is a polar curve for an ideal material for which reflection takes place according to Lambert's cosine law, which says that a light beam which meets a completely matt surface is reflected in each direction with an intensity which is proportional to the cosine of the angle between this direction and the normal through the point of incidence:

$$I(\varphi) = \cos\varphi \tag{8.28}$$

- A surface which gives a perfectly diffuse reflection according to this law is known as a Lambertian surface.
- *Indicatrix 2*, shows the directional dependence for a highly matt paper with only low gloss.
- *Indicatrix 3*, has been obtained for drawing paper. This almost elliptical shape is a typical example of mixed reflectance.
- *Indicatrix 4*, is for a supercalendered coated art paper. The indicatrix is here spherical as for a matt surface, but it is superposed with a peak relating to the specular reflectance. This peak does not however lie exactly in the direction of specular reflection, but in a direction which forms an angle to the normal which is slightly larger than the incident angle.

8.11.3 Measurement of Gloss

A conventional gloss meter is a reflectometer with incident illumination at a given angle and with measurement in the direction of specular reflection. The reflectance is measured with the use of a filter which corresponds to the sensitivity of the human eye to light, i.e. the $V(\lambda)$ function referred to in Chapter 5.

Measurements are commonly made with an angle of 75° between the incident ray and the normal through the point of incidence, and with measurement at the corresponding specular angle of 75°. Two methods are available as International Standards: These give different results primarily because the the methods have different angles of acceptance around the specular angle. The importance of this can be seen in *Figure 8.14* (3). A wider angle of acceptance means that more of the light scattered around the specular angle will reach the detector and will be included in the measurement.

It is not however satisfactory to measure all kinds of paper at the same angle and measurement at a smaller incidence angle is recommended for highly glossy papers. Two methods are used, one having an incidence angle of 20° and another having an incidence angle of 45°.

The result of a measurement is given not in reflectance units, but in gloss units which are defined in relation to the reflection by a smooth black glass reference with a defined refractive index, as indicated in *Table 8.1*. This glass is given a gloss value of 100 in each method. It can be seen in the table that the reference refractive index is not the same in all cases. The reasons for this are historical rather than optical.

Table 8.1. Gloss scale references in different Standards.

Standard	Angle	Reference refractive index	Reference wavelength (nm)	Spectral line
ISO 8254-1	75°	1,540	589,3	Na D-line
ISO 8254-2	75°	1,567	587,6	He D-line
ISO 8254-3	20°	1,540	587,6	He D-line
EN 14086	45°	1,567	587,6	He D-line

8.11.4 Gloss Variations

The standard type of gloss measurement gives a mean value for a fairly large area of paper, but there is a growing interest in measuring not only the average gloss but also small-scale variations in gloss, since these can be a disturbing factor in the visual appreciation of a printed surface. For this purpose, image analysis techniques are usually required, where the area studied can be broken down into a large number of picture elements (pixels) and the matrix of pixels can be analysed.

In the measurement of the colour, brightness of a paper etc, it is considered desirable that the gloss effect be eliminated. As a result, the sphere in the instruments used for this purpose is fitted with a so-called gloss trap, which prevents any direct specular reflection from the sample from reaching the measurement cell. In a similar manner, it has been argued that diffuse reflection from the material can interfere with a proper assessment of gloss, and a method has been proposed for the determination of what is called visual lustre. In this method, the paper is illuminated at an angle of 45° to the normal and measurements of the reflectance are made at both the specular 45° angle and at 0°. The theory is that the latter measurement provides an assessment of the diffuse reflection, as indicated in the indicatrix shown in Figure 8.14 (4). This value is then subtracted from the value obtained at 45° to leave the true gloss value.

8.12 The Creation of the Surface

The papermaker must devote considerable attention to the task of achieving a good surface. At his disposal, he has a number of variables and processes. These are dealt with in detail in other chapters. Here a few general comments are appended.

Components of the stock

The furnish strongly influences the nature of the surface. A furnish consisting of long coarse fibres is not likely to give a very smooth surface. Instead both shorter fibres (birch) and fillers are added to improve the surface. Here also, sizing agents which affect the wettability and absorptivity, and additives which improve the surface strength are also important.

Sheet-forming

The surface is born in the sheet-forming process. It has already been mentioned that, for instance, a twin-wire forming system may be important way of minimising the risk of an asymmetric two-sidedness in the paper.

Wet pressing

The wet pressing of the fibre web is usually discussed in relation to the consolidation of the web, but its importance for the structure of the surface should not be forgotten. An important consideration in the application of the extended shoe type of pressing is the extent to which this improves the surface or may in fact harden the surface and make it less amenable to subsequent calendering.

Drying

A special type of drying which imparts a high gloss to one side of the paper but has no effect on the other side is drying on a hot so-called „yankee" cylinder. The paper web is held in contact with the surface of this large rotating highly glazed cylinder for a part of its passage through the paper machine.

Surface sizing

A feature of many paper machines making paper or lithographic offset printing is a surface-sizing unit after the main dryer section. Here a CMC or starch solution is applied to the surface to impart a greater surface strength.

Calendering

A process designed especially to improve the surface, to reduce the roughness and to impart gloss is a treatment between rollers known as calendering.

In cases where the paper is subjected to treatment in a supercalender, the choice of which side of the paper is exposed to the metal roller in the first soft nip may determine the extent to which the two-sidedness of the uncalendered sheet is reduced or enhanced.

Coating

A coating is often applied to the coating to improve the surface. This is dealt with in another chapter. The coating must be porous in order to absorb certain components of the ink in printing process, and in order to provide a light-scattering layer which covers the substrate and gives a white appearance. The coating must also have sufficient strength to resist linting (piling) in a lithographic offset press. The strength of the coating and the strength with which it adheres to the substrate are governed by the type and amount of binder used in the coating formulation. This means that the concept of „binder requirement" may be important property of a coating formulation, i.e. the amount of binder required to resist a given stress. To achieve the highest gloss, cast coating is adopted.

9 Paper and Printing

Göran Ström
Innventia AB

9.1 Introduction 234

9.2 Conventional Printing Techniques 234

9.2.1 Offset Lithography 234
9.2.2 Flexography 235
9.2.3 Rotogravure 236

9.3 Digital Printing Techniques 238

9.3.1 Electrophotography 238
9.3.2 Inkjet 241

9.4 Short About Prepress 242

9.4.1 Gray Scale and Resolution 242
9.4.2 Colour Printing and Colour Separation 245

9.5 Printing Papers 245

9.5.1 Newsprint 245
9.5.2 Magazine Paper 246
9.5.3 Book Paper 246
9.5.4 Office Paper 246
9.5.5 Graphic papers 247
9.5.6 High Quality Inkjet Paper 247

9.6 Printing Inks and Ink Drying 248

9.6.1 Offset Inks 248
9.6.2 Rotogravure Inks 249
9.6.3 Flexographic Inks 249
9.6.4 Ultraviolet Curable Inks 250
9.6.5 Inkjet Inks 251

9.7 Print Quality 252

9.7.1 Print Density 252
9.7.2 Print Gloss 252
9.7.3 Unevenness in Print Density and Gloss 253

9.8 Ink Drying in Offset Printing 253

9.8.1 Physical Drying 254
9.8.2 Chemical Drying 255

9.1 Introduction

The production of a printed material can be divided into three main stages viz. *prepress, printing* and *finishing*. Prepress involves all the operations needed before the material can be printed. Today most of those operations are carried out with computers, but before the revolution of prepress most operations were done manually. Printing is accomplished in conventional and digital processes. Print finishing operations depend on types of printed work. The most common operations are cutting, trimming, folding and binding. In cutting, the material is cut to enable other finishing operations, while trimming is the term used for cutting the material into its finished size.

Each printing process requires specially designed ink to enable transport in the press and efficient drying. Although the inks differ strongly in composition and properties, the principal components are pigment, binder and solvent/oil. The most common printing media is paper and board. General requirements on the media are proper physical appearance, good transport through the press, a balanced interaction with the ink and an attractive print quality.

9.2 Conventional Printing Techniques

Conventional printing processes all have a non-variable printing plate that can be made off-press or in-press. The printing process is named after the characteristics of the plate.

The most important conventional techniques are *offset, flexography* and *gravure*. Other less important are *letterpress*, since this is an old and outgoing technique and *screen printing* which mainly uses other media than paper e.g. bottles, plastics, fabrics, etc.

The printing evolution started around 1440 when Johan Gutenberg invented the letterpress process. Today this process has a small and decreasing market share mainly in production of books. The market share of flexography increase while that of rotogravure is fairly constant. Offset has a large market share, about 50.5 % but it decreases due to the expanding digital printing business.

9.2.1 Offset Lithography

Offset lithography or simply offset is a *planographic* process, meaning that the areas on the printing plate that carry ink (*image areas*) are in the same plane as those that do not carry ink (*non-image areas*). The non-image areas are processed in such a way that the adhesion between ink and this area becomes very weak. This hinders transfer of ink from the form roller (an ink-carrying roller) to the non-image areas.

In traditional offset the non-image areas are hydrophilic (water loving). They consist of porous aluminium oxide onto which a hydrophilic polymer (e.g. gum arabic) is adsorbed. The area is then coated by water (*fountain solution*) and this „water-surface" does not pick up ink in the form roller nip. In *waterless offset* the non-image areas consist of silicone rubber. The adhesion between the ink and this non-image area is very low since silicone has a very low surface energy. Moreover, it is believed that inks for waterless offset contain small amounts of silicone oil, which concentrates at the boundary between ink and non-image area, and this further decreases the adhesion.

Since heat is generated during printing and fountain solution cools the press, waterless offset presses are equipped with a cooling system. The main advantages with waterless offset over conventional offset is lower waste paper since there will be no need for adjustment of ink-water balance, a process in conventional offset that can create considerably amounts of waste. Waterless offset also eliminates one of the reasons for dot gain namely emulsification of dampening solution and this sharpens the image. The disadvantage with waterless offset is the more expensive and the shorter lifetime of the printing plate since the silicone rubber layer is easily worn down.

The image area is always a hydrophobic polymer, in fact it is a photo polymer which polymerises when it is exposed to light during development of the image. Areas which have not been polymerised carry unpolymerised substance which is washed away and the underlying surface becomes the non-image area. *Figure 9.1* shows a traditional offset plate with ink and fountain solution, and a plate for waterless offset with ink.

Figure 9.2 shows a schematic drawing of an offset printing unit. One unit is needed for each colour. The dampening system with the fountain solution contacts the plate before the ink. Ink will not adhere to the non-image area since the split in the exit side of the nip will take place in the weakest layer which is the water layer. The image is then transferred to the paper via a cylinder covered by a rubber blanket. This cylinder is called *blanket cylinder* or *offset cylinder*. Since water is used in the process and this is transferred to various surfaces it is important that the ink can incorporate water. This is accomplished by emulsification.

A *sheet-fed offset press* prints on individual sheets, while a *web offset press* prints on a continues web of paper. The printing speed increases as the presses develop. The impression per hour is in the range of 15 000 for modern sheet-fed presses and 40 000 for modern web presses.

A web press uses *cold-set inks* when printing on newsprint and *heat-set inks* when printing on higher quality media such as magazine and fine papers. If the press uses heat-set ink, a hot-air dryer to remove ink solvents will be included after the last printing unit. Chill rolls are then located after the dryer. A sheet-fed press uses ink which dries through absorption of ink solvents into the paper surface followed by a polymerisation of binders. These presses may also utilize a dryer in order to speed up the drying process. Other drying mechanisms besides evaporation and absorption are involved in the process where the ink film becomes solidified. These will be discussed later.

9.2.2 Flexography

Flexography is a *relief process* since the image on the printing plate (or form) is raised over the non-image area. The form can be manufactured from a flexible rubber material, which enable printing on rough media such as packaging boards, or from a polymer to give high print quality on media with smooth surfaces. A flexographic printing unit is illustrated in *Figure 9.3*

The image area receives ink from an *anilox roller*. Its outer surface most often consists of a ceramic material with engraved small and uniform cells. Excess ink is removed from the roller by a doctor blade or by the fountain roller, which applies the ink to the anilox roller. The amount of ink transferred to the media is very much determined by cell geometry and volume. The impression cylinder generates a light pressure, just enough to transfer the ink. The pressure is much lower than that in offset printing.

Figure 9.1. Illustration of an offset plate with ink and fountain solution (left) and a plate for waterless printing containing silicon rubber as non-image area (right).

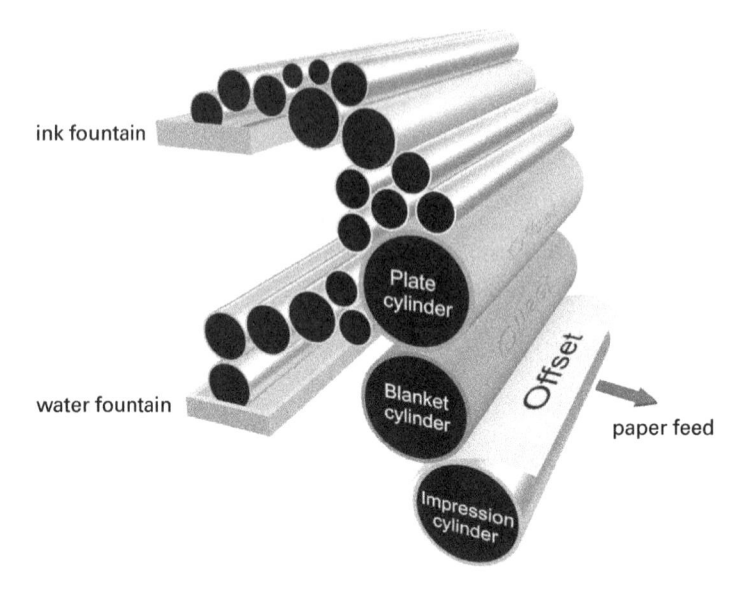

Figure 9.2. Configuration of a print unit of a sheet-fed offset press.

In contrast to offset both flexography and gravure uses *liquid inks*, which have low viscosity, somewhere between water and thin oil. Offset inks are termed *past inks* and are more viscous, close to syrup. Flexo ink contains high amounts of solvents and the image is dried after each printing unit in contrast to web offset where the drying unit is placed after the last printing units.

9.2.3 Rotogravure

Rotogravure is an *intaglio process* since the printing cylinder holds the ink in recessed areas (see *Figure 9.4*). The ink is feed to the printing cylinder by various techniques e.g. spraying, by immersing in the ink fountain, from an intermediate inking roller. Ink excess is removed by a doctor blade and the scraped off areas become the non-image areas. A dryer is located after each printing unit. Toluene is a common solvent in roto inks when printing on publication paper where fast drying is strongly requested. The drier has an efficient system to take care of the evaporated solvent. Water-alcohol based inks can be used when printing on packaging board because the printing speed is quite low which puts less demand on fast ink drying. Rotogravure presses are large and expensive. The most serious disadvantage with these presses is the expensive printing cylinders. However the print quality is superior to flexo and offset although offset print quality of today is almost as good as rotogravure print quality.

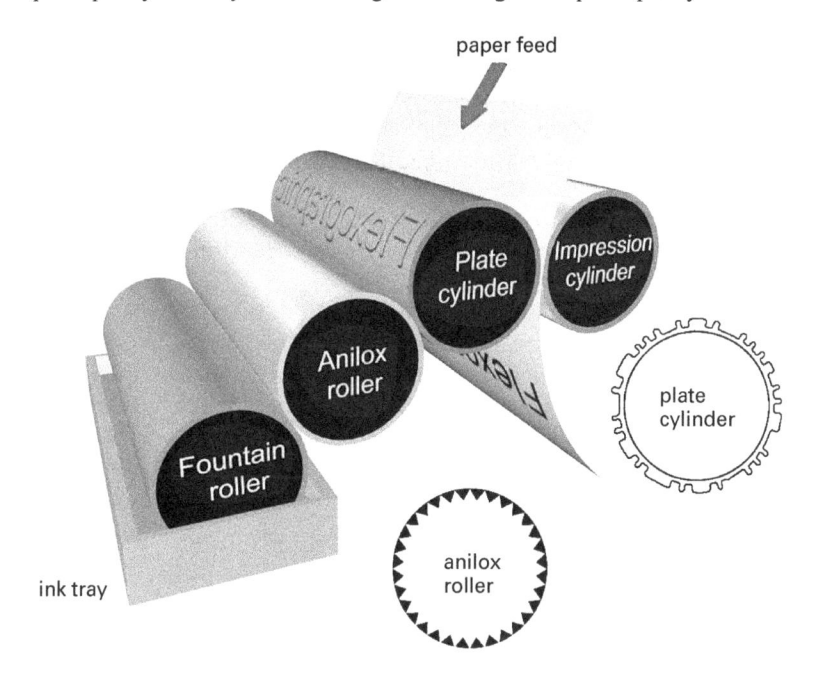

Figure 9.3. A Conventional Flexographic Unit.

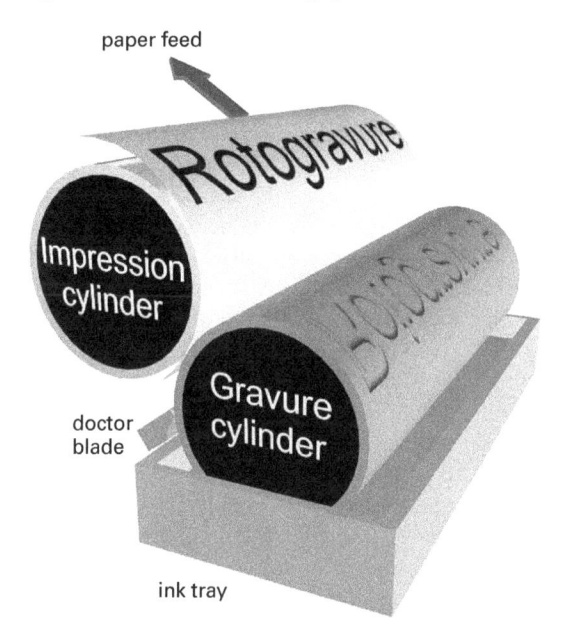

Figure 9.4. Illustration of ink transfer in rotogravure.

9.3 Digital Printing Techniques

Electronic or digital printing refers to printing where the complete workflow from manuscript to final product is handled from the computer. Digital printing also requires a printing system where the image produced can vary from sheet to sheet. This may be accomplished by a variable printing cylinder as in electrophotography or by the software as in inkjet printing. Variable printing enable production of personalized materials for direct mail, conferences, catalogues, etc.

Digital printing enables „*printing on demand*" (PoD), which means printing what you need, when you need it and where the customer needs it. This is an advantage when very short lead-time and/or small editions are required.

One important advantage of digital printing over conventional printing is the low cost per copy for small editions. In offset, for instance, the start up (set-up time) of a print job is time consuming. Most often it requires off-press manufacture of printing plates, mounting the plates in the press and start up, where the first 30–50 copies are rejected. Digital printing requires no printing plates and the first copy is as good as the rest. However, the traditional printing industry has recognized this shortcoming and made strong improvements. For instance certain offset presses (e.g. Heidelberg Quickmaster DI) have been developed where the workflow is handled from the computer and the printing plate is developed on the press. These presses have not a variable image but the work is done digital and therefore they are often referred to as digital offset.

The disadvantage of digital printing over conventional printing (e.g. offset) is lower print speed, lower print quality and higher production costs at high editions. The brake even point is at around 1000 copies.

The two most important digital printing processes are *electrophotography* and *inkjet*. Electrophotography is sometimes called laser electrophotography or just laser printing. This originates from the earlier way to use laser to create an image on the photo conducting drum. Today the laser, to a large extent, is replaced by light emitting diodes (LED).

When printing on plain, uncoated paper electrophotography gives better print quality than inkjet, but the hardware of electrophotography is very complex, costly, unreliable and creates high maintenance costs. When printing on expensive special paper, inkjet can reach photo quality although at a low print speed. Therefore inkjet is more common as home desktop printer while electrophotography has a large market as office printers and industrial digital print presses. High-speed inkjet printing can be achieved using special print head techniques and will be discussed below. In addition inkjet has a market in wide format printing of posters although special papers are used and printing speed is low.

9.3.1 Electrophotography

Electrophotography is today the most common process in digital printing. It uses dry or liquid toner. The liquid toner is a suspension of toner particles. The size of liquid toner particles is around 1–2 µm, which is much smaller than that of dry toner. The size of dry toner particles is around 6–10 µm but development of smaller toner particles is in progress. The smaller the toner particles the higher potential for print quality. The toner transfer process is illustrated in *Figure 9.5* for dry toner.

The image is created on a drum coated with a *photoconductor*, which may be selen based such as As_2Se_3 or, what is more common nowadays, an organic photo conducting material. The photoconductor is charged by a corona[1] at stage 1. The image is then exposed to the photoconductor by means of a tightly focused light beam generated from a laser or light emitting diodes (LEDs). The light beam focuses on the non-image areas and dissipates the charge in this area. The areas of the drum that still carry charge attract the toner in stage 3 developer and a toner image is formed onto the drum. The toner has developed a tribo-electric charge opposite that of the drum by means of friction during thorough mixing with magnetic carrier particles in the station and is transferred to the drum as a so-called „*magnetic brush*". At station 4 the toner is transferred to the paper by electrostatic interaction with the paper surface. The electrical charge on the paper has been induced by a second corona unit located on the backside of the paper. In the next stage the toner is fixed to the paper by a fuser. This is most often done by the combination of heat and pressure. The next stage of the drum is to remove any remaining electrostatic charge and to clean it from toner particles.

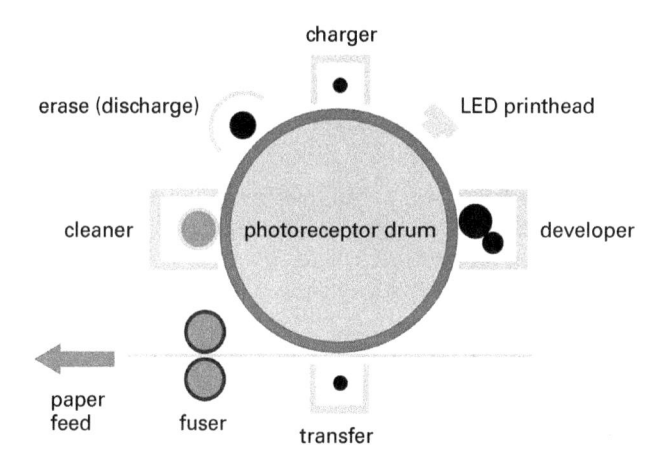

Figure 9.5. The principle of dry toner transfer in electrophotography.

During the end of 1993 Xeikon launched a web-based digital printing machine. This was the first electrophotographic printer to print in full colour on both sides of the paper (duplex) in a single pass. A scheme of the machine is shown in *Figure 9.6*. Eight electrophotographic printing stations are included in the tower to give four colors (CMYK) on each side. The fusing unit is located after the print tower meaning that the whole print on both sides is fused at the same time. Station 5 is a gloss unit where the print is calendered in a heated polymer nip and then cut into desirable size. Electrical and heat conductivity are important paper properties. A unit to dry the paper to the right electrical properties is therefore placed before the tower. The press uses dry toner and reaches a resolution of 600 dpi. The printing speed in A4 sheets per minute is 70 for Xeikon DCP 32 D and 100 for Xeikon DCP/50 D, a later member of this family.

---

[1] A corona is a thin gold-coated tungsten wire which is subjected to a voltage of thousand volts. The wire can impart electrostatic charge to objects close by.

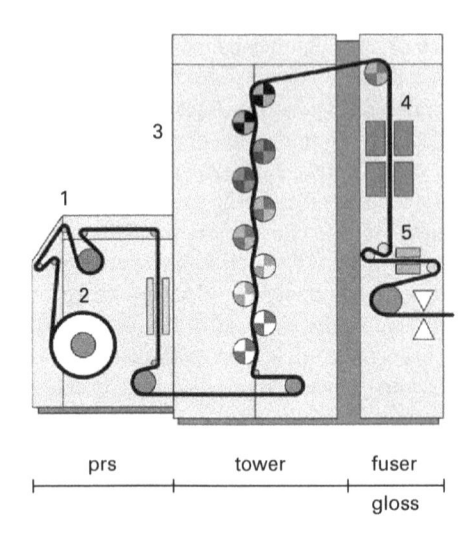

Figure 9.6. Scheme of the Xeikon DCP 32D press.

While the Xeikon press work with a direct transfer of toner from the drum to the paper the Xerox press is equipped with a belt to which the toner is first transferred. When the whole image has been transferred to the belt it is deposited to the paper. This indirect toner transfer improves print quality and put less demand on the paper. The Xerox machine is shown in *Figure 9.7*. The resolution is 600 dpi and print speed is 60 A 4 sheets per minutes which is similar to that of the Xeikon press.

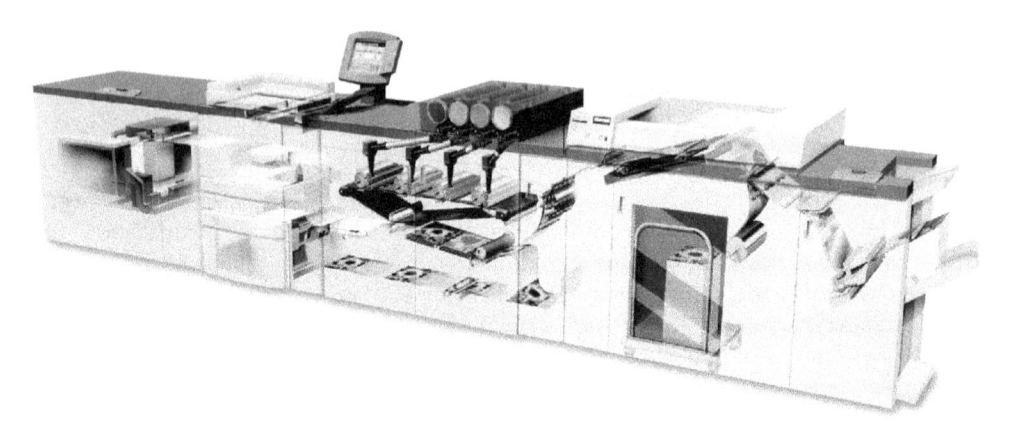

Figure 9.7. The Xerox DocuColor 2060 digital printing machine using indirect toner transfer.

9.3.2 Inkjet

Inkjet technology is widely spread in personal desk printers for homes and offices. The printers are inexpensive although the ink cartridges are quite expensive. Inexpensive plain papers e.g. ordinary uncoated copy papers can be used if the print quality demand is moderate or if only text is to be reproduced. Inkjet printing can yield very high print quality, in the same range as photographs, but this requires special photo quality paper and those are quite expensive. Inkjet has a unique potential for high print quality since the intensity of an ink spot on the paper can be varied by adding several ink droplets on top of each other. Printers with very high resolution are on the market. Epson Stylus C 80 has a resolution of 2880 dpi using variable size droplets from a volume of 3 pl.

The market for inkjet technology is within small printers for homes and small offices, industrial printing of wide format and industrial printing in general. The last segment is today quite limited but is believed to grow. The wide format (typically 1 to 1.5 m) and super wide format (> 2 m) color printing are used for short run poster work. The print quality is high and the normal resolution is 360–720 dpi, although high quality coated papers are required.

The ink droplets are generated as a continuous flow or as a „*drop-on-demand*" (DOD) In the continuous flow technique (see *Figure 9.8*) ink droplets are generated by piezo and the nozzle. The drops are charged and can then be directed to the image. The unwanted drops are deflected into a recycling reservoir. In DOD a drop is generated only when a dot on the paper is required.

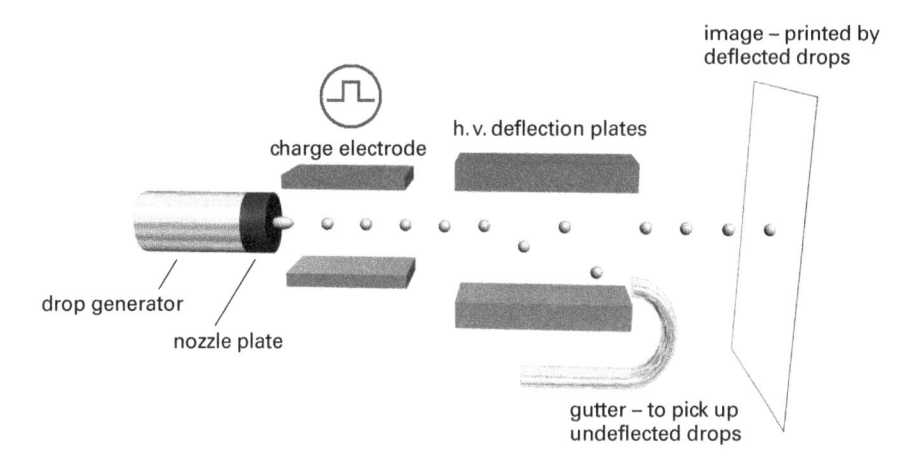

Figure 9.8. Scheme of a continuous inkjet head.

In DOD, ink droplets are formed either by the piezo technology (*piezo inkjet*, PIJ) or by heat pulses, which vaporizes some of the ink, which in turn induces a pressure. The latter is referred to as *thermal inkjet* (TIJ) or *bubblejet*. It is general believed that *continuous inkjet* (CIJ) gives higher print quality than DOD.

The first technical application of inkjet was disclosed by Siemens-Elema AB in Stockholm 1952. It was a chart recorder (Mingograf) based on a continuous jet with primary use in medical instrumentation. The inkjet was formed in a glass capillary 3 cm long and a diameter of 0.1 mm ending in a 15 µm nozzle opening. In 1976 IBM launched an inkjet system based on CIJ that achieved near letter quality, 240 dpi. The first piezo DOD inkjet printer to reach the market was introduced by Siemens in 1977.

The invention of thermal inkjet fundamentally changed the inkjet picture. This technique is both cheaper and the print heads are much smaller compared to PIJ. The print head is equipped with a transistor which heats the adjacent ink to > 290 °C for 2–3 µs. The ink vaporizes quickly and creates a pressure pulse which ejects the ink drop. Also the condensation of vaporized ink is quickly. The total cycle is 100 µs, which results in a drop rate of 10 thousand per second. It was invented by Cannon in the late 1970th and given the name *bubblejet*.

So far only one printing press for industrial use has reached the market. It is manufactured by Scitex and based on CIJ. The printing speed is 300 m/minutes or 2000 A4 colour sheets per minutes at a resolution of 300 dpi. The speed is much higher (> 10 times) than in electrophotography but the resolution is lower and more expensive papers are needed when high print quality is requested.

Advantages and disadvantages of the different inkjet technologies can be summarized as:

- CIJ can generate high drop rate by manipulating the ink supply pressure (high printing speed). CIJ is more expensive than TIJ and PIJ. Pigment based inks can not be used in CIJ due to clogging during recirculation.
- TIJ has become the most commercially successful technology approach to DOD design due to smaller size and lower production cost. This enables disposable print heads with a lifetime of 5 years. TIJ is limited to vaporizable inks. The bubble collapse process is hard to control. PIJ can use all types of inks, also the non-vaporizable ones like oil and wax based inks.

As guidelines we may say that inkjet drops have a volume of 40 to 80 pl (pico litres) although volumes down to 3 pl are reality today. A drop of 3 pl has a diameter of about 18 µm and the dot diameter on the paper becomes 30 to 50 µm depending on the paper. The drop velocity is typically 2 m/s for DOD and 10–20 m/s for CIJ. The distance from print head to paper is 1 mm or just above. A drop flight time of 50 µs is common.

9.4 Short About Prepress

Prepress is the common expression for operations that need to be done before the actual printing (presswork). Examples of operations are design, layout, typesetting, graphic art photography, colour separation, proofing, image assembly, filmmaking and plate making. In conventional prepress most or at least some of these operations are done manually, while in electronic prepress, which also is called *desktop publishing* these operations are done from the desktop. The advantages of electronic prepress are lower production cost and decrease in production time. For instance manual film image assembly and plate making may be eliminated.

9.4.1 Gray Scale and Resolution

Most printing systems deposit a certain amount of ink on a certain area of the paper. The printed image consists of inked areas and non-inked areas, and the amount of ink per area unit is the same for all inked area. Thus, it is either a certain amount of ink or nothing at all. A few exceptions from this exist. In rotogravure the ink amount can be varied through the depth of the cells. In inkjet certain systems can fire a few drops in a fast sequence and the drops merge to a large one just before it contacts the paper. But we will consider the general situation where we have ink in a fixed amount per area or no ink on the paper. The actual amount can be controlled by the printer through measurements of print density on full tone areas. In fact the ink amount is highly dependent on paper quality. A rough paper requires more ink than a smooth paper to reach the same print density.

Pictures taken by a conventional camera have a continuous gray scale meaning that the strength of the colour varies over the picture. In order to reproduce this with a fixed ink amount per surface area, *halftone dots* are produced. The grey scale is constructed by printing small dots, which vary in size. Small dots give light coloured areas and large dots give darker area. On very dark areas the ink forms a network with dots of non-inked area. Thus, instead of varying the ink film thickness, the area covered by ink is varied. Since the dots are very small the human eye perceive the area as light or darker without observing the individual dots. The dots can take different shapes like round, square, oval, etc.

Halftoning can be done with a special camera equipped with a halftone screen that is in contact with the unexposed film. The screen is constructed in such a way that high light intensity creates a larger halftone dot than lower light intensity. A more modern way to covert a continuous tone photograph into a halftone picture is by the use of a scanner. The picture is then converted into a digital form and can be stored and further treated in the computer.

Grey scale refers to the number of different levels of grey a picture can display. It goes from completely white, i.e. no ink at all to an area completely covered by ink, i.e. a full tone area. In offset, for instance, and by using a halftone screen, a very high number of grey levels can be obtained since the halftone dot size can be varied almost continuously. *Figure 9.9* shows an example with ten grey levels from 0 % to 90 %. The rectangles show the grey levels and the right part of the picture shows an enlargement to visualize the single halftone dots. When the grey levels are produced by camera and films the number of grey levels is limitless since the size of the halftone dot (or area that will carry ink) can be varied continuously. When using a scanner to reproduce a picture, or in digital printing, the situation is different since the area that carries ink only can be varied step wise. This will be discussed in the following section on halftone resolution. It is generally believed that the human eye can recognize 64 different grey levels but the eye is more sensitive in certain ranges and thus it is normally recommended to use at least 100 levels.

Halftone resolution is a measure of the screen frequency i.e. how many lines there are per length unit, i.e. one over raster width (see *Figure 9.10*). Most often this is given as lines per inch (lpi) but sometimes also as lines per cm or mm. In digital printing using electrophotography or inkjet the *printer resolution* is given in dots per inch (dpi). There is an important difference between halftone resolution given in lines per inch and printer resolution given in dots per inch. The latter is used for scanners in reproduction of images and in digital printing where the dot or more correctly the *exposure dot* is the smallest produced ink unit. Therefore exposure dots are used for creating *both* grey scale and halftone resolution. A certain number of exposure dots are needed to produce the grey scale. These dots fill up a *raster cell* and form a halftone dot. As mentioned above, at least 64 levels of grey are needed and this is up to 64 exposure dots, i.e. 8x8 dots, see *Figure 9.10*. Then a 240 dpi printer with 64 grey levels would give a halftone resolution of 30 lpi (240/8) and a 600 dpi printer would yield 75 lpi. Newspaper prints usually are screened at 65–85 lpi, while magazines are screened at 100–150 lpi and art prints at > 200 dpi.

244

To be more accurate, an 8×8 raster cell gives 65 grey levels since one adds the situation with no exposure dot in the cell. Thus we always have nxm+1 grey levels.

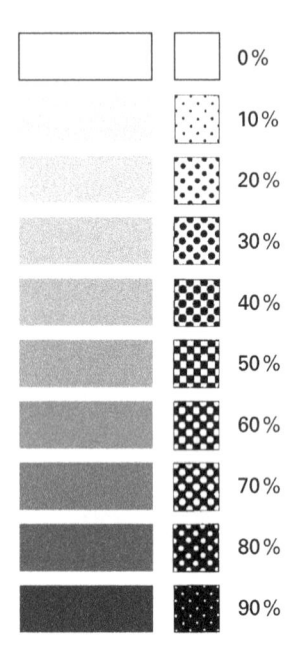

Figure 9.9. Grey scale obtained from changing the size of the half tone dots. Nine grey scale levels are shown. 64 levels plus the zero one are needed for an acceptable print.

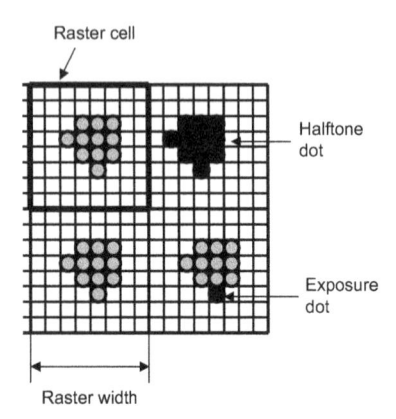

Figure 9.10. Generation of grey scale in a halftone (raster) cell. Raster cells that can house 8×8 exposure dots are needed to obtain a grey scale of 1+64 levels where the first level is without any dots.

9.4.2 Colour Printing and Colour Separation

Only three *colours, cyan, magenta and yellow* are needed to reproduce a print with all possible colours. When two of these *primary colours* are mixed *secondary colours* are obtained. For instance yellow and magenta gives red, while yellow and cyan results in green. Black is obtained if all three primary colours are mixed in equal proportions. Therefore black is not needed but used in practice to yield higher contrast. A full colour printing press uses four printing stations, one for each primary colour and one for black. It is quite common that a printing press has more than four printing units (and then more than four colours) in order to reach higher colour reproduction and special effects. The sequence in sheet-fed offset is generally to first print black and then cyan, magenta and yellow. One printing plate is needed for each colour. A colour image needs to be separated into four images, one for each primary colour and one for black. Today this is done by a scanner. Colour reproduction involves a large number of operations which can not be included in this text.

9.5 Printing Papers

Printing papers are best classified from the field of its use. Since quality and price go hand in hand a very high number of paper grades exist on the market, and it is important to choose the right paper grade for the product. Newspapers and magazines require paper of lower quality than exclusive books with illustrating pictures like books about art, gardening, cooking etc. The important paper quality features differ among the grades but smoothness, brightness, bulk and printability are important for all. With printability we refer to the paper ability to facilitate ink drying and yield a high print quality. Smoothness and bulk is most often a compromise. Smoothness can be highly improved by calendering but this reduces bulk, which in turn reduces stiffness.

9.5.1 Newsprint

The largest end product for this paper grade is of course newspapers. The paper is either made from deinked pulp or from a mechanical pulp (groundwood or thermomechanical) or a combination of deinked and mechanical pulp. Small amounts of kraft pulp may be added to improve paper strength. Common grammages are 42, 45 and 49 g/m^2, 45 g/m^2 as the most common.

Important paper properties are:

- *Strength.* Newspapers are most often printed in web offset and the grammages is low in order to keep cost down. Web brakes occur although they are not common, but very costly. Surface strength is required in order to minimize dusting.
- *Opacity.* High opacity reduces print through i.e. an unwanted appearance of print from images on the reverse side of the paper. Opacity is important for low grammages products.
- *Surface roughness.* High surface roughness reduces print quality such as print gloss, but may also reduce smearing of ink from printed images to adjacent paper or the reader.

Improved newsprint is used in inserts, magazines, etc. This product may contain bleached pulp and various special pigments to improve opacity. It is also produced in higher grammages. Common grammages lie between 49 and 56 g/m^2.

9.5.2 Magazine Paper

This paper is used in production of magazines, journals, catalogues, etc. Several different paper grades are manufactured. The two most common are *supercaledered* (SC) and *light weight coated* (LWC). SC paper is made from a furnish containing high amounts of mechanical (groundwood, thermomechanical) or deinked pulps, some reinforcement pulp (sulphate) and high amounts of filler, normally clay. The paper is heavily calendered to reach a smooth and glossy surface. LWC paper is made from chemical pulp and mechanical pulp, typically in a 1 to 1 ratio. Filler is present from broke. The paper is coated with low coat weight 8–10 g/m^2 and side. The paper is finished by calendering. The grammages for SC and LWC papers is in the region 49 to 60 g/m^2. The most common grammages are 52 and 56 g/m^2 for SC and somewhat higher for LWC.

Important paper properties are:

- *Strength*. Magazine paper is most often printed in web offset or rotogravure and web brakes must be avoided. The grammages is low in order to keep production and mailing cost down. These products are often mailed to the consumer.
- *Surface strength* is needed in offset printing due to the tackiness of the heat-set ink.
- *Opacity*. High opacity is requested to reduce print through for these thin papers.
- *Surface roughness and paper gloss*. Smoothness is an important feature since it increases paper gloss and print quality such as print gloss and missing dots in rotogravure. This paper is heavily calendered and as a consequence bulk and stiffness is reduced.

9.5.3 Book Paper

Book paper is used in production of ordinary books with text only, pocket books, novels, etc. It is made from mechanical pulp with some chemical pulp to improve strength. It is uncoated but sometimes pigmented. Normal grammages are 60–90 g/m^2 and the most important paper property is bulk.

9.5.4 Office Paper

Normal papers for use in copy machines and office/home printers are uncoated and made from a furnish containing high amounts of short fibres (e.g. birch), some long fibres (e.g. pine) and high amounts of filler. Calcium carbonate is common due to its low price and high brightness. Precipitated calcium carbonate is quite popular since it also improves bulk. The product is internal and surface sized to control wetting and absorbency of aqueous inks. It is important that these papers can be used in many printing processes like offset, electrophotography, and inkjet. Papers for documents and letters are often pre-printed in offset to mark it with logo, addresses, etc. It is then used in electrophotography or inkjet. These papers are often termed multipurpose papers. They contain no or very low amount of mechanical pulp and are often referred to as wood free uncoated fine papers. Grammages are in the range of 75 to 100 g/m².

Higher quality papers, often coated, for colour copying is entering the market but used in small amounts today.

Important paper properties are:

- *Brightness, opacity.* The optical appearance of the paper is important.
- *Bulk.* A high bulk is wanted in this product as in almost all paper products.
- *Electrical conductivity.* Charge transfer mechanisms are involved when toner is transferred to the paper in electrophotography.
- *Heat transfer.* The fusing process in electrophotography involves heating the paper from the backside. The heat is transferred through the paper to melt the toner.
- *Ink absorptivity and wettability.* These properties are important for ink drying and print quality in inkjet printing.

9.5.5 Graphic Papers

Graphic paper is used for products like annual reports, commercials, excusive magazines, illustrated books like art books, etc. The furnish is quite similar to the one used for office paper, i.e. chemical pulps of birch or other short fibre, some long fibres and calcium carbonate as filler. These papers are coated in one single pass on line, or in several passes in an off-line coater. Matt grades are often single-coated while multi-coated papers are used for silk and gloss grades after appropriate calendering. The pigment coating composition is normally a mixture of calcium carbonate and clay with calcium carbonate being the dominant pigment due to cost and brightness advantages. Plastic pigment are often used in the colour of glossy grades to improve gloss which often is > 70 %. Common grammages are 100–170 g/m2, but higher exists.

Important paper properties are:

- *Brightness, bulk.* This is a very exclusive paper and must have a good appearance and stiffness.
- *Smoothness.* This property is very important for print quality like print gloss. The paper needs to be uniform in physical properties in order to yield evenness in print quality parameters like print gloss and print density.

9.5.6 High Quality Inkjet Paper

Uncoated office paper is used in inkjet printers when printing only text or the requirement on print quality is moderate. These papers are multipurpose papers and also used in offset printing presses and laser printer. When high quality inkjet printing is wanted, papers with special coatings are required (*Figure 9.11*). These coatings may contain several specific layers, for instance a dye fixation layer and an ink-sorbent layer. The dye fixation layer contains chemicals that interact with the dye molecules and anchor the dye at the surface. This improves print density, colour saturation, and water fastness. Chemicals for dye fixation are cationic polymers such as polydadmac and polyvinyl pyrolidon. The ink-sorbent layer may consist of porous particles with high specific area such as silica gels, or of a swellable polymer e.g. gelatine. The ink absorption into the micro porous layer is fast and governed by capillary forces (capillary absorption) while the absorption into the swellable layer is slow and controlled by molecular diffusion. Glossy photo quality inkjet papers most often have a swellable layer and print quality is extremely high although the drying time is long.

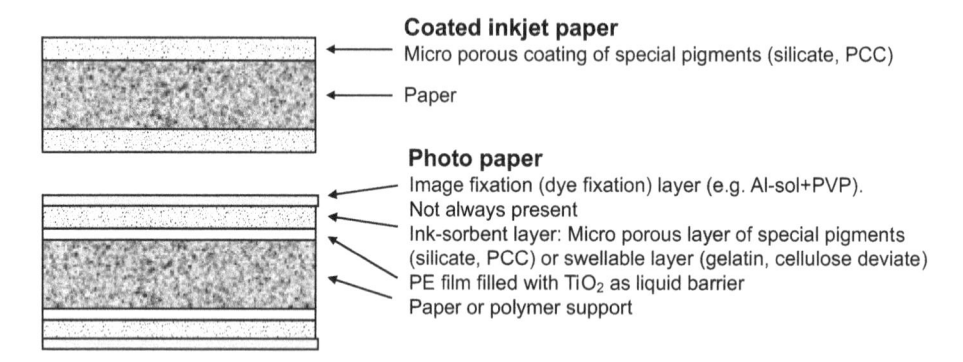

Coated inkjet paper
← Micro porous coating of special pigments (silicate, PCC)
← Paper

Photo paper
← Image fixation (dye fixation) layer (e.g. Al-sol+PVP). Not always present
← Ink-sorbent layer: Micro porous layer of special pigments (silicate, PCC) or swellable layer (gelatin, cellulose deviate)
← PE film filled with TiO_2 as liquid barrier
Paper or polymer support

Figure 9.11. Structure of high-quality inkjet papers.

9.6 Printing Inks and Ink Drying

All printing inks consist of three main components, viz. colorant, binder and solvent/oil. The colorant is most often a pigment but dyes occur, for instance in inkjet inks. The solvents range from pure hydrocarbons through ketones, esters and alcohols to water. Water and alcohols are frequent in flexo and inkjet inks. Type of binder depends very much on solvent. For instance acrylate latex binder is frequent in water born inks, while oil born inks often uses alkyd and hard resin binders. UV curable inks occur in offset, flexo and inkjet.

9.6.1 Offset Inks

Sheet-fed offset inks are manufactured by dispersing the pigment into a *vehicle*, which is a mixture of binder and oil. Normal binders are *alkyd* and *hard resin*. The latter is solid and amorphous, while the alkyd is a liquid. Additional oils and various additives are added to the ink to reach the right properties. A typical ink formulation is given in the *Table 9.1*.

Table 9.1. A typical sheet-fed offset ink formulation.

Ink constituent	Physical state	Content (%)
Pigment and filler	dispersed particles	15–20
Hard resin (binder)	dissolved polymer	20–30
Alkyd resin (drying oil, binder)	liquid polymer	8–12
Triglyceride (drying and semi-drying oil)	high viscous oil	10–20
Mineral or vegetable oil esters (non-drying oils)	low viscous oil	15–25
Additives e.g. wax, driers, antioxidants		3–5

Pigments are most often organic compounds but inorganic pigments occur. The particle size is typically 0.1–0.5 µm. Carbon black is used as pigments for black inks. The primary particles of carbon black are very small, in the range of ten to twenty manometers, but they form clusters that are larger.

Hard resins used in lithographic inks are normally an ester of a polyhydric alcohol such as glycerol or pentaerythritol and a rosin acid or a derivative of rosin acid. Such derivatives may be dimerised or polymerized rosin acids, maleic or fumaric fortified rosin or a *rosin modified phenolic resin*. Since the rosin derivative is a polybasic acid the hard resin normally has a high molar mass. It is solid and amorphous with a softening point well above 100 °C often above 150 °C.

Alkyd resins are also esters of a polyhydric alcohol but the acids are a mixture of monobasic and dibasic carboxylic acids. The mono basic acids are unsaturated C_{18} acids while the dibasic acid is phatalic acid or isomers of phatalic acid. The alkyd resin is a liquid at room temperature.

Drying oils are triglycerides with high amounts of highly unsaturated fatty acids like linolic and linoleic acid. They solidify (dries) through an oxygen induced polymerisation. Linseed oil is a drying oil and quite often alkyd resin is referred to as a drying oil.

Sheet-fed offset inks are used to print on coated paper. It dries through absorption of ink oil into the coating and polymerisation of the alkyd binder and drying oil, i.e. the part of the drying oil that still remain in the ink film. Most of the oil deplete from the ink film and are absorbed by the coating. Catalytic driers based on soaps of cobalt and manage are used to increase the drying rate, while antioxidants are used to slow down drying rate on the press.

Heat-set offset inks are used in web offset printing, quite often on coated paper. This printing press is equipped with a hot air drying unit to evaporate the oils from the print. The oils are depleted from the ink film not only through evaporation but also through absorption into the coating. Oils with low boiling point are used to facilitate oil evaporation. Resins are specially selected for their ability to release ink oil rapidly and completely upon heating. Drying oils are added in small amounts, not as binders but to act as pigment wetting agents and as plasticizers.

Cold-set offset inks are used in web offset printing on newsprint. These inks contain only small amount of drying oil, mainly to facilitate pigment wetting and dispersion. Consequently, there is no oxidative (polymerisation) drying. This ink dries only through absorption of ink oil into the paper and film formation of binder, which mainly is hard resin. This ink usually contains somewhat more pigment than sheet-fed and heat-set offset inks.

9.6.2 Rotogravure Inks

Offset inks have a very high viscosity and are classified as *past inks*. Rotogravure inks, on the other hand have a low viscosity and thus belongs to the group *liquid inks*. Roto inks dry through evaporation of the solvent, which most often is toluene. Metal resenates are common binders in these inks. Water born roto inks are used although they are much less common. Nitrocellulose is a common binder in those inks.

9.6.3 Flexographic Inks

Flexo inks are also liquid inks. Solvent and water based as well as UV curable inks exist. Water based inks are common for printing on paper and board while UV inks are used for printing on plastic films. Nitrocellulose is the most common resin in solvent born inks. It has good pigment wettability and wide compatibility with other resins. Acrylic resin is also commonly used both in solvent and water born inks. When used in water it can either be formulated as a latex or be used at a high pH. Acrylic resin has a good compatibility with nitrocellulose. Other resins are rosin esters and polyamide resins both are used in alcohol born inks. Common solvents are ethyl alcohol, propyl alcohol, propyl acetate, ethyl methyl ketone and water in combination with ammonia to provide high pH for dissolution of acrylic resin. *Tables 9.2 and 3* give some examples of compositions of flexographic inks.

Table 9.2. Example of solvent borne flexographic inks.

	For PE films	For folding cartons
Colorants	12 % organic pigment	14 % organic pigment and 6 % titanium dioxide as filler
Binder	22 % polyamide	11.5 % nitrocellulose
	4 % nitrocellulose	8 % maleic resin
Solvents	34 % n-propyl alcohol	25 % ethyl alcohol
	13 % ethyl alcohol	25 % n-propyl alcohol
	12 % n-propyl acetate	10 % n-propyl acetate
Additives	2 % PE wax	5 % plasticizer
	1 % amide wax	3.5 % PE wax

Table 9.3. Examples of water born flexographic inks for printing on non-absorbent and absorbent substrates.

	Constituents	For non absobent substrates	For absorbent substrates
Colorant	35 % pigment dispersion	50	40
Binders	Acrylic soluble polymer	10	30
	Acrylic emulsion (latex)	30	12.5
Solvents	Water	5	13
	Organic amine	1	1
Additives	Polyethene wax	3	3
	Organic antifoam agent	0.5	0.5
	Surfactant	0.5	0

9.6.4 Ultraviolet Curable Inks

These inks are special designed to go from a liquid state to a solid state when exposed to ultraviolet (UV) radiation. UV curable inks are available for most printing techniques e.g. offset, flexography, inkjet. Both non-aqueous and aqueous systems exist. The ink vehicle contains prepolymers, oligomers and monomers which polymerise in UV light by the aid of a photoinitiator e.g. benzophenone. The prepolymers are used to provide the ink with its resin component and to serve for proper rheological properties. The viscosity of an offset ink is much higher than that of an ink for flexography or inkjet. The time required for the ink film to polymerise depends among other on ink film thickness and number of UV lamps. The polymerisation process is fast and the film can be solidified within less than 10 ms.

9.6.5 Inkjet Inks

The first requirement on inkjet inks is low viscosity, preferably < 10 cp, but at least < 20 cp. Many inkjet inks have a viscosity of only a few cp. The inks may me classified as:

- *Oil based inks*, where the oil is aliphatic hydrocarbons. The ink dries by absorption of oil into the paper. The oil does not evaporate.
- *Solvent based inks*. Examples of solvents are methanol and methyl ethyl ketone. The ink dries by absorption into the paper and evaporation.
- *Aqueous inks*. The solvent is water but co-solvents (e.g. alcohol, glycerol, diethylene glycol monobutyl ether) and surfactants are required. The ink dries by absorption into the paper and evaporation.
- *Hot melt inks*. Mixtures of waxes and amides are typical. The material has a sharp glass transition temperature in the range 70 °C–120 °C and is jetted when it is in its liquid state. It cools down on the paper and solidifies.

Printers used in homes and small offices are equipped with aqueous or hot melt inks. Oil and solvent based inks are for industrial use. These inks must be handled correctly and care must be taken to provide a safe indoor environment and to minimize air pollution. The colorant is either a dye or a pigment. Dye colorants have high brightness and saturation when just printed but the colour fades rapidly. Pigments are believed to dominate in the near future due to its higher light stability and water fastness. The inks contain low amounts of binders, if any. Thus, fixation agents are needed in the paper. Inkjet ink formulations are given in *Tables 9.4* and *9.5*

Table 9.4. A formulation of a solvent borne ink for CIJ labelling.

Component	Content, wt.-%	Function
Methanol	42	solvent
Methyl ethyl ketone	30	co-solvent
Water	1.5	co-solvent
Ethyleneglycol methyl ether	9	evaporation retardant
Methyl ester of rosin	1.4	binder
Styrene-acrylic acid co-polymer	13	binder
Dye	2	colorant
Nonyl-phenoxpolyethoxy ethanol	4	surfactant

Table 9.5. An example of a water borne inkjet ink. Ink properties are: surface tension: 36.5 mJ/m^2, viscosity: 2.7 cp and pH 7.6.

Component	Content, wt.-%
Water (solvent)	73.5
Glycerine (co-solvent)	18
Dye	2.8
Biocide	0.2
Surfactant	5.5

9.7 Print Quality

Two of the most important general print quality parameters are *print density* and *print gloss*. These are important in all printing processes. There are certain quality defects that are important for a specific printing process, for instance missing dot in rotogravure and bleeding in inkjet. Missing dot means that the paper did not pick up the ink from the cell of the roto cylinder. The most frequent reason for this is the roughness and compressibility of the paper. If the paper has a shallow pit at the location where the dot was to be placed, there will be no contact between ink and paper, and the ink dot will not be transferred to the paper. Bleeding occurs in inkjet when drops of different colours bleed into each other upon contact during absorption.

9.7.1 Print Density

Print density (D) is the optical density of the print. It is defined as the logarithm of the ratio of the reflectivity of the paper (R_∞) and the reflectance of the print (R_p) when this is placed on top of a pad of unprinted paper sheets. The reflectance is the ratio between reflected light and incidence light.

$$D = \log(R_\infty / R_p)$$

Print densities as a function of ink amount of a few graphical papers (coated fine papers) are given in *Figure 9.12*. It is clear that print density is strongly determined by ink amount.

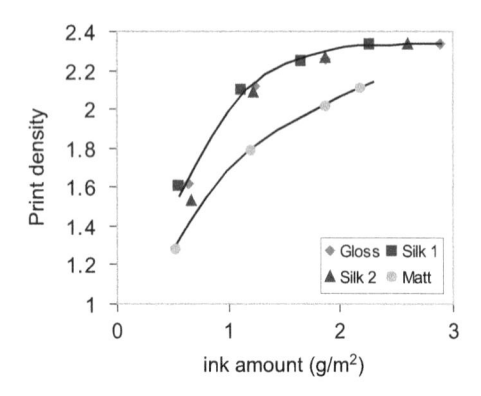

Figure 9.12. Print density is highly dependent on ink amount.

9.7.2 Print Gloss

Gloss is the ability of the surface to reflect light. Most often gloss is measured by illuminating the surface at a certain angle to the normal to the surface, and measure the reflected light at the same angle. Angles normally used are 20, 45 and 75 degrees. A glossy surface is measured at small angles while a matt surface like newsprint is measured at a high angle. Print gloss is determined by surface roughness, ink levelling and refractive index. Ink levelling takes place directly after print transfer and it relates to the flowing of ink lumps to an smooth surface. The effects of ink amount on print gloss for a few graphical papers are shown in *Figure 9.13*. The matt paper is the roughest one and the gloss paper is the smoothest one. It is seen that the smooth paper needs only a small amount to reach a high gloss while the matt paper needs higher amounts. While print density is more affected by ink amount than by surface smoothness, the opposite holds for print gloss.

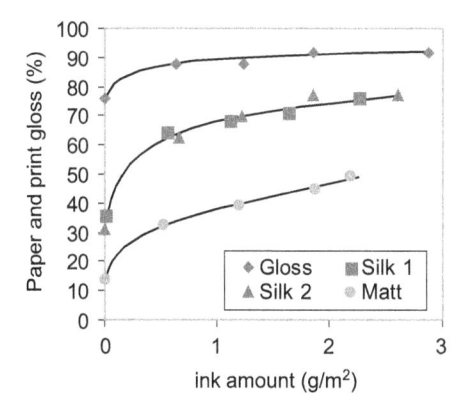

Figure 9.13. Print glosses very much depend on type of paper and the surface roughness is extremely important.

9.7.3 Unevenness in Print Density and Gloss

The mean values of print density and print gloss are important print quality parameters but the homogeneity in these features is as important or even more important. Nonuniformity in print density is referred to as *print mottle* or only mottle. There are several origins of this print defect. One is local variation in ink amount caused by local variation in ink setting rate, which in turn depends on local variation in the paper pore structure.

Nonuniformity in print gloss i.e. *print gloss variation* is a severe print defect. It depends to a large extent on surface roughness and is more pronounced on silk and matte coated papers than on glossy ones. *Evenness in print* is a term often used to describe print quality. This is determined subjectively by viewing the paper at different angles. There is no general definition of this quality property but print gloss variation plays an important roll.

9.8 Ink Drying in Offset Printing

The drying of offset prints on paper involves several phenomena which can be summarized into two main categories, namely:
- Removal (depletion) of ink oil from the ink film and

- Chemical curing[2] (mainly polymerisation) of residual drying oils e.g. alkyd resin and linseed oil.

The depletion of ink oil can take place through evaporation as in heat-set offset or through absorption into the substrate, which is the case for sheet-fed offset prints on coated substrates. The depletion of ink oils is referred to as the *physical drying* whereas the chemical curing of alkyd resin is referred to as the *chemical drying* or *oxidative drying*. The latter expression comes from the fact that the curing process is initiated by oxygen.

9.8.1 Physical Drying

The initial stage of physical drying is referred to as ink setting. After the ink setting stage the ink film does not smear when it is subjected to a light touch, which can be the hand of the printer or a paper surface in a set-off measuring device. It is quite obvious that ink setting time depends on the technique used to measure it. For coated papers and boards ink setting time varies from less than a minute to 10–20 minutes. Ink setting should be optimized. Too fast a setting reduces print gloss due to insufficient levelling of ink bomps from ink filaments and ink are building up on the rubber blanket while too slow setting may cause smearing in the stack. For coated papers ink setting is strongly determined by the coating porosity and pore size. Higher porosity and smaller pores gives faster setting. The effect of pore size on ink setting rate is shown in *Figure 9.14*. P&I slope is a measure of setting rate and a high value means fast setting.

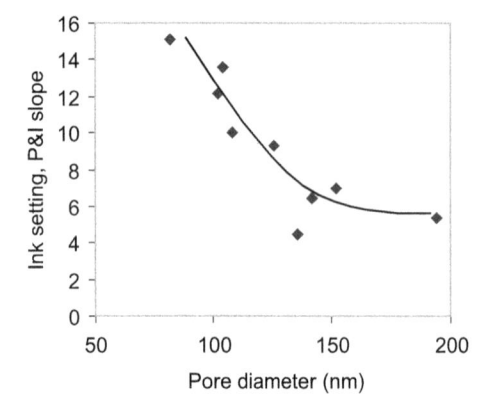

Figure 9.14. Ink setting rate decreases with increasing pore size of the coating.

The ink film is only partly dry when ink setting is completed. Once the ink is „set", the prints can be handled, although with care. At setting time roughly half of the oil is absorbed by the coating. The absorption of printing ink oils by the paper coating continues but at a much slower rate, and the mechanical stability of the ink film increases. Ink oil is not only absorbed by the pores of the coating but also by the latex used as binder in the coating. The oil molecules diffuse into the latex matrix which quite often is a co-polymer of butadiene and styrene.

---

[2] Another type of chemical curing takes place in UV curable inks. This will not be discussed here.

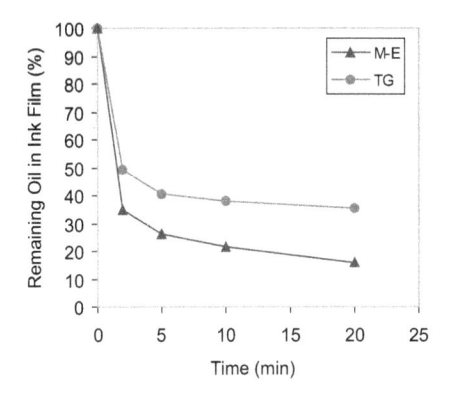

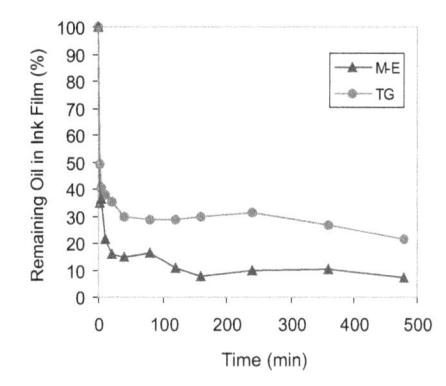

Figure 9.15. The initial absorption of ink oils into the coating (ink setting) is fast while the subsequent absorption is slow. The data shows the remaining oil in the ink film as a function of time after printing for a vegetable-oil ink. TG stands for triglycerides (mainly linseed oil) while M-E is the abbreviation for mono-ester oil.

Figure 9.15 shows the absorption of ink oils into the coating of a glossy coated paper.

9.8.2 Chemical Drying

The chemical drying of an offset inks (UV inks excluded) is an oxygen induced polymerization of alkyd resins and drying oils (e.g. linseed oil). The process is also called oxidative drying. The reactive part of the material is the unsaturation and the carbon next to the double bond of the fatty acid pendant groups. Oxygen molecules adhere to this site and a hydroperoxide is formed.

The second stage is the decomposition of the hydroperoxide into free radicals. This is a fairly slow process and catalysts based on soaps of cobalt and manganese is added to speed up the decomposition. The catalysts are also referred to as *driers* or *sicatives*. As a comparison we may stress that it may take 14 days for a film of linseed oil to dry in absence of driers but only a few hours in presence of a moderate amount of drier.

The third stage is a radical polymerisation of the binder. Also formed are oxygen containing functionalities like alcohol and aldehyde groups, as well as scissoring volatile compounds and loss of unsaturation.

Antioxidants are used in order to reduce polymerisation on the printing press during short stops or over-night stops. This material stabilizes the free radicals until it is consumed. Examples of antioxidants are hydroquinone, methyl ethyl ketoxime, cyklohexanone oxime and butylated hydroxy toluene. With the proper balance of catalytic driers and antioxidants, polymerisation will be delayed a certain time but once it starts it is completed within a relative short period of time. The *open time* is the time it takes for a relative thick ink film laid on a glass plate, to polymerise in open air. Inks with an open time of around 20 hours are referred to as *overnight inks* or press *stable inks*.

Figure 9.16 shows drying curves after pile drying of a vegetable-oil ink printed on a coated paper in sheet-fed offset. The three curves are for different temperatures and it is evident that drying temperature has a strong impact on ink drying. Ink drying was characterized by rubbing the print against an unprinted standard paper (uncoated) in one linear movement at a fairly high pressure. The amount of ink transferred from the print to the recipient paper is given by the rub-off value. The lower the value the dryer the ink film. The drying is very slow at low temperatures. Follow for instance the curve for 6 °C. The initial decrease is due to physical drying, i.e. depletion of ink oil from the ink film. This ends in a well-pronounced plateau. Then follows a sharp decrease to the final level. The sharp decrease corresponds to the final chemical drying i.e. polymerisation. The time for this to be completed is almost 20 days at 6 °C, about 4 days at 23 °C and only 16 h at 40 °C. The drying time is also very much dependent on excess to air. Drying with free excess to air is about 5–10 times faster than pile drying.

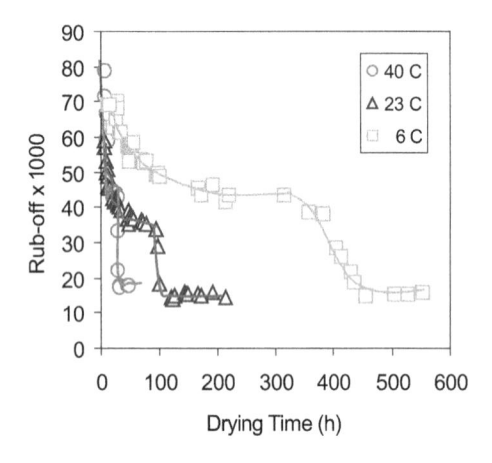

Figure 9.16. Ink drying after sheet-fed offset printing on coated paper using a vegetable-oil ink. The drying time is strongly affected by temperature. The print was dried in the stack. Drying with free excess to air is 5–10 times faster.

10 Packaging

Christer Söremark and Johan Tryding
Christer Söremark AB, Tetra Pak

10.1 Introduction 257

10.2 Packaging Materials and Converting 258

10.2.1 Packaging Materials 258
10.2.2 Corrugated Board 259
10.2.3 Converting 261
10.2.4 Barrier Properties 269

10.3 Requirements on Packaging Performance 270

10.3.1 Definitions 270
10.3.2 The Supply Chain 270
10.3.3 Packaging Requirements 273

10.4 Design of Packaging Performance 276

10.4.1 Technical Performance 277
10.4.2 Marketing Performance 284

10.5 Future Trends 284

10.5.1 Digital Printing and E-print 284
10.5.2 E-tags 284
10.5.3 Smart Packages 285
10.5.4 New Packaging Materials 286

10.6 Literature 286

10.1 Introduction

Packaging is a prerequisite for today's society with its demand for distribution of food and other products globally, nationally and regionally. From a business point of view the packaging system is part of the business successes. The packaging system should contribute to:
- An efficient logistic flow
- Marketing and selling the product
- A reduction of the environmental load in the goods flow

Product development in for example the foodstuff area will lead to a wider range of products and thus the development of new packaging systems. Changing consumer patterns, home shopping, environmental issues etc. will result in new distribution channels.

Worldwide the packaging industry is a major part of industry. In the industrialised countries it belongs to the top ten. The annual revenue is 550 billion € and the split between consumer and industrial packaging is 70 to 30. About 5 million persons are employed in 100 000 companies. By value paper and board is the largest material sector with 34 % followed by plastics with 30 %, metals with 25 % and glass with 6 %. Two-thirds of the world expenditure is in Europe, North America and Japan combined.

In Sweden the annual packaging production is 2 billion € and the industry employs directly more than 10 000 people.

To engineer packaging requires an understanding of all the requirements that exists and knowledge on how to match these with the packaging performance.

Chapter 10 will focus on packaging materials, packaging requirements and packaging performance for the paper and board sector.

First there will be a short introduction to packaging materials and converting. Then the packaging chain will be discussed followed by its generated requirements, direct or indirect, on the packaging. The process to create a packaging performance that meets the requirements will be analysed next. Finally future trends in packaging will be discussed.

10.2 Packaging Materials and Converting

In this first section we will briefly introduce paper based packaging materials and the structure and manufacturing of corrugated board. Finally the basics of converting packaging materials into packaging will be presented.

10.2.1 Packaging Materials

Packaging can be divided into two groups, Flexible Packaging and Rigid Packaging. As seen in *Table 10.1* the material for flexible packaging is paper. For rigid packaging the material consist of paper board or corrugated board, two materials that themselves have comparatively high bending stiffness. *Table 10.1* also gives the main applications for the two groups.

Table 10.1. Packaging and packaging materials.

	Flexible Packaging	Rigid Packaging	
Materials	Paper	Paper Board	Corrugated Board(liner & fluting)
Applications	Paper Wrapping Paper Sacks Paper Bags & Carrier Bags	Folding Cartons	Corrugated Boxes

For flexible packaging the largest application (by volume) is material for wrapping followed by sacks and bags (including carrier bags). The papers are mostly made from unbleached kraft pulp although the market share for bleached kraft pulp is increasing. To some extent also bleached sulphite pulp is used, primarily for smaller bags and envelopes. The bleached products can be coated with clay to improve the printing appearance. In addition special papers, e.g.

greaseproof and glassine are used for applications related to their specific properties. Greaseproof (grease-resistant) is for instance used for fatty food and industrial products protected by grease. Glassine (oil and grease-resistant) is used in much the same applications as greaseproof, but where the demands for protection are higher. Also tissue paper is used in packaging e.g. as soft wrapping for silverware, flowers etc.

The materials for rigid packaging consist of paper board or corrugated board. The manufacturing of the former and its different grades has been discussed elsewhere. The structure and manufacturing of corrugated board will be discussed below. A typical market for paper board is the food sector, e.g. cereals and snack foods, carriers for multi-packs (6-pack of beers), milk containers etc. The corrugated board is mostly used for transport packaging due to higher requirements on strength. However the development of corrugated board with lower basis weight and thickness has opened a „grey" market where both paper board and corrugated board compete.

10.2.2 Corrugated Board

Corrugated board is a structure that consists of a core (medium) which has been fluted to a wave-like form. To this core is glued one or two flat sheets of paper (facings) to form a single face or single wall corrugated board, respectively. Additional cores and facings can be added to give different structures, *see Figure 10.1*.

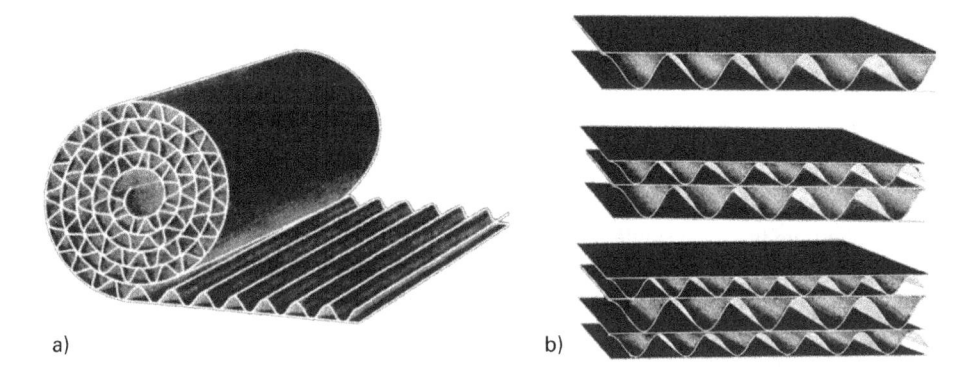

Figure 10.1. Different corrugated board structures. (a) Single face corrugated board. (b) Single wall (top), double wall (middle) and triple wall corrugated board (bottom).

The flat facing is called liner and the fluted medium is called fluting or corrugating medium. Both types of papers can be produced from virgin or recycled fibres. The structure of corrugated board is an adaptation of the engineering beam principle of flat load-bearing panels separated by a rigid core. In this way much more bending stiffness can be gained at the same basis weight compared to solid board.

The corrugated board is manufactured on a corrugator, *Figure 10.2*.

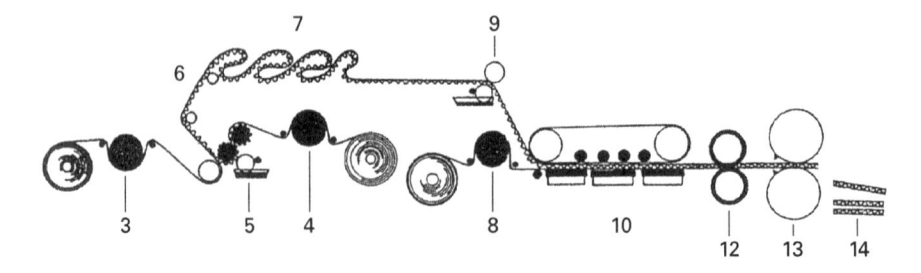

Figure 10.2. Manufacturing of corrugated board.

The process can be divided into six operations:

1. Unwinding and conditioning of the liners and medium
2. Corrugation of the medium
3. Gluing medium and liners together
4. Drying of the board
5. Cutting/slitting of the corrugated board. The board may also be creased in this process.
6. Stacking of sheets

The fluting is preheated (conditioned) and then passed through the hot (160 °C) corrugating rolls where the flute profile is generated. Immediately after the glue is applied to the flute tips and pressed together with the preheated liner. The introduction of heat is needed for the corrugation of the fluting, but also to achieve a fast cure of the glue. It is important that the glue develop a sufficiently strong tack in a short time so that the single face board can be transported to the bridge without separation of the two papers. This initial glue tack is called the green bond and develops within milliseconds. Besides temperature the adsorption of the glue to the papers governs this tack development. Normally the glue consists of 20 % modified starch dispersed in 80 % water.

The single face web is then glued to the second liner (double backer liner) after glue has been applied to the flute tips. The single wall board is then dried under pressure on hot plates in the dry end of the corrugator. Since the stiff corrugated board cannot be reeled it is slitted in lengthwise- and cut in crosswise direction to desired sheet dimensions before they are stacked at the end of the corrugator. In connection with the slitting operation the board can be creased lengthwise to facilitate later folding of the board.

The corrugated board is made with different structures that are defined by the top-to-bottom height of the fluting and the number of flute waves per unit length, see *Table 10.2*

Table 10.2. Different flute types, wave height and number of waves per meter.

Flute Type	Wave height, mm	No. of waves/m
A	4,8	110
C	3,6	130
B	2,4	150
E	1,2	290
F	0,7	350
G & N	0,5	550

Beside the structure of the board, papers with a wide range of properties can be used. Assume that we have 20 different liners and 10 different flutings to make boards from. From these papers it is possible to make 16 000 different boards of four different single wall flute types, or 1 600 000 different boards of two different types of double wall. These examples illustrate quite clearly the possibilities to create boards that satisfy a wide variety of requirements.

10.2.3 Converting

The runnability of packaging materials in packing lines is of greatest importance for the manufacturer. Knowledge and experience of the processes and issues affecting efficiency at each stage of paperboard conversion is the key factor for runnability. Materials that not run well in the converting machines will never come into the market.

The chain of conversion processes is shown in *Figure 10.3*.

print and varnish → cut and crease → folding and gluing → forming, filling and closing

Figure 10.3. The chain of conversion processes.

In all operations the packaging materials should be consistent in moisture content, stiffness, basis weight and thickness for best performance. They should also be free of debris and sheets must be supplied flat.

10.2.3.1 Print and Varnish

Printing is an essential step in the conversion process. The requirement on the print quality is high since consumers buying decision is influenced by the print quality.

The printing press consists schematically of the following units:

- Sheet or reel feed unit
- Printing units
- Sheet or reel collection unit

Sheet-to-sheet printing press is the most common used printing process for board and corrugated board. The reel-to-reel printing press is mostly used for liquid packaging board and preprinted liner. Reel-to-sheet printing press for board is usually used in combination with in-line cutting and creasing equipment.

The main purpose of the printing units in the printing press is to transfer the ink to the board. The ink has to be applied selectively to certain areas on the board in order to make an image. The transfer of the ink to the board surface is done under pressure which is necessary for the ink transfer to occur. The board is pressed between a cylinder with the ink and an impression cylinder. The printing units in a printing press consist typical of 4-6 units with vanishing.

The principles to transfer the ink to the board in the printing unit are:

- Flexography
- Offset lithography
- Rotogravure

In flexographic (flexo) printing ink transfer to the board is done by first transferring the ink via an ink roll to a printing plate mounted on a cylinder. The plate distributes the ink onto the board at the nip as schematically shown in *Figure 10.4(a)*. The plate on the cylinder is flexible and made of polymer or rubber. The plates are directly mounted to the cylinder by self-adhesives or indirect by the use of sleeves.

In offset lithographic printing the ink is transferred to the board via a separate transfer cylinder covered with a rubber blanket as shown in *Figure 10.4(b)*.

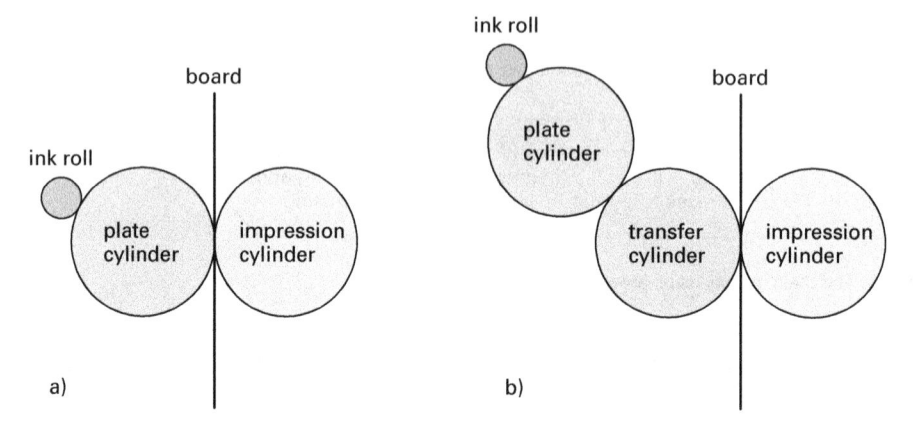

Figure 10.4. (a) Flexographic and (b) Offset lithographic principles.

In rotogravure printing the printing cylinder is made of chromed copper and the printing image is a negative relief engraved or etched on the copper surface. Rotogravure printing is expensive but can produce large number of prints at high speed.

The most common types of printing for paper board are sheet-fed offset (lithography) and rotogravure. Liquid packaging board is mostly printed in flexography. For general packaging the use of flexography has been limited, however it is gaining increasing acceptance due to technical improvements that give better print quality. Lithographic printing gives in general a better quality than flexographic printing. Varnishing of the printed board is done in order to protect the print against rubbing that may occur in the packaging chain.

Runnability requirements in the printing and varnishing operations are flatness and dimensional stability of the board, delamination strength in the thickness direction of the board and dust and debris free board. Dust and debris cause spots in the image areas of the board. Problems with ink are relatively rare. If ink problems arise they are linked to taint and odours.

When printing corrugated board (post-print) the flexographic technique is by far the most common. Also in pre-print (when the liner is printed separately before converted to corrugated board) flexographic printing is the prevailing technique.

10.2.3.2 Cutting and Creasing

The purpose of cutting and creasing is to convert a sheet of packaging material into a so-called blank, see *Figure 10.5*. After these operations, which are done at the same time, the blank has the right dimensions and can be folded into a box. The creasing is necessary to get a god fold and thus consistent dimensions of the boxes.

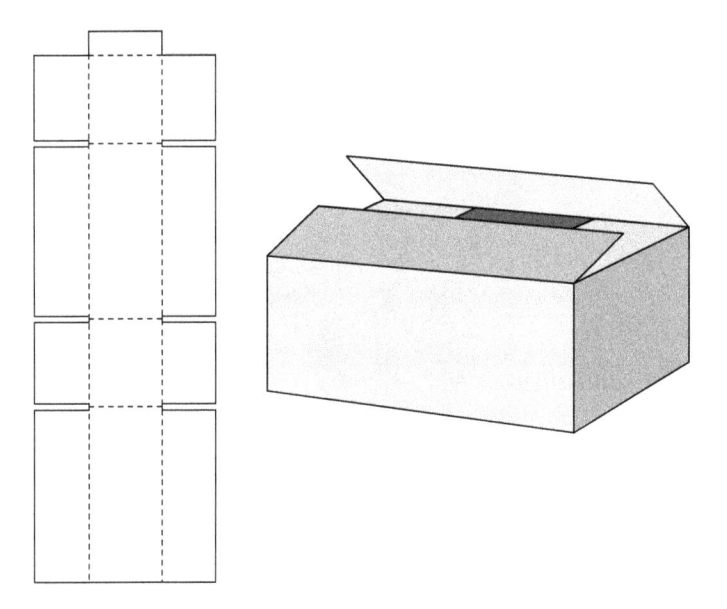

Figure 10.5. (a) Sheet of packaging material, i.e. blank and (b) blank folded into a box.

This process is the most critical when considering improvements in the line efficiency in the folding, gluing and packing operations. The creasing rule must be of the correct specification for the board type and quality. Changing the board specification without considering the crease rule width, depth and profile will result in less satisfactory creases. The converter must be aware of the correct cutter and creaser profiles and settings to use. This will affect the folding and gluing operations and the final erection and sealing of the material. This is one of the major causes of dimensional stability problems in board packaging.

Boards coated with polyethylene (PE), polypropylene (PP) and polyethyleneterephthalate (PET) need special creasing and cutting forms, to ensure the cut is clean and the film is not damaged.

There are two principal operations to cut and crease board and packaging materials,

- Flat bed and
- Rotary

The flat bed cutting and creasing principle is schematically shown in *Figure 10.6*. The cut and crease tools are parallel to each other and consist of a male die with creasing rule and cutting knife and a female die with scored channels. The sheet is placed between the dies. The male die is punched into the sheet and female die. During the punch operation the sheet is creased and cut by the ceasing rule and cutting knife, respectively. A flat bed cut and crease machine can be place off-line after a sheet-fed printing press or in line with a reel-fed printing press.

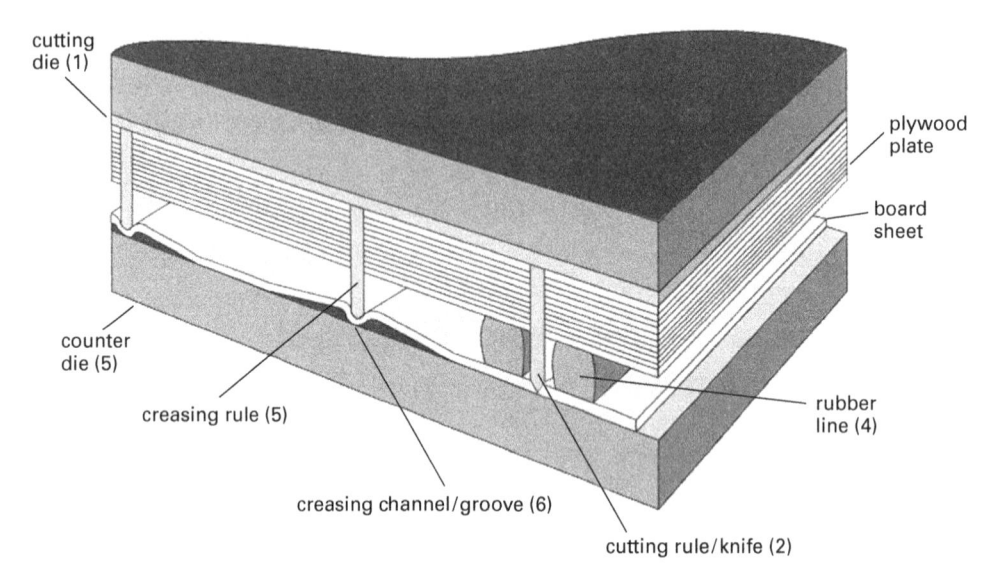

Figure 10.6. Principal sketch of flat bed cutting and creasing.

In the rotary cutting and creasing operation the male die and female die tools are mounted on cylinders. The creasing and cutting of the board or packaging material occurs in the nip between the cylinders. The benefits with rotary over flat bed cutting and creasing are the higher speed and longer runs. The drawback is the cost for the rotary die tools. A rotary cut and crease machine is in-line with a reel-fed printing press.

To obtain a cutting that meets the requirements of distinct cut edges the cutting knifes on the male die must be sharp. Over time the knife will wear out and blunt. This will lead to cuts that produce dust and hairy cut edges.

During the punch operation the board is pre-stressed by tension acting in the plane of the sheet. The male die is punched into the board and the board deforms under compression immediately beneath the male die, see *Figure 10.7 (a)*. Shear deformation occurs in the gap between the male die and female die on either side. The shear deformation causes damage that is primarily concentrated at the interfaces between the different layers in multi-ply board. The damage due to the shear deformation causes delamination, which governs the quality of the crease, between the layers. Upon unloading of the male die, most of the deformation is recovered. The remaining deformation of the board is shown in *Figure 10.7 (b)*.

Runnability requirements of the board in the cut and crease operations are dependent on the moisture content of the board. For flatbed cutting and creasing the flatness of the board is an important requirement. Board properties that influence the results of the cutting, creasing and folding are the board thickness, and strength and elongation properties in the in-plane and out-of-plane directions including fibre type and fibre orientation. Dust and debris also affects the runnability.

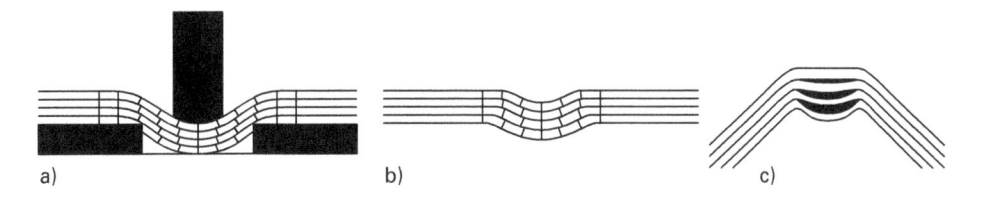

Figure 10.7. The behaviour of board during (a) creasing, (b) after creasing and (c) at folding.

For large carton blank formats it is especially important that the board is of the same consistent stiffness and that creases are formed correctly.

10.2.3.3 Folding and Gluing

Folding is done along the scored line as is shown in *Figure 10.7(c)*. The folding bends the board around the crease. The bending introduces compressive and tensile stresses in the creased zone. Due to the shear damage along the interfaces obtained during the indentation, the compression will cause the layers to buckle outwards separating the interfaces more and more at increasing folding angle.

Badly formed creases along the score line can after folding cause a skewness that varies from one side to another and gives a bad appearance of the folded scored line. Also surface cracks in the sheet plane give bad appearance and functionality. A symmetric crease will on the other hand give a scored line after folding with a neatly appearance and precisely located, and hence more attractive. *Figure 10.8* shows photographs of the appearance of the scored line after folding for one uncreased sample and four creased samples. The creased samples in *Figure 10.8* vary from badly to excellently creased samples.

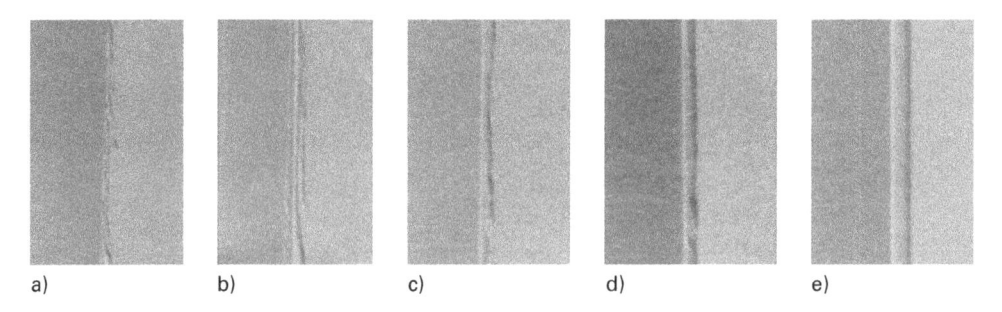

Figure 10.8. Folding for one uncreased (a) and four creased samples (b)-(e). The creased samples vary from (b) badly, (c) not well, (d) well and (e) excellently creased samples.

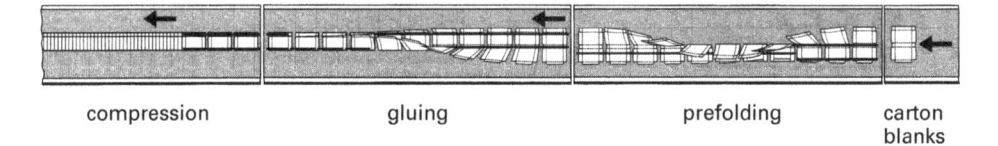

Figure 10.9. Side gluing.

Side gluing of carton blanks is the most frequently used pre-gluing of blanks before they are shipped to the goods manufacturer. At side gluing one side of the blank is glued and the other side is fold over and overlapping the glued side, as is shown in the gluing section in *Figure 10.9*.

The gluing machines used for side gluing consists of the operations:

- pre-folding
- application of adhesive
- folding
- sealing
- curing

In the gluing machine the blanks are fed into the pre-folding unit where the blanks are folded 90–180 degrees and thereafter folded back to the original position as shown in the pre-folding section in *Figure 10.9*. The adhesive is applied at the length side of the blank and thereafter the other side is folded to overlap the glued side and sealed under pressure, see the gluing section in *Figure 10.9*. The pre-glued blanks are then stacked for curing and finally packed for delivery as illustrated in the compression section in *Figure 10.9*.

The procedure of joining the two board sides together by gluing can be summarised in the following steps:

- application of the glue on the board substrate
- wetting of the glue to spontaneously flow out on the board
- penetration of the glue in to the board
- curing the sealed board, so that the glue consolidate, followed by drying or cooling

The adhesives that are typically used for gluing are water based dispersion glue and hot melt glue. The water based dispersion glue can be applied to the board directly at room temperature. After gluing it has to be dried. The hot melt glue has to be heated up before it can be applied onto the board.

If the board is coated with a thermoplastic polymer the sealing is done by melting the coating and then joining the opposite board surfaces.

Important parameters for the performance in the gluing machine are:

- open time, i.e. the time from application of the glue to the sealing
- closing time, i.e. the time the sealed blank is under pressure
- pressure on the sealed zone
- amount of glue
- temperature
- speed of the gluing machine

It is important that the board surface where the glue will be applied shall not be varnished in order to get good adhesion.

The consistency and quality of the print surface, crease profile, squareness, and the integrity of adhesive joints are all-important factors when maximum efficiencies are being sought. Other factors influencing improved efficiencies are the consistency of the board specification, especially stiffness and thickness, ambient conditions of the cartons – they must be stored at least 24 hrs in the packing area prior to use - and age of cartons - ideally not more than three months old after gluing.

10.2.3.4 Forming, Filling and Closing

The basic function of a packaging machine system is to perform:

- Forming of the package
- Filling the package with a product
- Closing the package

There are two principal ways of feeding the packaging machine system with the packaging material, i.e.

- Reel
- Blanks

At reel feeding of the packaging material into the packaging machine, the longitudinal form-fill-seal is done by forming the packaging material from the reel into a tube over a forming collar, see *Figure 10.10*. The next step is to form, fill and horizontally seal the package. This is done during the downward movement of the packaging material. Horizontal knives cut the tube to produce a bag-like package. Thereafter the package is either packed as it is or the package is feed into a final forming unit where its final form is obtained. A well-known package filled with this method is the Tetra Brik.

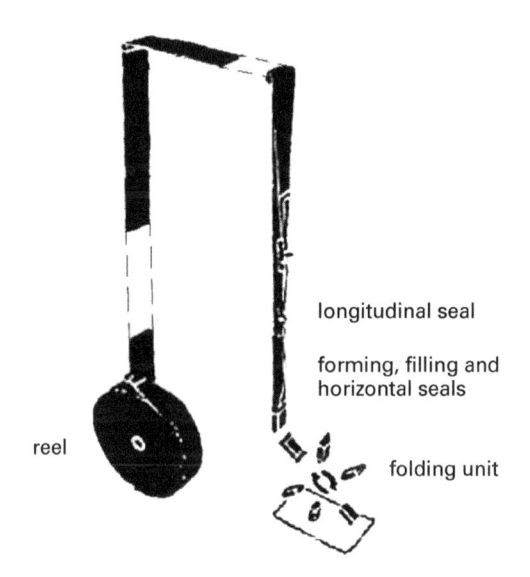

Figure 10.10. The packaging material in the packaging machine. The form-fill-seal is done by forming the packaging material from the reel into a tube over a forming collar.

The blanks that are fed into a packaging machine system are either flat pre-glued blanks or blanks. The blanks are erected to a rectangular or a square form. The product is filled into the erected blanks and thereafter the package is closed and sealed, see *Figure 10.11*. There are two ways of filling the container. One is the top-loading system were the container is bottom sealed after the erecting and the top is opened for the product to be filled into the container. The other is the end-loaded system, see *Figure 10.11*, were the product is first filled and then the ends are closed and sealed.

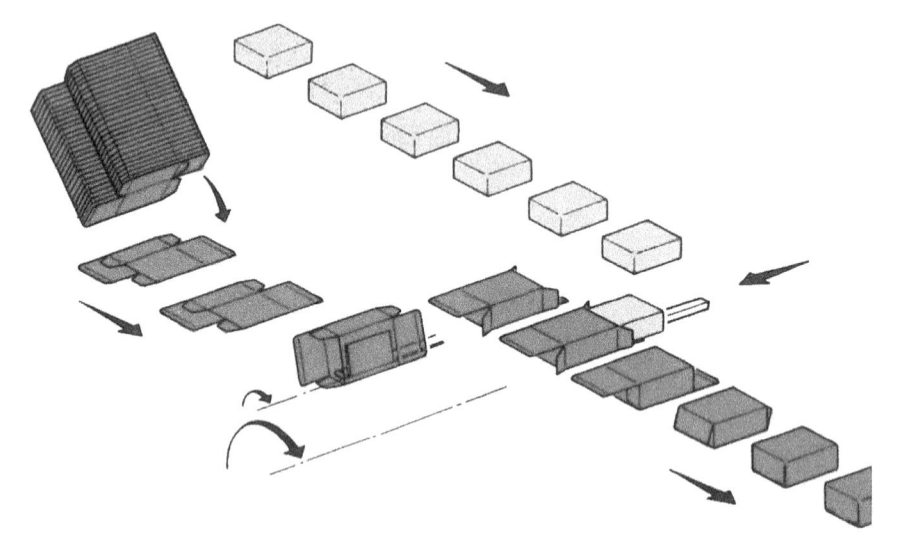

Figure 10.11. The product is filled into the erected blanks and thereafter the package is closed and sealed.

The products that are filled into a container can be divided into three different groups of goods:

1. Liquid and viscous products
2. Granulates and powder products
3. Solid products

The process of filling the package consists of three basic steps. The first step is transportation of the product to the filling equipment, the second step is to control that the amount of product that is to be filled in the package is correct and the final step is transfer of the right amount of the product from the filling equipment to the package.

An additional process function in the packaging machine system is the condition under which a product is filled into a package, i.e. if the product should be filled under vacuum, gas injection, or sterilisation.

Factors to consider when selecting a packaging machine system are:

- What kind of product should be packed
- Which package material and package method should be used
- Which productivity demands do the market have on the product
- What package formats are demanded from the market

10.2.4 Barrier Properties

Paper and board are porous and consists of fibres linked together to form a network. The voids in the network make paper and board porous and hence sensible to gas and liquid mass transport. A barrier on the paper or board is then used to protect and restrain the package material from mass transport of gases and liquids. The requirement on boards with barriers is that the barrier should fulfil the demands a product put on the package barrier properties.

Barriers on paper and board are applied on the surface of paper and board as extrusion coating, dispersion coating or lamination. At extrusion coating a molten high molar mass polymer from an extruder is applied onto the surface of a moving paper or board web. Dispersion coating is a technique, were the water-soluble barrier substrate is coated onto the surface of the board. The dispersion coating yields a package material that is easy to re-pulp, thus favouring recycling.

The most frequently used polymers at extrusion coating are low and high-density polyethylene (PE) and polypropylene (PP). Low density PE (LDPE) is the most used barrier for frozen food and it resists all oils and greases in short term. High density PE (HDPE) is a more dens material than LDPE and has better barrier properties than LDPE. At long term exposure the HDPE has a higher resistance to cracking then LDPE. Higher demands on barrier properties of packed products, than PE can offer, are fulfilled with PP. Typical end-use applications for PP-coating are for example ready meals and pet food.

An overview of polymer coating, often PE, on both sides of the board and lamination of an aluminium foil is schematically shown in *Figure 10.12*. At the left chill roll in the figure the molten plastic is extrusion coated onto the surface of the board and an aluminium foil is laminated onto the molten polymer. When the coated board leaves the chill roll the temperature in the coated board is about 20–30 °C. At the left chill roll in *Figure 10.12* the uncoated board side is extrusion coated with a polymer. Thereafter the package material is rewound into a package material roll. The procedure shown in *Figure 10.12* is typical for converting of liquid board materials.

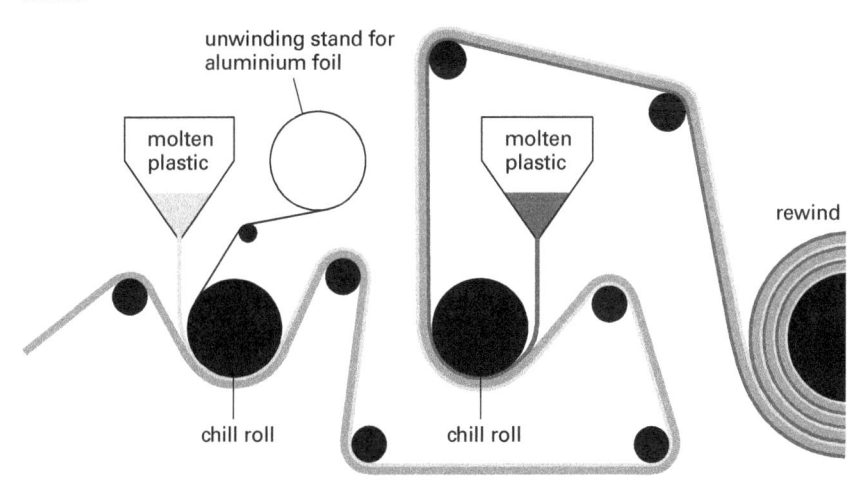

Figure 10.12. Extrusion coating and lamination of barriers on liquid board.

Lamination barriers that are used are aluminium foils, metallised polyester and metallised oriented polypropylene. The metallised polyester and metallised oriented polypropylene are laminated directly onto the surface of the board. Metallised foils have very high resistance against gas mass transportation.

10.3 Requirements on Packaging Performance

10.3.1 Definitions

For the following a definition of the different packaging levels is needed. The definitions according to the EU Directive on Packaging and Packaging waste define primary, secondary and tertiary packaging as:

• Sales or primary packaging:	Packaging conceived so as to constitute a sales unit to the final user or consumer at the point of purchase.
• Grouped or secondary packaging:	Packaging conceived so as to constitute at the point of purchase a grouping of a certain number of sales units whether the latter are sold as such to the final user or consumer or it serves only as a means to replenish the shelves at the point of sales. It can be removed from the product without affecting its characteristics.
• Transport or tertiary packaging:	Packaging conceived so as to facilitate handling and transport of a number of sales units or grouped packages in order to prevent physical handling and transport damage. Transport packaging does not include road, rail, ship and air containers.
• Packaging system:	The combination of primary, secondary and/or tertiary packaging that is used in the supply chain for a certain product.

In addition to these definitions of packaging levels a load carrier is often used. The most common example is here the standardised pallet. The pallet is made of either wood or plastic materials. The pallets constitute the unit load and are easy to handle with for example a fork lift.

The packaging system has to fulfil a number of different requirements throughout the supply chain. All different requirements shall be met by the performance of the packaging. The performance, on the other hand, is related to one ore more properties of the packaging, e.g. stiffness. Finally, the material and structural properties constituting the packaging give the packaging properties.

10.3.2 The Supply Chain

To understand the packaging requirements, expressed by the actors in the distribution system, calls for a description of the complete packaging flow. The right box in *Figure 10.13* illustrates the flow of products from the manufacturer/filler to the end users. In order to facilitate the distribution and protection of the products different aspects on the requirements are set by the different actors in this supply chain. The left box in *Figure 10.13* illustrates the suppliers of packaging materials that deliver engineered packaging with adapted performance. As an example a manufacturer of toothpaste has to contact suppliers of plastics, carton board and corrugated board.

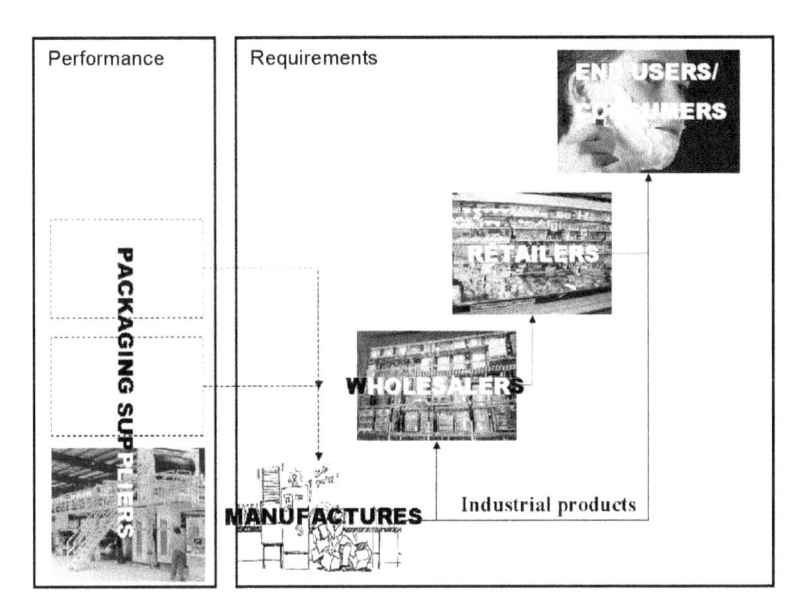

Figure 10.13. The supply chain.

The operations in different parts of the supply chain that result in the requirements on the packaging are described below.

- *The manufacturer:* The product is normally first packed into sales or primary packaging, which thereafter is packed into a grouped or secondary packaging. The secondary packag-

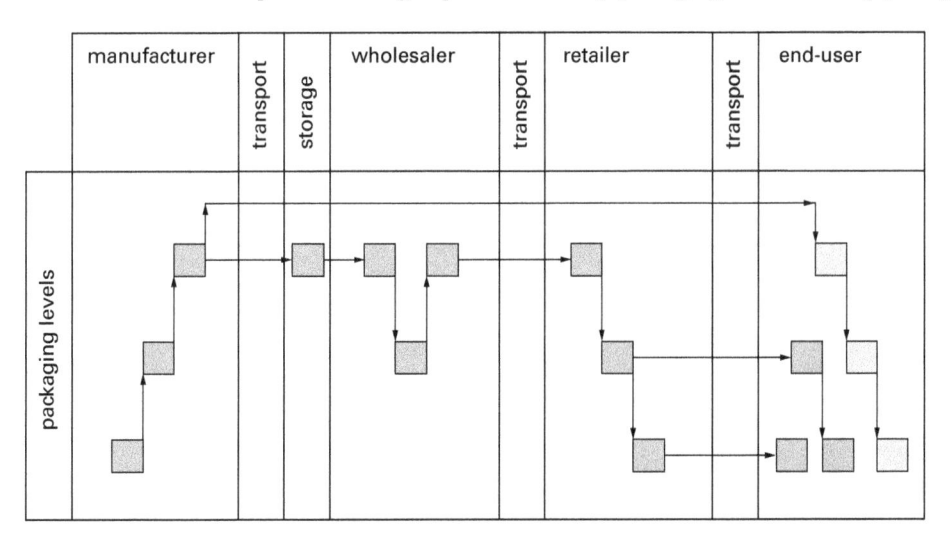

Figure 10.14. Illustration of material flow from manufacturer to end-user.

ing facilitates the handling at both wholesalers and retailers. To simplify the transport from the manufacturer to the wholesalers the secondary packaging is packed into a transport/tertiary packaging. Normally the tertiary packaging is stacked on a pallet (load carrier). The packaging levels are schematically illustrated in *Figure 10.14*.

- *The Wholesaler*: The tertiary packaging is opened and the secondary packaging is reloaded into new transport packaging (i.e. roll cages, *Figure 10.15*) for distribution to retailers, see *Figure 10.14*. Waste handling of packaging materials is an important task for the wholesalers. It should be noticed that most of these handling operations are performed manually.

Figure 10.15. The most common transport system from wholesaler to retailer.

- *The retailer:* The tertiary packaging and most often the secondary packaging is unpacked. The primary or secondary packaging is then put on the shelves, see *Figure 10.16*. These

Figure 10.16. Point of sales at retailer.

processes are time consuming since all the handling is done manually. Waste handling of packaging materials is an important and time-consuming task.
- *Transport and Storage:* The packaging is transported between the manufacturer, wholesaler and retailer by different types of vehicles (truck, train, boat, air). Reloading occur a number of times along the transport chain and is mostly done mechanically by forklifts. During transport the packaging is subjected to both static and dynamic loads as well as changes in surrounding environment (humidity, temperature, light etc.) During intermediate storage the impact on the packaging consists of static loads and changes in environment.
- *The end user/consumer:* The primary packaging must clearly inform about the product. It must be easy to open the package and to get rid of the packaging material. The retailers offer both primary and secondary packaging, e.g. one bottle of soda and a six-pack of soda.

Some products, especially industrial products, are delivered directly from the manufacturer to the end-user, for example products from subcontractors to the car industry.

10.3.3 Packaging Requirements

The fundamentals for packaging is to transport, serve and sell a product. This should be done in the most economically efficient way and at the same time taking environmental aspects into consideration.

The product primarily gives the requirements on packaging. From the manufacturer's point of view the product should be given an optimal combination of primary, secondary and tertiary packaging – a so-called packaging system – has to be designed. In doing this all different requirements along the supply chain has to be considered.

A systematic illustration of the packaging requirements is shown in *Figure 10.17*. This system is based on interviews with employees at the different actors in the supply chain for food. The systematisation consists of five groups, i.e. product flow, protection, runnability, environment and communication. Each group has different requirements that are linked to the supply chain.

1.	product flow	– handling – transport – storage
2.	protection	– mechanical – barrier
3.	runnability	– stiffness – flatness
4.	environment	– material savings – environmental concern
5.	communication	– information – promotion

Figure 10.17. Systematisation of packaging requirements during the supply chain.

The terms technical and marketing requirements will be introduced to differentiate the requirements into two groups. Technical requirements refer to physical properties that protect and facilitate the distribution of products. Marketing requirements are related to the customer end-use and the marketing of the product. Marketing requirements are on the credit side of packaging.

10.3.3.1 Product Flow

Product flow includes all physical aspects of taking a product from the manufacturer to the end-user. During this process a number of handling operations take place at different packaging levels. At the manufacturer this handling is mostly done automatically due to the fact that large quantities of identical products packed in the same way are produced at one place. Most common is to use pallets for transportation to the wholesaler, where the pallets are stored before reloading and shipping of the packaging to retailers.

In order to make the transportation efficient the capacity of carriers has to be utilised to its maximum. In the food industry most often the volume limits the amount of goods to be transported in a carrier, rather than the weight. Volume efficiency is a measure of the utilisation of cubic capacity. Since the pallet (EU-pallet) is the most common load carrier, the dimensions of transport packaging are crucial. The outer dimensions must be adjusted so that the footprint of one or more packaging is the same as the dimension of the pallet. As a consequence, not only the dimensions of secondary and primary packaging, but also the product itself has to be considered for highest total volume efficiency. Also the height of the transport packaging must be adjusted in such a way that two or more pallets stacked on top of each other utilise the full height of a carrier. The dimensions at different packaging levels are not that crucial when the limiting factor for the carrier is weight.

At the wholesaler the incoming unit loads (pallets) are broken down, sometimes repacked, and reloaded onto a new load carrier. The new load carrier is adapted for the specific demands of supply of the individual retailers. This is often a pallet or in the retail business a roll cage, which is then shipped to the retailer. At the retailer the unit is broken down and tertiary and secondary packaging opened and the primary packaging (sales packaging) put on the shelves. At the wholesaler an extensive labour intensive manual handling starts, which then ends at the end-user who finally opens the primary packaging. All this handling introduces a number of different aspects on packaging requirements, not only on handleability, but also on management of packaging waste.

Handleability is a wide concept that includes a number of requirements on packaging such as:

- to grip
- to stack
- to open and fold
- waste handling
- recycling of material or energy re-use
- packaging re-use

10.3.3.2 Protection

Product protection includes the interaction between the product and the protective power of the packaging towards different external impacts in the packaging system. These impacts can be mechanical or related to the surrounding climate (temperature and humidity). Light and other electromagnetic radiation fields are other example that could harm the product. For other products chemical and/or biochemical impacts (oxygen, mould) are crucial. Another aspect of protection is that the packaging itself should not contaminate the product e.g. chemically or biochemically.

The protection of the product is essential in packaging design – both from an economical and an environmental point of view. The consequences of insufficient barrier properties may result in large waste of food (i.e. milk) – and the resources that were taken into account to produce the food. If the mechanical protection fails the product may even contaminate land areas (e.g. toxic chemicals).

10.2.3.3 Runnability

The runnability of packaging materials in the packing lines is of greatest importance for the manufacturer. Materials that do not run well in the machines will never enter the market.

10.2.3.4 Environmental

There are two aspects on environmental requirements on packaging. One is related to protection and the environmental consequences if the products are destroyed or contaminate the environment in case of inadequate protection. The other aspect deals with the packaging itself and is related to both the production of packaging and the use of it. Environmentally friendly and efficient use of processes, raw materials (including energy) and additives is expected in the production of packaging materials and packaging. In the design of a packaging system the goal is minimum use of packaging materials, which saves both resources and minimise the amount of waste. Furthermore the packaging should support re-use or recycling.

10.2.3.5 Communication

The requirements on communication of packaging can be divided into two aspects. One is related to the distribution and involves requirements on flow information. Here the packaging should carry information so that the right products are sent to the right place and that they are handled in the correct way. Different requirements prevail on different packaging levels.

The other aspect of communication requirements relates to marketing performance. These requirements can be divided into three different groups:

- Product Information
- Selling Capacity
- Safety

The first group includes requirements on description of both the product and the packaging. Examples are; declaration of content, how to use, last date of consumption, composition of packaging etc. The requirements on selling capacity are much differentiated between products. Most aspects are related to packaging design and graphic design with the purpose to increase the attractiveness of the product. Safety requirements may include aspects on protection against theft and childproof design and components.

10.4 Design of Packaging Performance

Packaging performance denotes the ability of a packaging to satisfactory manage the different tasks as they are defined by the requirements. Packaging performance is thus the tool to match all the requirements in the most efficient way.

Packaging performance depends on:

- The properties of a single packaging.
- The interplay between product and packaging and the interplay between packaging at different levels (primary, secondary, tertiary).

By properties it is understood the characteristics that contribute to the performance of a single packaging. These properties can often be measured and thus quantified. There are only two ways to influence the packaging properties and in the extension the performance of a packaging. These are:

- Material properties
- Structural design

Consider as an example a bridge. The structural design of the framework and the materials used (including joints) give the properties that govern the performance. Performances that shall match requirements such as maximum load bearing capacity and deflection, torsional stiffness, colour and external attractiveness. In packaging, material properties like stiffness, friction and barrier are examples that mainly impact the technical performance and colour and print have their main impact on marketing performance. A package must be assembled in such a way that the joint itself is not the weak point. Structural design influences the volume efficiency but also to a high degree marketing performance.

To optimise packaging performance is a fairly complex process, which calls for a holistic approach, i.e. an approach that involves both the product itself and the distribution chain. Sometimes different requirements result in contradictory demands for packaging properties. Examples are: less packaging material vs. high stacking strength, excellent print vs. low material cost, easy-to-open vs. high packaging stiffness etc. From this it follows that the design involves different knowledges and skills and that it involves a number of compromises in order to achieve the best cost/performance for a packaging system.

In the following sections technical and marketing performances will be discussed separately.

10.4.1 Technical Performance

Technical performance is the combined packaging properties that sufficiently manage the technical requirements, see *Figure 10.18*. By changing the material properties and the structural design it is possible to influence the packaging properties and thus the performance of the packaging. By considering the interaction between the product and the primary, secondary and tertiary packages it is possible to design a packaging system that is optimal for the technical requirements. A successful optimisation of a packaging system is dependent on how well the technical requirements are defined quantitatively and how well the packaging performance can match these requirements.

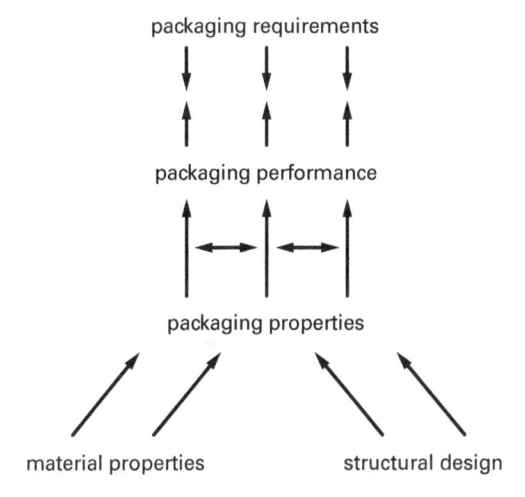

Figure 10.18. Packaging requirements versus material properties and structural design.

The basic procedure for design of the technical performances includes:

- Theoretical predictions (if possible)
- Verification of performance in laboratory
- Full-scale tests

The process is most often iterative and contains elements of „Design – Build – Test". To avoid sub-optimisation a final evaluation of the performance versus all requirements must be made, with the purpose to find a packaging solution that satisfactory fulfils these requirements. This step involves compromises between contradictory requirements.

These procedures for packaging design do not in principle differ from those used in aircraft and car industries. It starts with identification of requirements and then the task is to find a solution for performance that satisfy these requirements. The engineering follows the same routes and the design variables are material properties and structural design. Laboratory and full scale testing follow classical methods for mechanical analysis, i.e. the finite element method. During the last ten years an increasing amount of the engineering is done by the help of simulation methods run on computers that speeds up the development phase considerably. This evolution is due to better software and constitutive (material) models. The same trend can be followed in the packaging business although this industry is some years behind the automotive industry primarily due to the lack of advanced constitutive models for paper and board.

To illustrate the procedure for optimisation of the technical performance the stacking performance during transport and storage will be discussed, i.e. the resistance against damage due to compression loads. For this general discussion we assume a package on any level of packaging (primary, secondary or tertiary). The packages are transported and stored stacked on a standardised pallet. The packages are staked on top of each other either as tower stacking or interlock stacking, see *Figure 10.19*. The interlocked stack gives a more stable stack since the individual boxes are constrained by their neighbours.

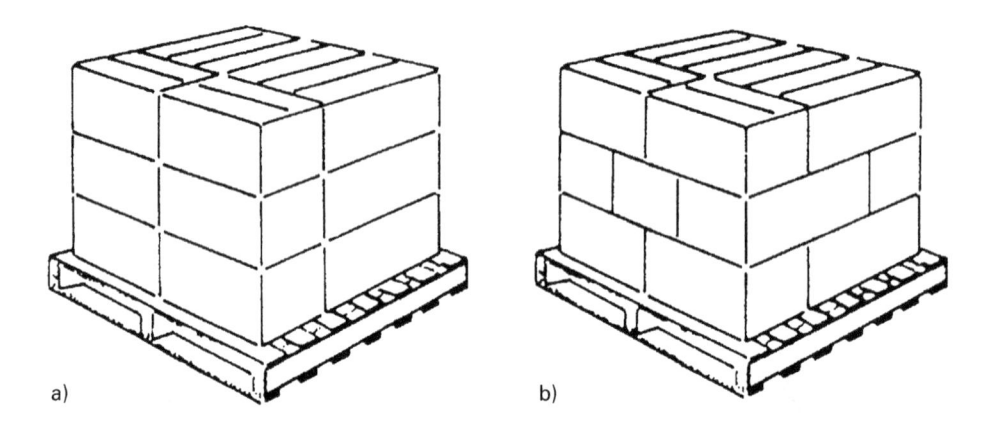

Figure 10.19. (a) Tower and (b) Interlocked stack pattern.

During transport and storage the packaging will be subjected to static loads, vibrations, shocks, and environmental changes. Static loads are given by the weight of the packages stacked on top of the bottom package, and dynamic loads are given by a combination of vertical and horizontal accelerations. The static and dynamic loads affect the choice of both the structural design and material properties whereas changes in relative humidity affect primarily the material properties.

To illustrate the procedure for packaging design only static compression loads will be discussed in the following. For a more complete design it is necessary to include the dynamic forces during transport and a separate analysis must be made for these requirements. If the magnitude of the shocks and vibrations are known a new estimation of the requirements are done, which then should be matched by the stacking performance of the packaging.

10.4.1.1 Static Compression Loads

The packaging must be engineered so that it can handle the static top-load compression during transport and storage. The top-load compression is estimated by answering the question:
- What is the top-load compression of the most stressed package in the pallet?

The most stressed packages on a pallet are those in the bottom layer. The number of layers and the weight of each packaging then give the top-load compression. To account for uncertainties due to impact of relative humidity, imperfections in stacking or shock loads a safety factor is used. The safety factor, which is most often based on experience, is multiplied with the calculated load to give the final requirement on compressive strength.

From the estimated top-load compression strength value and the packaging geometry the material needed can be calculated. The theoretical calculations can be divided into:

- Empirical analytical calculations
- Numerical simulations based on the finite element method

The empirical analytical calculation is often based on the McKee formulae. The McKee formulae are calibrated to laboratory compression tests of packaging. In recent years analysis based on the finite element method has been used more and more as an alternative to empirical analytical calculations.

Investigations where results from finite element analysis are compared with experimental observation show that use of formulae based on buckling theories, e.g. the McKee formulae, are not able to predict neither the stiffness nor the maximum box compression strength of packaging with short panels. Short panels are panels with a height (H) to width (W) ratio less than one, i.e. $H/W < 1$, if the top-load compression is in the height direction.

Below we will give an example of the design procedure for a specific corrugated transport packaging. For good reasons it is expected that the most severe requirements on compression strength prevail during transport or storage. The assumed boxes including the product, with a mass of say 10 kg, are stacked on a pallet with a height of six layers. This gives a load on the bottom box of five boxes equal to 50 kg. During transport two pallets are stacked on top of each other, which give a maximum load on the lowest box of 110 kg, i.e. 11 boxes multiplied with the weight of the package 10 kg. (The weight of the pallets is here neglected.) During storage, however, three pallets are stacked on top of each other, which correspond to a maximum load of 170 kg. This may lead to the conclusion that the compression performance should be 170 kg – at minimum. However, during transport the boxes are subjected not only to a static load, as in storage, but also to vibrations and shocks. These dynamic impacts in the vertical direction may result in a requirement for compression strength that is higher than 170 kg although the static load is only 110 kg during transport.

Next will follow a more general discussion on compression requirements on packaging during transport and storage, see *Figure 10.20*. The solid line represents the stacking requirement during different situations along the distribution chain. In position A the requirements represent the compressive load of three pallets stacked on top of each other. Position B refers to two stacked pallets in a high humidity storage. After some time they are rearranged and thereby subjected to vertical vibration loads. Finally, in position C one pallet is handled carelessly and subjected to a vertical shook. The dotted lines in *Figure 10.20* (1, 2 and 3, respectively) represent three different boxes engineered to have three different levels of strength.

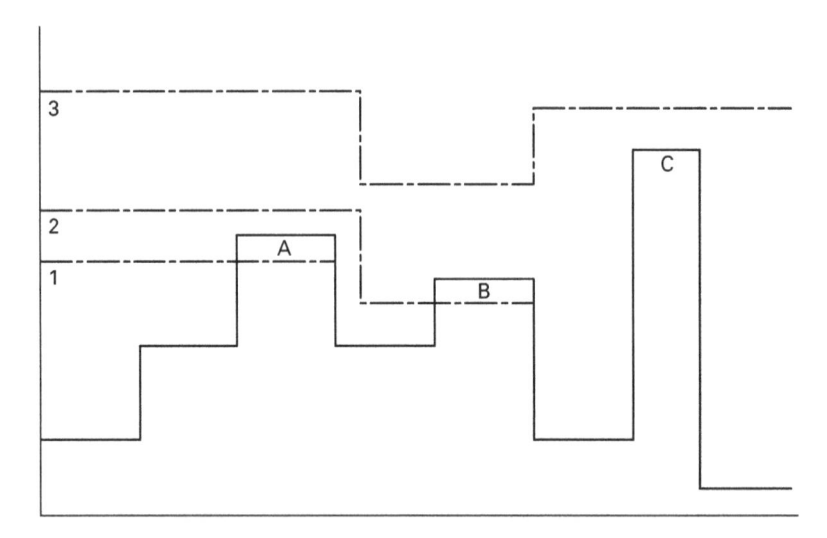

Figure 10.20. Compression requirements (solid line) and performance (dotted lines). The (1), (2) and (3) dash-dotted lines correspond to different boxes.

Let us first consider the corrugated box represented by the dash-dotted line (1) in *Figure 10.20*. In position A the impact of three stacked pallets exceeds the property of the boxes in the lowest layer on the lowest pallet and a collapse of the box will occur. A new box with higher box strength is then evaluated, i.e. the box denoted by the line (2). In a position after A on the solid black line the relative humidity in the environment is temporarily increased, which results in a decrease in the compressive properties of the box, the step down of the line (2). This will cause a box collapse in position B. A box with even higher strength has to be considered to manage the requirements in position B. The box denoted by the dash-dotted (3) line is engineered. This box withstands the impacts in position B. Thereafter the relative humidity is decreased to its initial value, which results in an increase in the compression properties of the box. The box recovers more or less its original strength. As a result of this recovery the box also manages the requirements from the shock loading at position C.

The same design procedure shall be used for other requirements along the supply chain, i.e. to find the dimensioning requirements for every specific property of the packaging. This way to design the packaging solution is the fastest way but the requirements must be known quantitatively. In addition to this the different properties of the packaging must be possible to calculate from material properties and structural design.

In most situations neither the requirements are known nor is it possible to predict all requirements from material properties and structural design. Consequently other procedures than the one described above must be used. These involve both laboratory testing and full-scale transportation tests.

To finally verify the design a full-scale transport test of box (3) in *Figure 10.20* has to be performed. This is the normal procedure because large values may be lost if the boxes fail. For this reason „dummies" of the same weight as the product sometimes substitute the product in these tests.

In the manufacturing of packaging there is always a distribution in the different properties due to variations in the process and the material. This is also true for the requirements. Consider for instance the impact of change in relative humidity or dynamic loads. These may vary due to time of day/year or traffic situation. If these distributions are Gaussian (see *Figure 10.21*) it is possible to estimate the percentage of box failures from statistical considerations. If for instance the standard deviation is σ % for both the requirement and property, and the mean of the property exceeds the mean of the requirement by four, i.e. 5.2σ see *Figure 10.21*, the frequency of failure will be 2 % (+/− 2,6σ includes 99 % of all values). If the mean value of the requirement and the property is the same, 50 % of the packaging will by definition fail („underpacking"). If the performance exceeds the requirement enough no failures will occur („overpacking"). Overpacking, to use more packaging material than necessary, implies expensive packaging, increased costs for transportation and higher environmental impact. Underpacking may result in damaged or wasted products, which results in delayed deliveries, bad will and in the end disappointed customers. Underpacking is most often very expensive for the producer. Also from an environmental point of view underpacking may be harmful due to waste of production resources for the product, and eventually contamination of land (in case of harmful products).

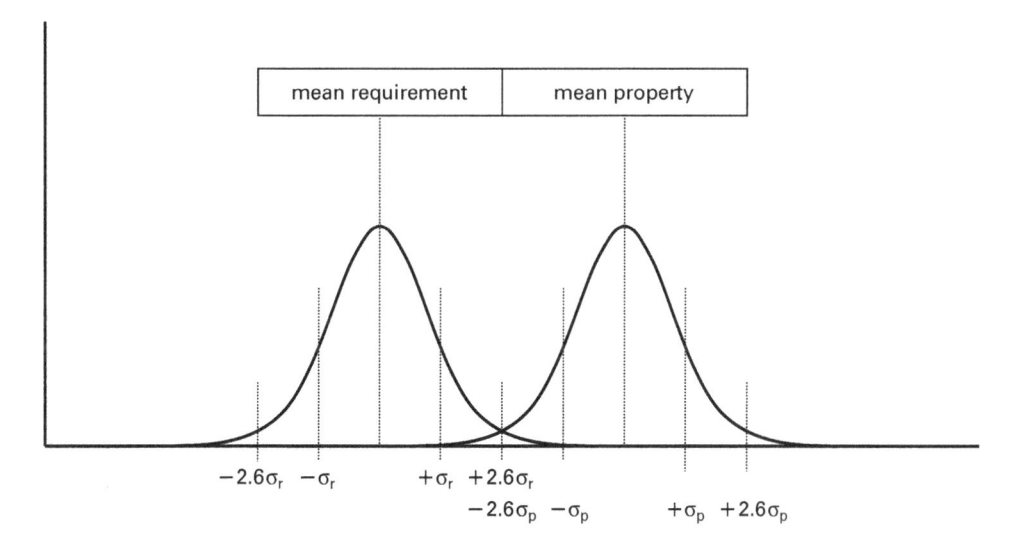

Figure 10.21. Gaussian distribution of possible box failures.

The balance between over- and underpacking is an important consideration in design of packaging systems. These considerations can be summarised as shown in *Figure 10.22*, where the relation between total cost and product protection (red line) is outlined. Increased protection asks for more packaging material and thus higher costs (blue line). On the other hand less protection increases the cost for damaged products (black line). This increase is normally more pronounced than the increase in cost for packaging.

Life Cycle Analysis (LCA) is a tool by which the environmental consequences systematically can be analysed. LCA takes into account the consumption of renewable and non-renewable resources and emissions to air and water. It is used to compare the consequences of the production of different packaging solutions or to evaluate the consequences of different packaging systems.

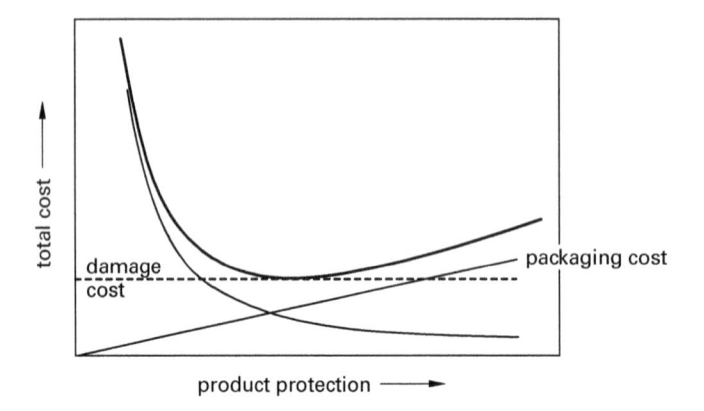

Figure 10.22. Environmental exposures versus package performance.

10.4.1.2 Interaction Between Packages

Let us consider two different products, canned tomatoes and potato chips, respectively. The can is stiff and has high compression strength and can thus fulfil all compression requirements from stacking. The packaging solution can then be an open try with a wrap-around in plastic that facilitates the handling of e.g. 12 cans. This packaging constitutes the grouped/secondary packaging as well as the transport/tertiary packaging. At the retailer the wrap-around is easily removed and the sales packaging can be put on the shelves.

In the other example, potato chips in a paper bag, no compressive load can be taken by the sales/primary packaging. A secondary packaging is needed to make it easy to handle the bags and satisfy the compression requirement that exists from producer to retailer. These demands create a need for a separate tertiary packaging. In the first mentioned case the cans take 100 % of the load and the transport packaging (wrap-around) 0 %. In the second case the reverse is valid, the transport packaging takes 100 % and the primary packaging 0 %. However in many situations packaging at more than one level contribute to the overall compression strength.

Figure 10.23 illustrates schematically a laboratory compression test where the compression load versus the deformation of the package is recorded for some primary package and secondary corrugated package.

In *Figure 10.24(a)* the typical load-deformation curves for the primary and the secondary packaging are shown in the same graph. The figure shows that the entire compression load-deformation curve for the primary package is completed before the secondary package carries any substantial load. Thus, we have a situation where the primary packaging is damaged at a load Qa, see *Figure 10.24(a)*.

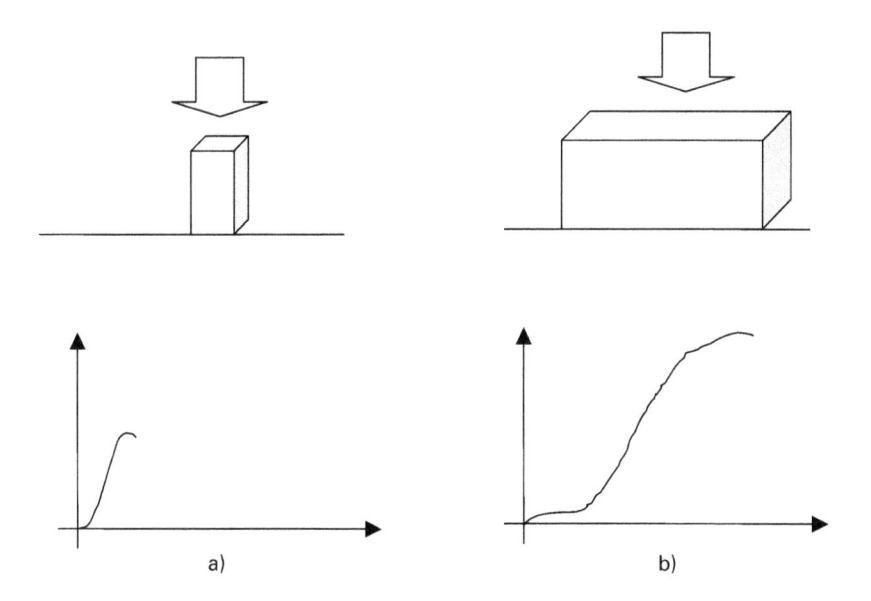

Figure 10.23. Load versus deformation during compression. (a) Primary and (b) secondary packaging.

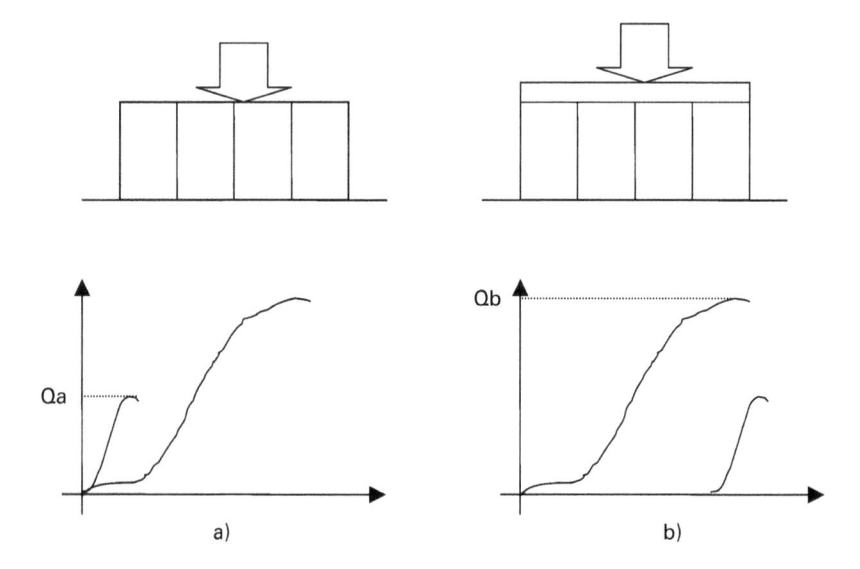

Figure 10.24. Influence of head space on box collapse loading. (a) Zero head space between the primary and secondary packages. (b) Head space between the primary and secondary packages.

However, if we during packing leave a space between the primary and the secondary packaging this is equivalent to moving the load-deformation curve for the primary packaging in *Figure 10.24(b)* to the right. The space between the primary package and the secondary package is called headspace. During loading the packaging system interacts, i.e. first the secondary package will carry the load and experience the deformation, and when the deformation has exceeded the headspace the primary package will start to deform. In the example given in *Figure 10.24(b)* a load Qb can be achieved, which is substantially higher than Qa, even before the primary packaging is deformed at all.

10.4.2 Marketing Performance

Marketing performance is the combined packaging properties that sufficiently meet the marketing requirements (c.f. *Figure 10.17*).

Packaging is often the only tangle point of difference between products. This is particularly true for many food products, beverages and personal health-care products. Thus, product development is often limited to a new design of packaging.

Packaging is a very cost-effective marketing tool. In a self-service environment the packaging has a very strong impact on buying decisions. After purchase, the packaging can also be used to inform about and promote other products and stimulate a repetitive purchase. The packaging is the „Silent Salesman"!

The tools available to affect packaging properties and change the performance are packaging design, material properties and graphic design. The process to create good marketing performance is very much a dialogue between manufacturer and packaging supplier. The evaluation of a new or changed marketing performance is not easy to perform; it takes time and is most often related to sales volumes.

The requirements for high quality print demand specific properties of the packaging surfaces. These properties are dependent on the printing method used.

10.5 Future Trends

10.5.1 Digital Printing and E-print

A future trend, in the development of the existing printing technologies with ink on paperboard, is the use of computers in the process. By digital printing technologies the design image is printed directly onto the package. The process is similar to the process of printing a document from a PC to a desk-top printer. The trend in the market is that digital printing technologies increase their market shares. This technology enables the possibility to make every printed package unique.

New technologies to display images on packages emerge and today some very interesting new technologies have been developed, for example metallic ink printed onto paper to form circuits and conductive polymers that are laminated onto paper.

10.5.2 E-tags

In the supply chain the products physically flows in the direction from the producer to the end-user. The product information flow is in the reverse direction, i.e. from the end-user to the producer. The information of the products is done by uniquely identifying each product in the packaging chain at each packaging level.

Today the market uses the EAN-UCC system for product identification. The development of Internet has triggered standardisation organisations, companies and research teams to link the product identification to the Internet. The research today is concentrating around the product identity code that will link the product to stored information on the Internet and development of a mark-up language for describing the physical objects on Internet.

The physical medium that can be used for identification is divided into the following groups

- Printed barcodes
- Radio Frequency Identification (RFID)
- Magnetic strips

The barcodes is printed on a label or directly onto the package, and the key feature of barcodes is that the barcode is a standardised and very inexpensive system. The drawback of the printed barcode is that it is sensible to dirty or wet environments.

The RFID-system and magnetic strips are not so often used as barcodes in identification of products. This is mainly due to the cost.

The RFID-system consists of a transponder unit and a read/write unit. The transponder unit is often applied onto a label that is put onto the product. Both the transponder and the write/read units have internal intelligence and an antenna that receives, sends and stores data.

The key features of RFID-technology are:

- Data can be stored on the object and changed.
- The transponder unit can be read through water, dirt and other non-metallic surfaces.
- Identification of several objects simultaneously is possible.

The drawback of the RFID-system is today that the cost for the transponder unit, with a chip and an antenna, is much higher than the cost for a printed barcode.

Magnetic strips will in the future probably be replaced by the RFID-technology in combination with barcodes.

10.5.3 Smart Packages

A smart package is defined as packaging that has build-in intelligence. The new possibility, with cheap computer chips that are placed on packages, allows a new range of consumer driven applications. For example one can design chips that

- measure moisture and temperature and from this information can display in real time the shelf life of the product
- adjust the microwave oven to automatically prepare the cocking

This form of interaction between the package and its surrounding will enhance the performance of the package for the consumer.

10.5.4 New Packaging Materials

The performance of packages depends on both the material properties and the formability of the packaging material.

Today, paperboard is folded into a package. The sides of the package consist of flat panels. There is an increasing market demand for a change in geometrical shape of packages. One way to increase the flexibility for the designers of packaging systems is to form the sides into more curved panels. To be able to do this with paper and board the permanent elongation of paper and board, without damaging the material, needs to be increased.

A sandwich construction consists of two stiff outer skins that are separated by a more compliant core. The stiffness of the board relies on the stiffness of the outer skins and the distance between the skins. Sandwich constructions of board materials are multi-ply board and corrugated board.

In corrugated board the stiffness is primarily given by the distance between the two liners on each side of the corrugated fluting. During the last hundred years a wave pattern of fluting has separated the liners. A more suitable pattern consists of a honeycomb pattern. At the Catholic University in Leuven, Belgium a continuous process to form the core into a honeycomb type of pattern for corrugated board has been developed. By gluing liners on each side of the honeycomb pattern a very stiff corrugated board is obtained, not exhibiting some of the drawbacks of the wave pattern.

10.6 Literature

For further studies of packaging material in the supply chain the following literature is recommended.

Förpackningslogistik, andra utgåvan, Packforsk, 2000.
Fundamentals of Packaging, Ed. F. A. Paine, Institute of Packaging, Brookside Press Ltd, Leicester, 1985.
Packaging Foresight, Packa Futura 2001, Carl Olsmats, Packforsk 2001.
Packaging in the 21st Century, A. Stirling-Roberts, J. Prebble and P. Page, Pira International Ltd., 2001.
Papermaking Science and Technology, Book 12, Paper and Paperboard Converting. A. Salvolainen (ed.), Fapet Oy, 1998.
Papermaking Science and Technology, Book 18, Paper and Board Grades. H. Paulapuro (ed.), Fapet Oy, 2000.

11 Laminate Theory for Papermakers

Christer Fellers
Innventia AB

11.1 Introduction 288

11.2 Definition of a Laminate 289

11.3 Stress and Strain for a Layer in a Laminate 292

11.4 Matrices 294

11.5 Comparison between Orthotropic and Isotropic Materials 295

11.6 Strain in the Plane due to Moisture Absorption 296
11.6.1 Strain in the Plane due to an External Mechanical Force 296
11.6.2 Total Strain in a Layer in the Plane due to both Moisture Sorption and Mechanical Forces 297
11.6.3 Strain due to Bending 297

11.7 Total Strain in a given Layer due to Moisture Changes, Mechanical Influences and Bending 299

11.8 Forces and Moments 299

11.9 Laminate Theory where the Layers are Oriented with an Arbitrary Angle in Relation to the 1-Direction (MD) 301

11.10 Transformation of Stresses and Strains 301

11.11 Transformation of the Hygroexpansion Coefficient 302

11.12 Transformation of the Stiffness Matrix, Q 302

11.13 Forces and Moments 303

11.14 Bending Stiffness and Tensile Stiffness 303

11.15 Simplified Expression for Bending Stiffness 305

11.16 Calculation Example for Bending Stiffness 306

11.17 Curl and Twist, Strain, Shear, as a Result of a Change in Moisture 307

11.18 Curl and Twist 308

11.19 Leaning Stacks of Folded Printing Papers 308

11.20 Tensile Stiffness and Bending Stiffness Studied in a Polar Diagram 309

11.21 Hygroexpansion, Curl and Twist Studied in Polar Diagrams 309

11.22 Hygroexpansion Orientation and Tensile Stiffness Orientation 310

11.23 Shear-free Angle for Hygroexpansion 311

11.24 Twist Free Angle 311

11.25 Finding the Height at a Point on a Bent Paper 313

11.26 Stress and Strains in the Thickness Direction of the Laminate due to Hygroexpansion 314

11.1 Introduction

In order to be able to calculate the bending stiffness of a paper and its tendency to curl and twist, knowledge of the paper's elastic and hygroscopic properties is required. These properties are discussed in this chapter.

When the moisture content in paper or carton board changes, the sheet often does not remain flat. The term curl is used if the sheet bends in the machine-direction or cross direction of the sheet, MD or CD, and the term twist is used to indicate that the sheet becomes twisted like a propeller. In corrugated fibreboard, the term warp is usually used to describe this type of bending. The reason why the sheet bends when the moisture content changes is that the hygroexpansion is different in the different layers of the sheet. Twist or warp occurs if one or more of the layers has an oblique symmetry with respect to the machine direction.

Curl and twist in paper, carton board and corrugated fibreboard can cause great problems in the converting and use of these materials, and can lead to great costs as the result of complaints. The most important mechanical property of a paper, board or corrugated fibreboard is the bending stiffness of the sheet.

With the help of laminate theory, it is possible to calculate how a sheet material such as paper is deformed as a result of external forces and of the internal forces caused by hygroexpansion in the different layers of the sheet. A paper is regarded as a layered structure with different properties in the different layers.

With the help of laminate theory, it is possible to do several things:

- Make sensitivity investigations of the causes of curl and twist
- Calculate bending stiffness.

A separate computer program has been developed to make it possible to carry out the calculations rationally.

Examples of problems where the moisture has an influence:

- Multi-colour offset printing
- Continuous stationery for computers
- The twist and curl of corrugated fibreboard
- Twist and curl and bending stiffness can be calculated with laminate programs.

11.2 Definition of a Laminate

- A laminate consists of a number of uniformly thick layers which do not move in relation to each other.
- The laminate is ordered in a coordinate system 1-2, with the 3-axis, the Z-direction, directed downwards, *Figure 11.1*.
- Each layer is assumed to be orthotropic, i.e. to have different properties in any two mutually perpendicular directions.
- Each layer is ordered in a coordinate system *x-y* (on-axis), which lies parallel with the 1-2 system (off-axis).
- The coordinate system *x-y* of each layer can be displaced at an angle relative to the coordinate system 1-2.
- The positive angle is measured clockwise from the 1-system to the *x*-system.

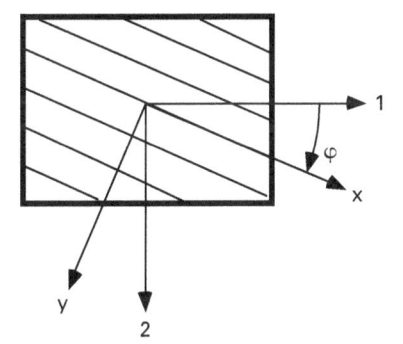

Figure 11.1. Each layer lies in an x-*y*-coordinate system. The layer is placed in a 1-2 coordinate system according to. In the case of a paper, the sheet is placed so that 1 = MD and 2 = CD.

Figure 11.2 shows a laminate consisting of *n* layers. The origin lies at the geometrical midpoint of the laminate. The layers are numbered from the top and downwards. Note that the thickness coordinates are calculated from the middle to the lower interface of the layer.

The thickness coordinates are calculated as follows.

$$t = \sum_{k=1}^{n} t_k \tag{11.1}$$

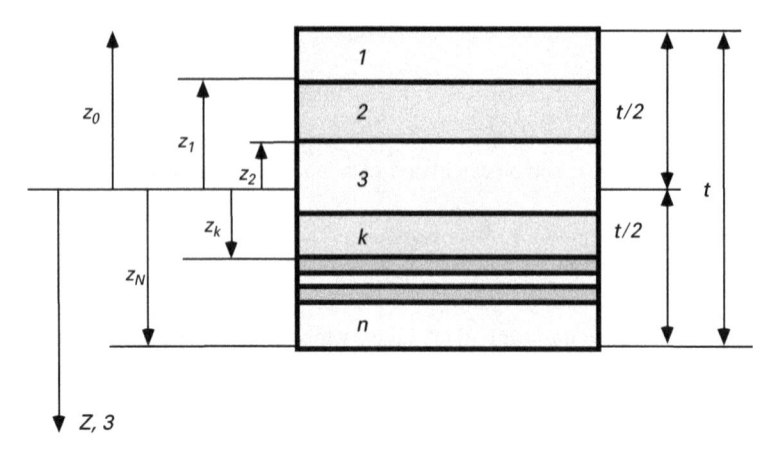

Figure 11.2. A laminate consisting of n layers. The origin lies at the geometrical mid-point of the laminate. The layers are numbered from the top and downwards.

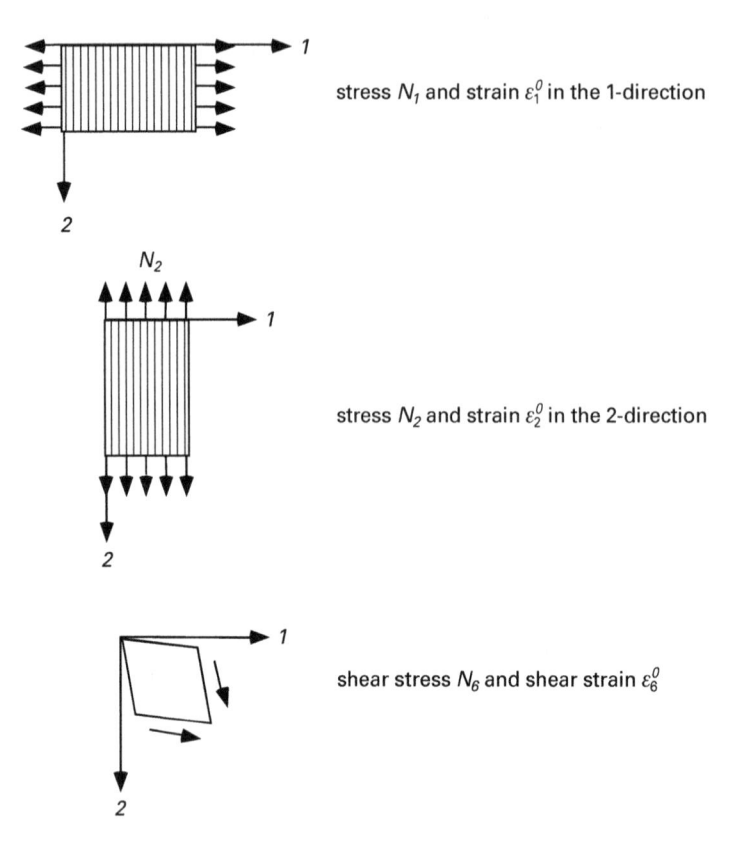

stress N_1 and strain ε_1^0 in the 1-direction

stress N_2 and strain ε_2^0 in the 2-direction

shear stress N_6 and shear strain ε_6^0

Figure 11.3. Coordinate system and directions.

The coordinates of the layer are given by:

$$z_0 = -\frac{t}{2} \tag{11.2}$$

$$z_k = z_{k-1} + t_k \tag{11.3}$$

where
t_k = the thickness of the layer k
$k = 1, 2, \ldots, n$
Positive stress N and positive strain ε are defined in *Figure 11.3*.
Any arbitrary curl of a paper sheet can be divided into three components, κ_1, κ_2 and κ_6.
Positive bendings (κ) and positive moments (M) are shown in *Figure 11.4*.

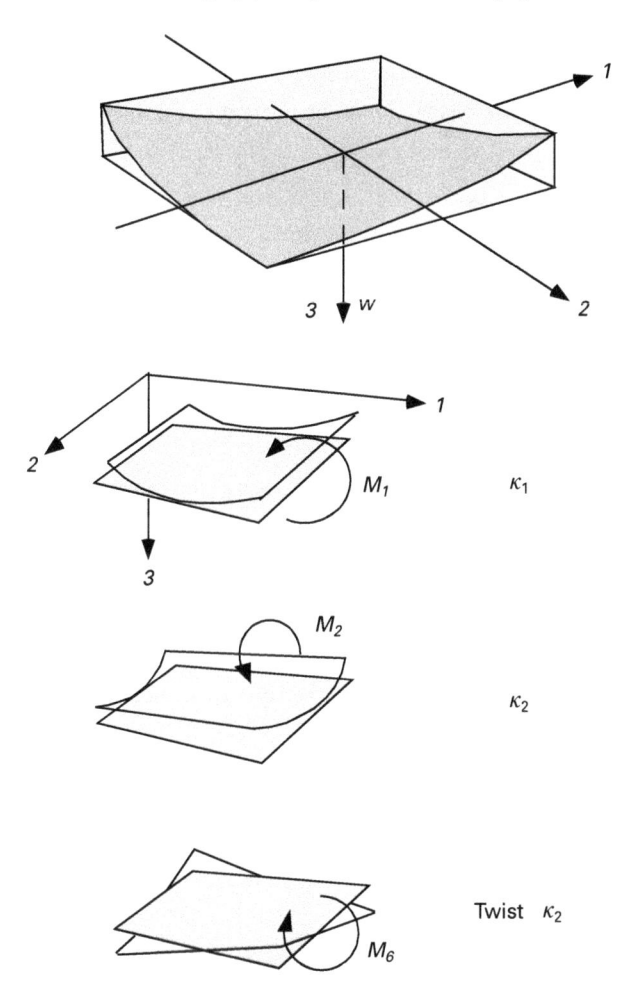

Figure 11.4. A curl of a paper sheet can be divided into three components, κ_1, κ_2 and κ_6.

Curl in direction 1

$$\kappa_1 = -\frac{\partial^2 w}{\partial x_1^2} \qquad \kappa_1 = \frac{1}{R_1} \qquad \left[\frac{1}{m}\right] \tag{11.4}$$

Curl in direction 2

$$\kappa_2 = -\frac{\partial^2 w}{\partial x_2^2} \qquad \kappa_2 = \frac{1}{R_2} \qquad \left[\frac{1}{m}\right] \tag{11.5}$$

where R is the radius of curvature in the appropriate direction.
Twist:

$$\kappa_6 = -2\frac{\partial^2 w}{\partial x_1 x_2} \qquad \kappa_6 = -2\frac{(h/b)}{l} \qquad \left[\frac{1}{m}\right] \tag{11.6}$$

The twist, κ_6, can be illustrated as twice the change in the inclination of the surface from point A to point B in *Figure 11.5*.

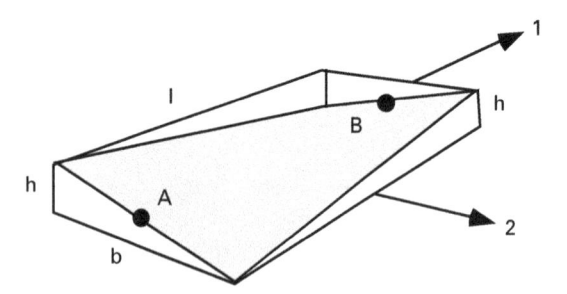

Figure 11.5. The twist, κ_6, can be illustrated as twice the change in the inclination of the surface from point A to point B.

11.3 Stress and Strain for a Layer in a Laminate

Consider one of the orthotropic layers in the laminate. The layer is oriented with its symmetry directions in the *x*- and *y*-directions according to *Figure 11.6*. If the layer is strained ε_x in the *x*-direction, the layer contracts in the *y*-direction (ε_y in the *y*-direction is negative).

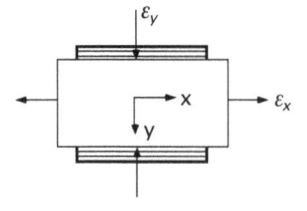

Figure 11.6. Consider one of the orthotropic layers in the laminate.

Hooke's law gives:

$$\varepsilon_x = \frac{\sigma_x}{E_x} \tag{11.7}$$

The contraction in the y-direction perpendicular to the loading direction is given by the equation:

$$\varepsilon_y = -v_{xy} \cdot \varepsilon_x \tag{11.8}$$

where
σ_x = stress in the x-direction
ε_x = strain in the x-direction
ε_y = strain in the y-direction
v_{xy} = Poisson's ratio (strain in the y-direction because of loading in the x-direction)
E_x = Elastic modulus in the x-direction
If the layer is strained in the y-direction, we obtain by analogy:

$$\varepsilon_y = \frac{\sigma_y}{E_y} \tag{11.9}$$

The contraction in the x-direction perpendicular to the loading direction is given by the equation:

$$\varepsilon_x = -v_{yx} \cdot \varepsilon_y \tag{11.10}$$

where
σ_y = stress in the y-direction
ε_x = strain in the x-direction
ε_y = strain in the y-direction
v_{xy} = Poisson's ratio (strain in the x-direction because of loading in the y-direction)
E_y = Elastic modulus in the y-direction
If the layer is loaded in both the x- and y-directions, the strain in the x-direction is:

$$\varepsilon_x = \frac{\sigma_x}{E_x} - v_{yx} \frac{\sigma_y}{E_y} \tag{11.11}$$

The corresponding strain in the y-direction is:

$$\varepsilon_y = \frac{\sigma_y}{E_y} - v_{xy} \frac{\sigma_x}{E_x} \tag{11.12}$$

The shear strain is defined as:

$$\gamma_{xy} = \frac{\partial v}{\partial x} + \frac{\partial u}{\partial y} \tag{11.13}$$

The relationship between shear stress τ_{xy} and shear strain γ_{xy} is

$$\tau_{xy} = G_{xy} \cdot \gamma_{xy} \tag{11.14}$$

where
G_{xy} is the shear modulus, N/m2

11.4 Matrices

The equations above can be expressed compactly in matrix form as:

$$\begin{bmatrix} \varepsilon_x \\ \varepsilon_y \\ \gamma_{xy} \end{bmatrix} = \begin{bmatrix} S_{xx} & S_{xy} & 0 \\ S_{yx} & S_{yy} & 0 \\ 0 & 0 & S_{ss} \end{bmatrix} \cdot \begin{bmatrix} \sigma_x \\ \sigma_y \\ \tau_{xy} \end{bmatrix} \tag{11.15}$$

where [S] is the compliance matrix with the following components

$$S_{xx} = \frac{1}{E_x} \quad S_{xy} = -\frac{v_{yx}}{E_y}$$

$$S_{yx} = -\frac{v_{xy}}{E_x} \quad S_{yy} = \frac{1}{E_y}$$

$$S_{ss} = \frac{1}{G_{xy}}$$

The matrix is symmetrical, $S_{xy} = S_{yx}$, from which it follows that

$$\frac{E_x}{E_y} = \frac{v_{xy}}{v_{yx}} \tag{11.16}$$

If we assume that the material has its highest modulus of elasticity in the *x*-direction and if we assume that we strain the material equally in both directions, the contraction will be greatest in the *y*-direction when a load is applied in the *x*-direction, as shown in *Figure 11.7*.

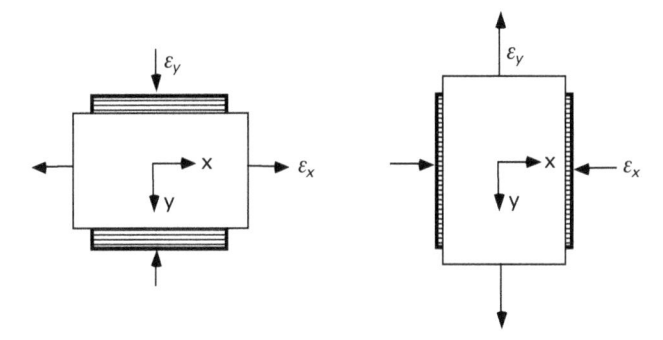

Figure 11.7. Contraction in *x*-direction and *y*-direction as a function of applied equal strain.

The following equations indicates which strain is obtained as a function of the external stress. If one wishes to know which stresses are obtained as a function of the external strain, the compliance matrix is inverted and we obtain:

$$\begin{bmatrix} \sigma_x \\ \sigma_y \\ \tau_{xy} \end{bmatrix} = \begin{bmatrix} Q_{xx} & Q_{xy} & 0 \\ Q_{yx} & Q_{yy} & 0 \\ 0 & 0 & Q_{ss} \end{bmatrix} \cdot \begin{bmatrix} \varepsilon_x \\ \varepsilon_y \\ \gamma_{xy} \end{bmatrix} \qquad (11.17)$$

where [*Q*] is the stiffness matrix with the following components.

$$Q_{xx} = \frac{E_x}{1 - \nu_{xy} \cdot \nu_{yx}} \qquad Q_{xy} = \frac{\nu_{xy} \cdot E_y}{1 - \nu_{xy} \cdot \nu_{yx}}$$

$$Q_{yx} = \frac{\nu_{yx} \cdot E_x}{1 - \nu_{xy} \cdot \nu_{yx}} \qquad Q_{yy} = \frac{E_y}{1 - \nu_{xy} \cdot \nu_{yx}} \qquad Q_{ss} = G_{xy}$$

For reasons of symmetry, $Q_{yx} = Q_{xy}$.

The stiffness matrix given in these equations is the starting-point in calculations according to laminate theory.

11.5 Comparison between Orthotropic and Isotropic Materials

In the case of isotropic materials, the following relationship applies between elasticity modulus; shear modulus and Poisson's ratio.

$$G = \frac{E}{2(1+\nu)} \qquad (11.18)$$

In the case of an orthotropic material such as paper, the following approximate relationship applies, according to Baum:

$$G_{xy} = \frac{\sqrt{E_x \cdot E_y}}{2\left[1+\sqrt{v_{xy} \cdot v_{yx}}\right]} \qquad (11.19)$$

$$\sqrt{v_{xy} \cdot v_{yx}} = 0.293 \qquad (11.20)$$

$$G_{xy} = 0.387\sqrt{E_x \cdot E_y} \qquad (11.21)$$

Previously equation (11.16) shows

$$\frac{E_x}{E_y} = \frac{v_{xy}}{v_{yx}}$$

Combination of equations (11.16) and (11.20) gives:

$$v_{xy} = 0.293\sqrt{\frac{E_x}{E_y}} \qquad (11.22)$$

11.6 Strain in the Plane due to Moisture Absorption

We assume that the layer also suffers a strain ε^H because of moisture absorption

$$\varepsilon^H = \beta^H \cdot \Delta H \qquad (11.23)$$

where β^H = hygroexpansion coefficient
ΔH = the change in the moisture content of the paper
In the laminate equations, the index H is in general used, but unfortunately this usage covers two different cases.
Index $H = RH$ (Relative Humidity), and we obtain.

$$\varepsilon^H = \beta^{RH} \cdot \Delta RH \qquad (11.24)$$

Index $H = mc$ (moisture content), and we obtain.

$$\varepsilon^H = \beta^{mc} \cdot \Delta mc \qquad (11.25)$$

11.6.1 Strain in the Plane due to an External Mechanical Force

We assume that a layer is exposed to a strain ε^M due to the application of an external mechanical force.

11.6.2 Total Strain in a Layer in the Plane due to both Moisture Sorption and Mechanical Forces

$$\varepsilon = \varepsilon^M + \varepsilon^H \tag{11.26}$$

11.6.3 Strain due to Bending

If a beam is subjected to a bending moment, M, the beam is formed into an arc with a certain radius of curvature, R, Plane cross sections in the beam remain plane after bending. Longitudinal elements on the convex side of the beam are stretched, and those on the concave side are compressed. The region in the beam's cross-section, which is neither stretched nor compressed, is called the neutral surface. It need not lie in the middle of the beam. *Figure 11.8a* shows the strain distribution in a bent beam.

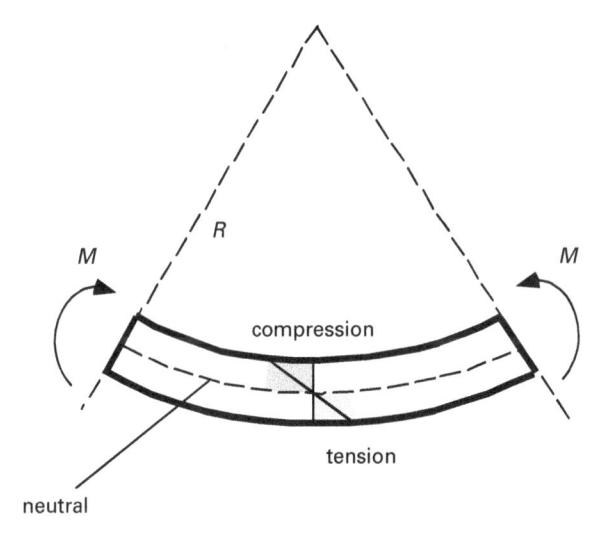

Figure 11.8a. The strain distribution in a bent beam.

Let us consider a region of the bent beam in greater detail, as shown in *Figure 11.8b*. Note that in laminate theory the origin of the z-coordinate is placed in the geometrical middle plane of the laminate.

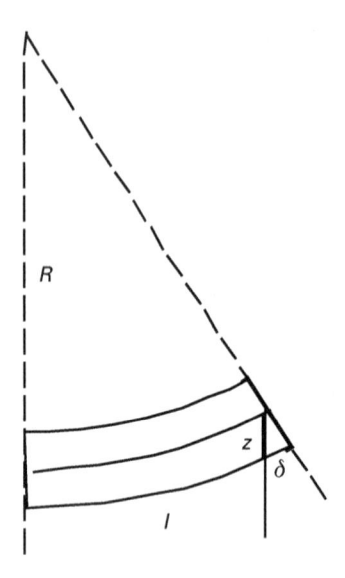

Figure 11.8b. Bent beam at a larger scale.

Similar triangles give:

$$\frac{\delta}{z} = \frac{l}{R}$$

The strain is given by:

$$\varepsilon_B = \frac{\delta}{l} = \frac{z}{R} \qquad (11.27)$$

or

$$\varepsilon_B = z \cdot \kappa \qquad (11.28)$$

where

$$\kappa = \frac{1}{R} \quad \kappa = \text{the curvature, m}^{-1}$$

The strain of the mean surface is ε°. The total strain in a given layer during bending is then:

$$\varepsilon = \varepsilon^\circ + \varepsilon_B \qquad (11.29)$$

Figure 11.9 illustrates the strains in a bent laminate, where z is the coordinate from the middle plane and κ is the curvature.

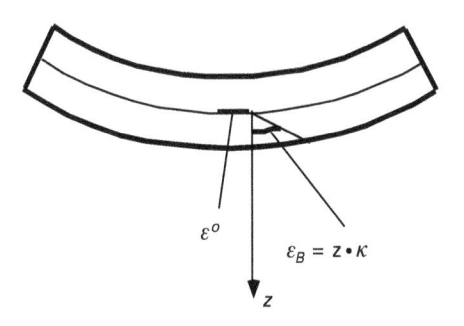

Figure 11.9. The strains in a bent laminate.

11.7 Total Strain in a given Layer due to Moisture Changes, Mechanical Influences and Bending

In a given layer, equations (11.26) and (11.29) apply:

$$\varepsilon^M + \varepsilon^H = \varepsilon^\circ + \varepsilon^B \tag{11.30}$$

Equations (11.24), (11.28) and (11.30) give:

$$\varepsilon^M = \varepsilon^\circ + z \cdot k - \beta^H \cdot \Delta H \tag{11.31}$$

11.8 Forces and Moments

In the uniaxial case, where the layers are oriented symmetrically in the 1-direction (MD) and the mechanical strain is ε^M, the stress in a given layer is, according to the equation equal to:

$$\sigma = Q \cdot \varepsilon^M \tag{11.32}$$

Equations (11.31) and (11.32) give

$$\sigma = Q \cdot \left(\varepsilon^\circ + z \cdot \kappa - \beta^H \cdot \Delta H \right) \tag{11.33}$$

The forces and moments in all the layers, with the z-axis directed downwards as in the previous figure, are integrated to give

$$N = \int \sigma \, dz \tag{11.34}$$

$$M = \int \sigma \cdot z \, dz \tag{11.35}$$

where
M = moment per unit width, Nm/m
N = force per unit width, N/m
Equations (11.33) – (11.35) give:

$$N = \int Q\left(\varepsilon^\circ + z \cdot k - \beta^H \cdot \Delta H\right) dz \tag{11.36}$$

$$M = \int Q \cdot z\left(\varepsilon^\circ + z \cdot k - \beta^H \cdot \Delta H\right) dz \tag{11.37}$$

These expressions are developed with the integration limits, where i goes from 1 to n

$$N = \sum_{i=1}^{n}\left(Q \cdot \varepsilon^\circ(z_i - z_{i-1}) + \tfrac{1}{2}Q \cdot k\left(z_i^2 - z_{i-1}^2\right) - Q \cdot \beta^H \cdot \Delta H(z_i - z_{i-1})\right) \tag{11.38}$$

$$M = \sum_{i=1}^{n}\left(\tfrac{1}{2}Q \cdot \varepsilon^\circ\left(z_i^2 - z_{i-1}^2\right) + \tfrac{1}{3}Q \cdot k\left(z_i^3 - z_{i-1}^3\right) - \tfrac{1}{2}Q \cdot \beta^H \cdot \Delta H\left(z_i^2 - z_{i-1}^2\right)\right) \tag{11.39}$$

This equation can be written compactly as:

$$\begin{bmatrix} N \\ \hline M \end{bmatrix} = \begin{bmatrix} A & B \\ \hline B & D \end{bmatrix}\begin{bmatrix} \varepsilon^\circ \\ \hline \kappa \end{bmatrix} - \begin{bmatrix} N^H \\ \hline M^H \end{bmatrix} \tag{11.40}$$

or

$$\begin{bmatrix} N \\ \hline M \end{bmatrix} + \begin{bmatrix} N^H \\ \hline M^H \end{bmatrix} = \begin{bmatrix} A & B \\ \hline B & D \end{bmatrix}\begin{bmatrix} \varepsilon^\circ \\ \hline \kappa \end{bmatrix} \tag{11.41}$$

where

$$A = \sum_{i=1}^{n} Q_i\left[z_i - z_{i-1}\right] \tag{11.42}$$

$$B = \frac{1}{2}\sum_{i=1}^{n} Q_i\left[z_i^2 - z_{i-1}^2\right] \tag{11.43}$$

$$D = \frac{1}{3}\sum_{i=1}^{n} Q_i\left[z_i^3 - z_{i-1}^3\right] \tag{11.44}$$

The internal forces and moments caused by the hygroexpansion are:

$$N^H = \sum_{i=1}^{n} \Delta H_i \cdot \beta_i^H \cdot Q_i [z_i - z_{i-1}] \qquad (11.45)$$

$$M^H = \frac{1}{2}\sum_{i=1}^{n} \Delta H_i \cdot \beta_i^H \cdot Q_i \left[z_i^2 - z_{i-1}^2\right] \qquad (11.46)$$

Each internal force and moment has three components.

$$\begin{bmatrix} N_1^H \\ N_2^H \\ N_6^H \end{bmatrix} = \text{internal forces per unit width, N/m} \qquad (11.47)$$

$$\begin{bmatrix} M_1^H \\ M_2^H \\ M_6^H \end{bmatrix} = \text{internal moments per unit width, Nm/m} \qquad (11.48)$$

The same applies to external forces N, moments M, strains ε° and curvatures κ.

11.9 Laminate Theory where the Layers are Oriented with an Arbitrary Angle in Relation to the 1-Direction (MD)

When a layer is oriented at a certain angle to the 1-2 system, the expressions above are in principle the same, although more complicated. To be able to make calculations, the following properties must be transformed to the 1-2 system (*off-axis*) from the *x-y* system (*on-axis*).

1. Stresses and strains
2. Stiffnesses Q
3. Hygroexpansions

11.10 Transformation of Stresses and Strains

Transform *x-y* strains (on-axis) to the 1-2 system (**off-axis**). The angle is defined in Figure 11.1.

$$\begin{bmatrix} \varepsilon_1 \\ \varepsilon_2 \\ \tfrac{1}{2}\gamma_{12} \end{bmatrix} = [T]^{-1} \begin{bmatrix} \varepsilon_x \\ \varepsilon_y \\ \tfrac{1}{2}\gamma_{xy} \end{bmatrix} \qquad (11.49)$$

where

$$[T]^{-1} = \begin{bmatrix} c^2 & s^2 & -2cs \\ s^2 & c^2 & 2cs \\ cs & -cs & c^2-s^2 \end{bmatrix} \qquad \begin{cases} c = \cos\phi \\ s = \sin\phi \end{cases} \tag{11.50}$$

Transform stresses to the 1-2 system (*off-axis*)

$$\begin{bmatrix} \sigma_1 \\ \sigma_2 \\ \tau_6 \end{bmatrix} = [T]^{-1} \begin{bmatrix} \sigma_x \\ \sigma_y \\ \tau_{xy} \end{bmatrix} \tag{11.51}$$

11.11 Transformation of the Hygroexpansion Coefficient

The hygroexpansion coefficient in the *x-y* direction (ON-axis) is transformed to the 1-2 direction (OFF-axis) as follows:

$$\begin{bmatrix} \beta_1^H \\ \beta_2^H \\ \tfrac{1}{2}\beta_{12}^H \end{bmatrix} = [T]^{-1} \begin{bmatrix} \beta_x^H \\ \beta_y^H \\ \tfrac{1}{2}\beta_{12}^H \end{bmatrix} \tag{11.52}$$

11.12 Transformation of the Stiffness Matrix, Q

$$\begin{bmatrix} \sigma_1 \\ \sigma_2 \\ \tau_{12} \end{bmatrix} = \begin{bmatrix} Q_{11} & Q_{12} & Q_{16} \\ Q_{12} & Q_{22} & Q_{26} \\ Q_{16} & Q_{26} & Q_{66} \end{bmatrix} \begin{bmatrix} \varepsilon_1 \\ \varepsilon_2 \\ \gamma_{12} \end{bmatrix} \tag{11.53}$$

The stiffness elements are.

$$\begin{bmatrix} Q_{11} \\ Q_{22} \\ Q_{12} \\ Q_{66} \\ Q_{16} \\ Q_{26} \end{bmatrix} = \begin{bmatrix} c^4 & s^4 & 2c^2s^2 & 4c^2s^2 \\ s^4 & c^4 & 2c^2s^2 & 4c^2s^2 \\ c^2s^2 & c^2s^2 & c^4+s^4 & -4c^2s^2 \\ c^2s^2 & c^2s^2 & -2c^2s^2 & (c^2-s^2) \\ c^3s & -cs^3 & cs^3-c^3s & 2(cs^3-c^3s) \\ cs^3 & -c^3s & c^3s-cs^3 & 2(c^3s-cs^3) \end{bmatrix} \begin{bmatrix} Q_{xx} \\ Q_{yy} \\ Q_{xy} \\ Q_{ss} \end{bmatrix} \tag{11.54}$$

11.13 Forces and Moments

The forces and moments become:

$$\begin{bmatrix} N_1 \\ N_2 \\ N_6 \end{bmatrix} + \begin{bmatrix} N_1^H \\ N_2^H \\ N_6^H \end{bmatrix} = \begin{bmatrix} A_{11} & A_{12} & A_{16} \\ A_{12} & A_{22} & A_{26} \\ A_{16} & A_{26} & A_{66} \end{bmatrix} \begin{bmatrix} \varepsilon_1^0 \\ \varepsilon_2^0 \\ \varepsilon_6^0 \end{bmatrix} + \begin{bmatrix} B_{11} & B_{12} & B_{16} \\ B_{12} & B_{22} & B_{26} \\ B_{16} & B_{26} & B_{66} \end{bmatrix} \begin{bmatrix} \kappa_1 \\ \kappa_2 \\ \kappa_6 \end{bmatrix} \qquad (11.55)$$

$$\begin{bmatrix} M_1 \\ M_2 \\ M_6 \end{bmatrix} + \begin{bmatrix} M_1^H \\ M_2^H \\ M_6^H \end{bmatrix} = \begin{bmatrix} B_{11} & B_{12} & B_{16} \\ B_{12} & B_{22} & B_{26} \\ B_{16} & B_{26} & B_{66} \end{bmatrix} \begin{bmatrix} \varepsilon_1^0 \\ \varepsilon_2^0 \\ \varepsilon_6^0 \end{bmatrix} + \begin{bmatrix} D_{11} & D_{12} & D_{16} \\ D_{12} & D_{22} & D_{26} \\ D_{16} & D_{26} & D_{66} \end{bmatrix} \begin{bmatrix} \kappa_1 \\ \kappa_2 \\ \kappa_6 \end{bmatrix} \qquad (11.56)$$

The equation is a more general form of the previous equations, see Equation (11.41)

$$\begin{bmatrix} N \\ \hline M \end{bmatrix} + \begin{bmatrix} N^H \\ \hline M^H \end{bmatrix} = \begin{bmatrix} A & | & B \\ \hline B & | & D \end{bmatrix} \begin{bmatrix} \varepsilon^0 \\ \hline \kappa \end{bmatrix}$$

11.14 Bending Stiffness and Tensile Stiffness

We now study the application of Equation (11.41). It is assumed that there is no hygroexpansion.

$N^H = M^H = 0$.

Equation (11.41) has in principle three different forms. The calculations refer to these three equations. The manner in which the inversion is carried out is not shown here.

A) Strains and curvatures give external forces and external moments

$$\begin{bmatrix} N \\ \hline M \end{bmatrix} = \begin{bmatrix} A & | & B \\ \hline B & | & D \end{bmatrix} \begin{bmatrix} \varepsilon^0 \\ \hline \kappa \end{bmatrix} \qquad (11.57)$$

B) Forces and curvatures give external strains and external moments

$$\begin{bmatrix} \varepsilon^0 \\ \hline M \end{bmatrix} = \begin{bmatrix} A^* & | & B^* \\ \hline C^* & | & D^* \end{bmatrix} \begin{bmatrix} N \\ \hline \kappa \end{bmatrix} \qquad (11.58)$$

C) Forces and moments give external strains and external curvatures

$$\left[\frac{\varepsilon^\circ}{\kappa}\right] = \left[\begin{array}{c|c} A' & B' \\ \hline C' & D' \end{array}\right]\left[\frac{N}{M}\right] \tag{11.59}$$

Equation (11.59) is expanded

$$\begin{bmatrix} \varepsilon_1^0 \\ \varepsilon_2^0 \\ \varepsilon_6^0 \end{bmatrix} = \begin{bmatrix} A'_{11} & A'_{12} & A'_{16} \\ A'_{12} & A'_{22} & A'_{26} \\ A'_{16} & A'_{26} & A'_{66} \end{bmatrix}\begin{bmatrix} N_1 \\ N_2 \\ N_6 \end{bmatrix} + \begin{bmatrix} B'_{11} & B'_{12} & B'_{16} \\ B'_{12} & B'_{22} & B'_{26} \\ B'_{16} & B'_{26} & B'_{66} \end{bmatrix}\begin{bmatrix} M_1 \\ M_2 \\ M_6 \end{bmatrix} \tag{11.60}$$

$$\begin{bmatrix} \kappa_1 \\ \kappa_2 \\ \kappa_6 \end{bmatrix} = \begin{bmatrix} C'_{11} & C'_{12} & C'_{16} \\ C'_{12} & C'_{22} & C'_{26} \\ C'_{16} & C'_{26} & C'_{66} \end{bmatrix}\begin{bmatrix} N_1 \\ N_2 \\ N_6 \end{bmatrix} + \begin{bmatrix} D'_{11} & D'_{12} & D'_{16} \\ D'_{12} & D'_{22} & D'_{26} \\ D'_{16} & D'_{26} & D'_{66} \end{bmatrix}\begin{bmatrix} M_1 \\ M_2 \\ M_6 \end{bmatrix} \tag{11.61}$$

Equation (11.61) is applied to *bending stiffness*. The bending stiffness is defined as the moment per unit curvature, *Figure 11.10*.

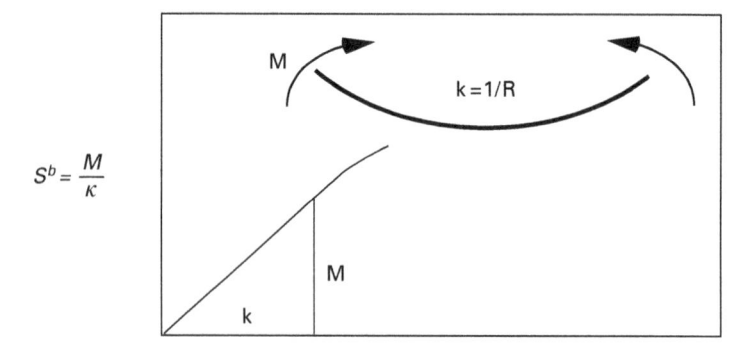

Figure 11.10. The definition of bending stiffness.

We add a moment in the 1-direction. No forces act on the laminate, i.e. $N_1 = N_2 = N_6 = 0$. From Equation (11.61) we obtain:

$$\kappa = D'_{11} \cdot M_1 \tag{11.62}$$

Through the definition of bending stiffness, we obtain the bending stiffness in the 1-direction, 2-direction and 6-direction (torsional stiffness).

$$\begin{cases} S_1^b = \dfrac{M_1}{\kappa_1} = \dfrac{1}{D'_{11}} \\ S_2^b = \dfrac{M_2}{\kappa_2} = \dfrac{1}{D'_{22}} \\ S_6^b = \dfrac{M_6}{\kappa_6} = \dfrac{1}{D'_{66}} \end{cases} \qquad (11.63)$$

Equation (8-48) applied to **tensile stiffness**. Tensile stiffness is defined as force per unit width per unit strain.

$$E^b = \dfrac{N}{\varepsilon^o} \qquad (11.64)$$

We apply a force per unit width N_1. From Equation (11.60) we obtain:

$$\varepsilon_1^o = A'_{11} \cdot N_1 \qquad (11.65)$$

and thus:

$$E_1^b = \dfrac{N_1}{\varepsilon_1^o} = \dfrac{1}{A'_{11}} \qquad (11.66)$$

$$E_2^b = \dfrac{N_2}{\varepsilon_2^o} = \dfrac{1}{A'_{22}} \qquad (11.67)$$

$$E_6^b = \dfrac{N_{61}}{\varepsilon_{16}^o} = \dfrac{1}{A'_{66}} \quad \text{shear stiffness Equation (11.68)}$$

11.15 Simplified Expression for Bending Stiffness

In the uniaxial situation, the bending stiffness can be calculated in the following way.
Consider the Equation (11.41)

$$\begin{bmatrix} N \\ M \end{bmatrix} + \begin{bmatrix} N^H \\ M^H \end{bmatrix} = \begin{bmatrix} A & B \\ \hline B & D \end{bmatrix} \begin{bmatrix} \varepsilon^o \\ \kappa \end{bmatrix}$$

If $N^H = M^H = 0$ there is no moisture change, and

$$N = A \cdot \varepsilon^o + B \cdot \kappa \qquad (11.69)$$

$$M = B \cdot \varepsilon^o + D \cdot \kappa \qquad (11.70)$$

If there is no external force, $N = 0$, so that $\varepsilon°$ is eliminated and we obtain:

$$M = \left[D - \frac{B^2}{A}\right] \cdot \kappa \qquad (11.71)$$

The definition of bending stiffness is the moment, M, per unit curvature, κ.

$$S^b = \frac{M}{\kappa} \qquad (11.72)$$

The final expression for bending stiffness is then:

$$S^b = D - \frac{B^2}{A} \qquad (11.73)$$

11.16 Calculation Example for Bending Stiffness

The following calculation example refers to a four-layer sheet with the properties given in the following table. The starting values are the grammage, w, the tensile stiffness index, E^w, and the density, ρ

Ply k	w g/m^2	$E^w = \frac{E}{\rho}$ MNm/kg	ρ kg/m^3	$E = E^w \cdot \rho$ N/mm^2 (MPa)	$t = \frac{w}{\rho}$ mm
1	80	7.5	800	6000	0.1
2	80	2.5	400	1000	0.2
3	50	4.0	500	2000	0.1
4	70	7.14	700	5000	0.1

The total thickness is calculated as:

$$t = \sum t_k = 0,1 + 0,2 + 0,1 + 0,1 = 0,5 \qquad (11.74)$$

As indicated in Figure 11.2 the Equations (11.2) and (11.3) are:

$$z_0 = -\frac{t}{2}$$

$z_k = z_{k-1} + t_k$ where t_k is the thickness of the layer $k = 1, 2, ..., n$

$$z_0 = -\frac{0.5}{2} = -0.25$$

From Equation (11.3), we obtain:

$z_1 = z_0 + t_1 = -0.25 + 0.1 = -0.15$

$z_2 = 0.05$

$z_3 = 0.15$

$z_4 = 0.25$

The following table gives squares and cubes of the z-coordinates.

Layer k	z_k mm	z_k^2 mm^2	z_k^3 mm^3
0	−0,25	0,0625	−15,625 · 10^{-3}
1	−0,15	0,0225	−3,375 · 10^{-3}
2	0,05	0,0025	0,125 · 10^{-3}
3	0,15	0,0225	3,374 · 10^{-3}
4	0,25	0,0625	15,625 · 10^{-3}

Equations (11.42) – (11.44) are used to calculate A, B and D

Layer k	$Q_k(z_k - z_{k-1}) \approx E_k(z_k - z_{k-1})$	$Q_k(z_k^2 - z_{k-1}^2) \approx E_k(z_k^2 - z_{k-1}^2)$	$Q_k(z_k^3 - z_{k-1}^3) \approx E_k(z_k^3 - z_{k-1}^3)$
1	6000 [−0,15− (−0,25)] = 600	6000 (0,0225−0,0625) = −240	6000 [−3.375−(15,625)] · 10−3 = 73,5
2	1000 [0,05− (−0,15)] = 200	1000 (0,0025−0,0225) = −20	1000 [0.125− (−3,375)] · 10−3 = 3,5
3	2000 [0,15−0,05] = 200	2000 (0,0225−0,0025)= 40	2000 [3,375−0.125] · 10−3 = 6,5
4	5000 [0,25−0,15] = 500	5000 (0,0625−0,0225) = 200	5000 [15,625−3.375] · 10−3 = 61,25
	$A = \Sigma = 1500$	$B = \frac{1}{2}\Sigma = -10$	$D = \frac{1}{3}\Sigma = 48,25$

The bending stiffness is calculated according to Equation (11.73).

$$S^b = 48,25 - \frac{(-10)^2}{1500} = 48,18 \quad \text{mNm}$$

11.17 Curl and Twist, Strain, Shear, as a Result of a Change in Moisture

In this case there are no external forces applied, but the hygroexpansion causes internal forces and moments. These can be very large. It is difficult to prevent expansion as a result of moisture absorption.

$N = M = 0$

We invert the matrices in the same way as equations (11.59) – (11.61). Internal forces and moments give strains and curvatures

$$\begin{bmatrix} \varepsilon_1^0 \\ \varepsilon_2^0 \\ \varepsilon_6^0 \end{bmatrix} = \begin{bmatrix} A'_{11} & A'_{12} & A'_{16} \\ A'_{12} & A'_{22} & A'_{26} \\ A'_{16} & A'_{26} & A'_{66} \end{bmatrix} \begin{bmatrix} N_1^H \\ N_2^H \\ N_6^H \end{bmatrix} + \begin{bmatrix} B'_{11} & B'_{12} & B'_{16} \\ B'_{12} & B'_{22} & B'_{26} \\ B'_{16} & B'_{26} & B'_{66} \end{bmatrix} \begin{bmatrix} M_1^H \\ M_2^H \\ M_6^H \end{bmatrix} \qquad (11.75)$$

$$\begin{bmatrix} \kappa_1 \\ \kappa_2 \\ \kappa_6 \end{bmatrix} = \begin{bmatrix} C'_{11} & C'_{12} & C'_{16} \\ C'_{12} & C'_{22} & C'_{26} \\ C'_{16} & C'_{26} & C'_{66} \end{bmatrix} \begin{bmatrix} N_1^H \\ N_2^H \\ N_6^H \end{bmatrix} + \begin{bmatrix} D'_{11} & D'_{12} & D'_{16} \\ D'_{12} & D'_{22} & D'_{26} \\ D'_{16} & D'_{26} & D'_{66} \end{bmatrix} \begin{bmatrix} M_1^H \\ M_2^H \\ M_6^H \end{bmatrix} \qquad (11.76)$$

These can be represented in compact form as:

$$\begin{bmatrix} \varepsilon^0 \\ \hline \kappa \end{bmatrix} = \begin{bmatrix} A' & | & B' \\ \hline C' & | & D' \end{bmatrix} \begin{bmatrix} N^H \\ \hline M^H \end{bmatrix} \qquad (11.77)$$

We can then extract the component, which we need from the equation.

11.18 Curl and Twist

With the help of Equation (11.76) **Curl and Twist** due to a moisture change can also be obtained. We can extract the component, which we need from the equation.

$$\begin{aligned}[\kappa_1] = [C'_{11}][N_1^H] + [C'_{12}][N_2^H] + [C'_{16}][N_6^H] + \\ [D'_{11}][M_1^H] + [D'_{12}][M_2^H] + [D'_{16}][M_6^H] \end{aligned} \qquad (11.78)$$

$$\begin{aligned}[\kappa_2] = [C'_{12}][N_1^H] + [C'_{22}][N_2^H] + [C'_{26}][N_6^H] + \\ [D'_{12}][M_1^H] + [D'_{22}][M_2^H] + [D'_{26}][M_6^H] \end{aligned}, \qquad (11.79)$$

$$\begin{aligned}[\kappa_6] = [C'_{16}][N_1^H] + [C'_{26}][N_2^H] + [C'_{66}][N_6^H] + \\ [D'_{16}][M_1^H] + [D'_{26}][M_2^H] + [D'_{66}][M_6^H] \end{aligned} \qquad (11.80)$$

11.19 Leaning Stacks of Folded Printing Papers

Folded printed papers are stacked in bundles. If the layers have an oblique symmetry and if the moisture content changes, each layer will shear slightly in relation to the 1-direction, MD. The shearing is then accumulated in the bundle and this develops a propeller-like appearance, as shown in *Figure 11.11*.

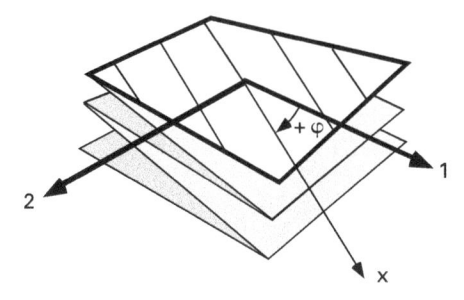

Figure 11.11. Stack lean.

With the help of Equation (11.75) we obtain *the shear* due to moisture.

$$\left[\varepsilon_6^0\right] = \left[A'_{16}\right]\left[N_1^H\right] + \left[A'_{26}\right]\left[N_2^H\right] + \left[A'_{66}\right]\left[N_6^H\right] + \\ \left[B'_{16}\right]\left[M_1^H\right] + \left[B'_{26}\right]\left[M_2^H\right] + \left[B'_{66}\right]\left[M_6^H\right]$$

(11.81)

11.20 Tensile Stiffness and Bending Stiffness Studied in a Polar Diagram

By coordinate transformation it is possible to obtain plots of bending stiffness and tensile stiffness in polar diagrams. The results may look as follows. By tensile stiffness orientation we mean the angle at which there is symmetry, *Figure 11.12*.

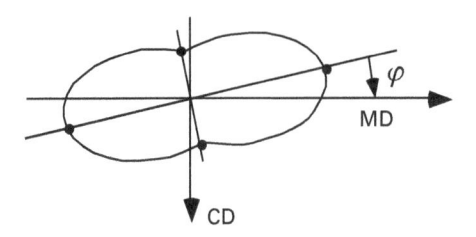

Figure 11.12. Tensile stiffness and bending stiffness in a polar diagram.

11.21 Hygroexpansion, Curl and Twist Studied in Polar Diagrams

By coordinate transformation it is possible to obtain plots of hygroexpansion, curl and twist in polar diagrams. The results may look as follows. By hygroexpansion orientation we mean the angle at which there is symmetry, *Figure 11.13*.

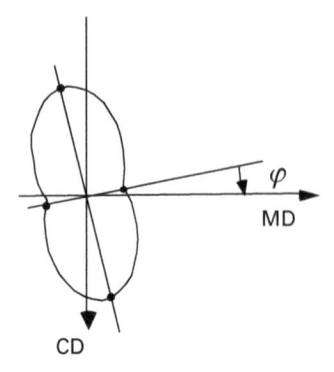

Figure 11.13. Hygroexpansion, curl and twist in a polar diagram.

11.22 Hygroexpansion Orientation and Tensile Stiffness Orientation

The tensile stiffness in different directions in the laminate can be measured using ultrasonic technique. Measurement of the hygroexpansion orientation, can be obtained with speckle photography. *Figure 11.14* shows the tensile stiffness orientation and hygroexpansion orientation for a copy paper at different distances from the edge. The two orientations seem to correspond well.

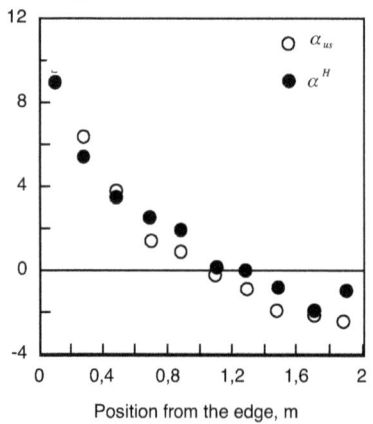

Figure 11.14. Tensile stiffness orientation and hygroexpansion orientation for a copy paper at different distances from the edge.

11.23 Shear-free Angle for Hygroexpansion

The hygroexpansion orientation can be calculated analytically, i.e. the shear-free angle for strain due to hygroexpansion
Transform 1-2 strains to the 1'-2' systems.

$$\begin{matrix} \text{unknown} & & \text{known} \end{matrix}$$

$$\begin{bmatrix} \varepsilon_1' \\ \varepsilon_2' \\ \tfrac{1}{2}\varepsilon_6' \end{bmatrix} = [T]^{-1} \begin{bmatrix} \varepsilon_1 \\ \varepsilon_2 \\ \tfrac{1}{2}\varepsilon_6 \end{bmatrix} \qquad (11.82)$$

where

$$[T]^{-1} = \begin{bmatrix} c^2 & s^2 & -2cs \\ s^2 & c^2 & 2cs \\ cs & -cs & c^2-s^2 \end{bmatrix} \qquad \begin{cases} c = \cos\phi \\ s = \sin\phi \end{cases} \qquad \text{see (11.50)}$$

We know the properties in the 1-2 system and require the properties in the 1'-2' system. The positive angle goes from the unknown to the known.

$$\begin{bmatrix} \varepsilon_1'^{\circ} \\ \varepsilon_2'^{\circ} \\ \tfrac{1}{2}\varepsilon_6'^{\circ} \end{bmatrix} = [T]^{-1} \begin{bmatrix} \varepsilon_1^{\circ} \\ \varepsilon_2^{\circ} \\ \tfrac{1}{2}\varepsilon_6^{\circ} \end{bmatrix} \qquad (11.83)$$

The shear-free angle is given by $\varepsilon_6^{\circ} = 0$ The angle of symmetry then becomes

$$\phi = -\frac{1}{2}\tan^{-1}\left(\frac{\varepsilon_6^{\circ}}{\varepsilon_1^{\circ} - \varepsilon_2^{\circ}}\right) \qquad (11.84)$$

In the case of hygroexpansion, it is desirable to indicate angles from the 1-axis to the symmetry axis. The hygroexpansion angle is defined as $\alpha H = -\varphi$

11.24 Twist Free Angle

Figure 11.15 shows a paper with curl in both MD and CD, and twist. The task is to find the twist-free angle, i.e. the curl orientation $\phi_{curl} \cdot \kappa_6' = 0$. **The Curl angle**, i.e. the twist-free angle for strain due to hygroexpansion, can be calculated.

$$\begin{bmatrix} \kappa_1' \\ \kappa_2' \\ \tfrac{1}{2}\kappa_6' \end{bmatrix} = [T]^{-1} \begin{bmatrix} \kappa_1 \\ \kappa_2 \\ \tfrac{1}{2}\kappa_6 \end{bmatrix} \tag{11.85}$$

$$[T]^{-1} = \begin{bmatrix} c^2 & s^2 & -2cs \\ s^2 & c^2 & 2cs \\ cs & -cs & c^2 - s^2 \end{bmatrix} \tag{11.86}$$

Equation (11.85) gives ($\kappa_2' = \kappa_\varphi$):

$$\kappa_\phi = \kappa_1 \cdot \sin^2 \phi + \kappa_2 \cdot \cos^2 \phi + \kappa_6 \cdot \sin\phi \cdot \cos\phi \tag{11.87}$$

$$\phi_{twist} = -\frac{1}{2} \cdot \tan^{-1}\left(\frac{\kappa_6}{\kappa_1 - \kappa_2}\right) \tag{11.88}$$

The curl orientation is indicated by the angle to axis 1. This gives the same sign of the angle as κ_6 = the twist, according to earlier definitions.

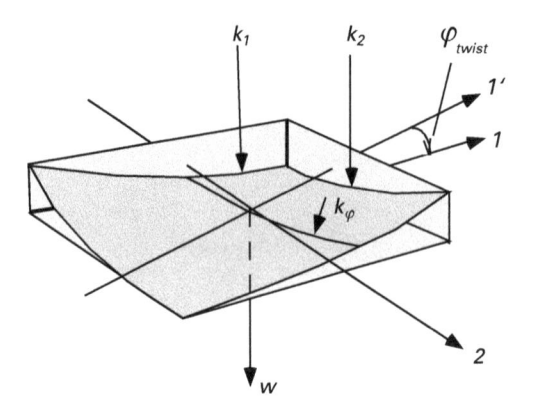

Figure 11.15. Twist free angle.

We know the curl orientation φ_{curl} and the curl $\kappa\varphi$ according to the above and wish to know the curvatures in the 1-2 directions. Equation (11.87) gives:

$$(\kappa_2 = \kappa_\phi) \tag{11.89}$$

The angle is defined as being positive from the unknown to the known, as shown in *Figure 11.16*.

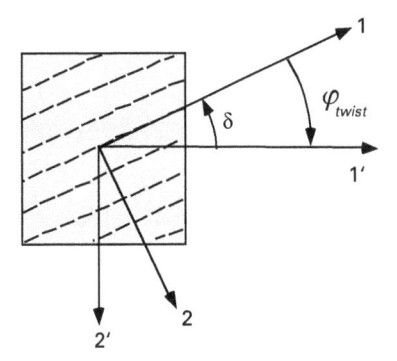

Figure 11.16. The angle is defined as being positive from the unknown to the known.

Insert the angle $\delta = -\varphi_{curl}$

$k_1' = k_\varphi \sin^2 \delta$
$k_2' = k_\varphi \cos^2 \delta$
$k_6' = -2 k_\varphi \sin \delta \cos \delta$

11.25 Finding the Height at a Point on a Bent Paper

The task is to find the height at a point on a bent paper, *Figure 11.17*.

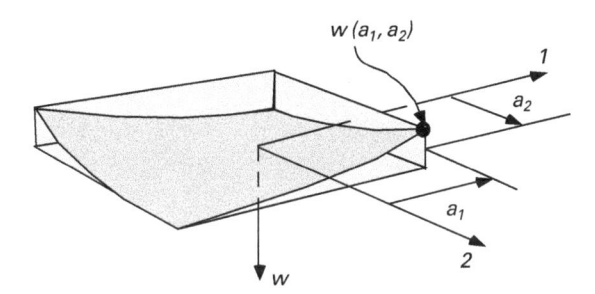

Figure 11.17. Finding the height at a point on a bent paper.

$$w = -\frac{1}{2}\left(a_1^2 \cdot k_1 + a_2^2 \cdot k_2 + a_1 a_2 k_{12}\right) \tag{11.90}$$

where
a_1 and a_2 are coordinates, (m)
$\kappa_1, \kappa_2, \kappa_{12}$ are curl and twist values, (m^{-1})

11.26 Stress and Strains in the Thickness Direction of the Laminate due to Hygroexpansion

For each interface in the laminate, the strains may be calculated according to Equation (11.31), and the stresses according to Equation (11.33). An example of a strain distribution is given in *Figure 11.18*.

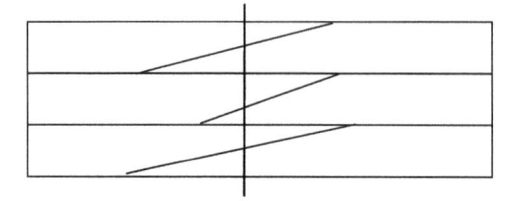

Figure 11.18. Stress and strains in the thickness direction of the laminate due to hygroexpansion.

12 Mathematical Modelling and Analysis of Converting and Enduse

Mikael Nygårds
Innventia AB
Sören Östlund
Royal Institute of Technology, KTH

12.1 Introduction 315

12.2 The Mechanical Modelling Process 317

12.3 Paper from a Mechanical Point of View 320

12.3.1 Length Scales 320
12.3.2 Continuum Models 321
12.3.3 Network Models 324

12.4 The Finite Element Method 325

12.4.1 Displacements 325
12.4.2 Stresses 327
12.4.3 Matrix Formulation 327
12.4.4 Principle of Virtual Work 328

12.5 Example: Creasing of Paperboard 329

12.5.1 Deformation Mechanisms 329
12.5.2 Model Generation 330
12.5.3 In-plane Model 330
12.5.4 Delamination Model 331
12.5.5 Modelling 333

12.6 References 333

12.1 Introduction

Numerical methods are becoming increasingly used to study the behaviour of materials and structures. For modelling of solids the finite element method (FEM) is the most popular approach. Among the reasons to do numerical modelling some can be mentioned:

- Different types of loadings can be investigated.
- The effect of different material properties can be investigated.
- More information about damage and deformation mechanisms can be gained.
- Material behaviour can be predicted.

- Properties that are important for the manufacturer can be linked to properties that are important for the user.

If the investigations above would be performed by experiments it would be expensive and time consuming, since trial materials and structures must be manufactured and tested. In a numerical model the parameters only need to be changed and a new simulation can be started. This can be used to better understand the problems before an experimental study is conducted. Then it becomes possible to learn what kind of experiments that will be of interest.

Ideally numerical simulations should be used together with experiments to learn more about the materials. Experimental investigations and visual inspections make it possible to detect the macroscopic behaviour of the materials. Numerical methods also enable studies of the macroscopic behaviour under different types of loading, but in addition the method also offers the possibility to study field properties such as stresses and strains within the material.

In the past experimental studies has been conducted to learn more about material behaviour. There it was shown how different properties such as elastic, plastic, viscoplastic etc can be measured by different test methods. In all methods presented a property is measured, and it is possible to predict how the material behaves under this particular loading. But if the loading condition change, it is not obvious how the mechanical response change. Therefore, a model is needed. The knowledge gained from an experimental investigations can been used to formulate theories that predict the material properties. It is however difficult to find analytical solutions to predict material responses if the boundary conditions are different from the experimental setup. With aid of numerical methods it is possible to solve systems of equations that are needed to do better and more complicated predictions than was possible earlier.

In the finite element method field properties are studied. The field properties of interest for paper are primarily;

- Stress,
- Strain,
- Temperature,
- Moisture.

Interpretations of these fields must be done in order to understand the problem solution. The next task (which probably is the most difficult one) is to understand how process and manufacturing parameters together with different loading conditions interact and affect the paper to create the field properties.

As an example consider the behaviour during converting and end-use. It is the process and manufacturing parameters that creates the actual material, but in order to fully understand how deformation and damage evolve knowledge of the underlying field properties, such as stress and strain, is required. Another example is curl. Curl is created during in the papermaking process since an unsymmetrical paper or paperboard is created. By changing process parameters it is possible to redesign the material in the thickness direction, or change the drying strategy etc. This will certainly change the curl behaviour of the material. It is however difficult to predict the change in curl a priori if there is no knowledge of how these process parameter changes affects the underlying stress field. It is an unbalanced stress field in the material that causes the out-of-plane deflection, and this can not easily be controlled directly by process parameters. Therefore it is important to gain knowledge about field properties in order to explain material behaviour.

12.2 The Mechanical Modelling Process

The aim with modelling is generally to create and simulate a condition that mimics reality. However, assumptions need to be made, because modelling needs to focus on certain aspects that are dominating in the analysis. If no assumptions would be made the models would become very large, and thus computer demanding to solve. Moreover, if modelling focus on important aspects it is easier to evaluate and understand the results. In general three types of assumptions can be made.

- **Geometry:** A complicated structure is normally idealized to a computable structure. The main features of the geometry need to be represented, but details such as e.g. surface roughness are normally not considered, basically because it does not influence the overall mechanical behaviour. The primary aspect that affects the phenomena of interest does however need to be incorporated.
- **Material behaviour:** Assumptions about material behaviour is always needed, because there are hardly any material models that exactly resemble the behaviour of real materials under all possible loading conditions. Assumption like this does however lead to satisfactory results in most cases.
- **Boundary conditions:** In many applications the load is not known or even measurable; it is then very difficult to get accurate boundary conditions for the model analysis. In other cases it is possible to model the total load on a structure, but it is difficult to exactly know the true distribution of the load, while the opposite is true in other situations.

Thus, what is created is a model of the real structure. It is important to be aware of the assumptions you do since that should influence the way you evaluate the results. In certain cases a coarse model is sufficient to detect principal behaviours of the structure, while other applications require large models with many details. Evaluation of what kind of model that is need for different applications require experience.

There are certain steps that are needed in the formulation of a mathematical model; a schematic description of the model generation is shown in *Figure 12.1*. To formulate a mechanical model you need to use the laws of mechanics directly or indirectly. It should be emphasised that it is the model properties that are evaluated when the problem is solved not the real behaviour. The interpretation of the model results in terms of the real physical behaviour has to be done by the user. If the criteria used to formulate the model are close to reality the interpretation of model result are straightforward. However, normally simplifying assumptions have been made. They are motivated by reductions in both modelling time and analysis (computer) time. Therefore, the accuracy of a model is often a tradeoff between available resources and time versus the complexity of the model. It is however important to be aware of how the response would change if the model had larger complexity.

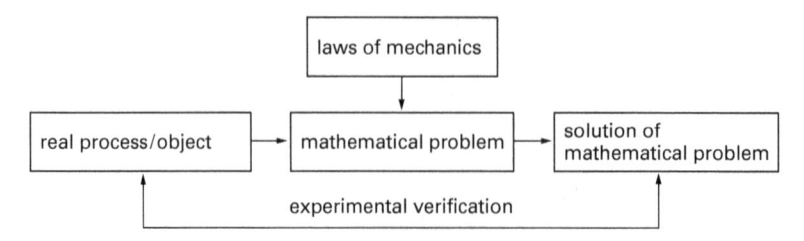

Figure 12.1. A schematic representation of the mechanical modelling process.

When the mathematical problem following from the modelling has been formulated it needs to be solved. Within mechanical modelling the resulting mathematical models often consist of partial differential equations (or in simpler cases ordinary differential equations) and corresponding boundary conditions. It is in general difficult to solve these differential equations analytically. Therefore efficient numerical methods that yield the result with engineering accuracy have been developed.

One of the most important steps in the analysis is to check if the model behaviour resembles that of the real phenomenon of interest. This needs to be done by experimental verification. If the comparison is not satisfactory the mathematical model needs to be reformulated, since important features have been left out of the model. This verification step is many times performed on a simplified problem because real problems might be too complex to study in detail experimentally.

Example

To illustrate the modelling generation concept consider a 10mm wide paper strip that is 100 mm long, e.g. cut from this page. Load this strip by a 1.0 kg weight. How much does the paper strip elongate due to this load?

If the paper has a thickness of 0.1mm, then stress in the paper strip can be calculated from

$$\sigma = \frac{F}{A} = \frac{1.0 \times 9.8}{10 \times 10^{-3} \times 0.1 \times 10^{-3}} N/m^2 = 10 \ MPa,$$

In this regime copy paper is in general elastic, with a possible elastic modulus of E = 3 GPa. Thus, Hooke's law can be used to approximate the elongation,

$$\Delta L = L \frac{\sigma}{E} = 0.1 \times \frac{10 \times 10^6}{3 \times 10^9} m = 0.3 mm.$$

In this example is it sufficient to consider paper as a continuum. This approach gives the total elongation that was needed in the problem. If more details were needed in the analysis it would have been incorporated in the model. For example, the fibrous structure does create locally high stresses and non-homogeneous strain fields. Moreover, it is assumed that the load due to the weight is evenly distributed. Thus, even though approximations are done the problem has been solved satisfactory. The elongation of the strip could be calculated without detailed knowledge of the fibrous structure and load transfer between fibres.

The level of complexity in a model should be different depending on what kind of modelling that are performed. A normal modelling approach in the industry is to incorporate more details if the scale is small. As an example of this approach it is possible to look at the different kind of length scales that are of interest in modelling of a diary product box, as illustrated in *Figure 12.2*. First, by experience of how the product behaves in normal use, it is possible to specify the loads that affect the box, i.e. the boundary conditions representing loads due to for example stacking or gripping. A box model can then be formulated to investigate how the model behaves under the specified load. From the results of this analysis it is possible to extract the loading

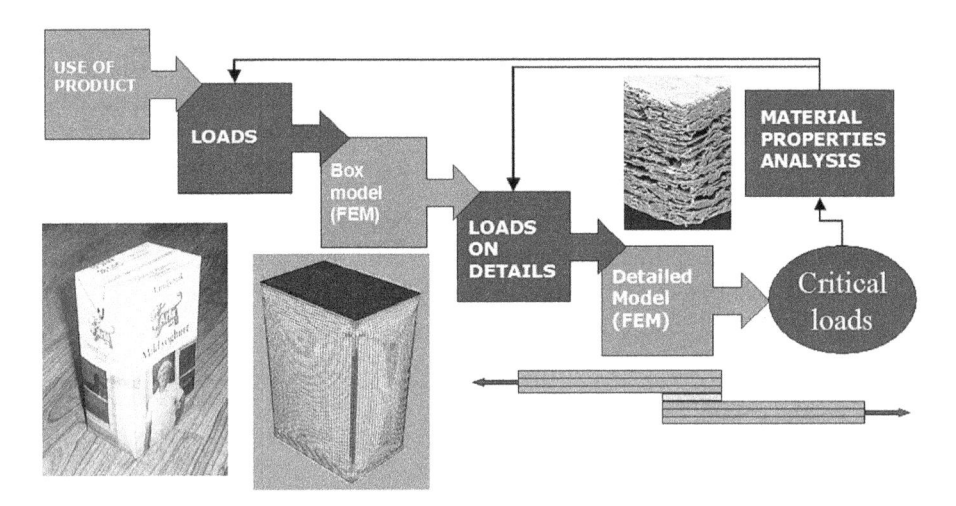

Figure 12.2. Modelling of different scales in a diary product box. Both the box model and studies of the creases can give information about critical material behaviour.

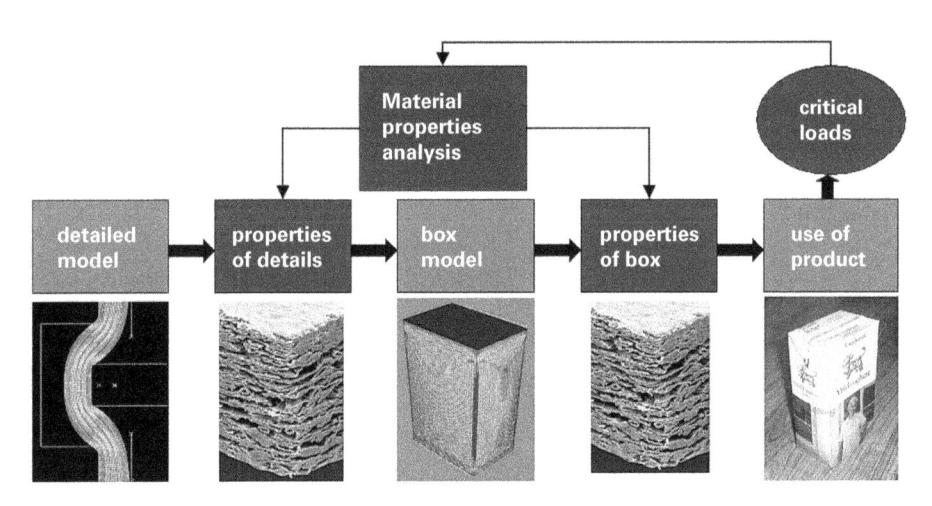

Figure 12.3. Modelling of different scales in a diary product box. Both the detailed model and the box model can give information about critical material behaviour for the product.

conditions at a corner of the box. On both the detailed scale and the box model it is possible to evaluate how the assumed material behaves, and from the analyses can critical loads be extracted where risk for failure arise. Thus, the material can be improved if the weakest areas are identified. Moreover, the same approach can be used in the opposite direction. If the loads on certain details are known a priori, then it is possible to calculate the criteria need needed on the product, which is schematically shown in *Figure 12.3*.

12.3 Paper from a Mechanical Point of View

Due to the complex structure of paper there are problems associated with modelling of paper. The paper fibres form a two-dimensional network that is loosely bonded in the out-of-plane direction. Therefore, the properties in the three principal directions MD (machine direction), CD (cross machine direction) and ZD (out-of-plane direction) differ considerably, as seen in *Figure 12.4*. This means that the paper has an anisotropic behaviour. The stiffest direction is MD, since this is the preferred direction of the fibres. Generally for machine made paper, the properties in the MD and CD direction can differ up to about a factor 5, while the stiffness in ZD on the other hand can be a factor 100 less than in the MD.

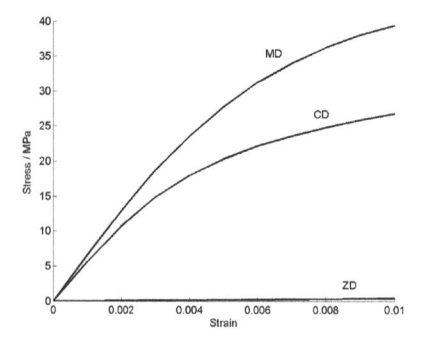

Figure 12.4. Tensile properties of paperboard in MD, CD and ZD.

12.3.1 Length Scales

Figure 12.5 shows examples of the different length scales that influence the mechanical behaviour of a paper product such as a box. Those are: box, web, fibre network and fibre with fibrils. In all constitutive modelling an appropriate length scale must be chosen. In modelling of mechanical behaviour of paper a box can be used to illustrate the end use products. Modelling of these require knowledge about the design as well as the material properties. On web or sheet scale there are continuum models that are used to model the material. On the fibre network scale details in the structure can be studied. Then appropriate models for the fibres and bonds must be known as input in the models. Very few models on the fibre level and below exist to predict mechanical behaviour. The interested reader is referred to the PhD- thesis by Persson (2000).

box　　　　　web　　　　fibre network　　fibre fibrils

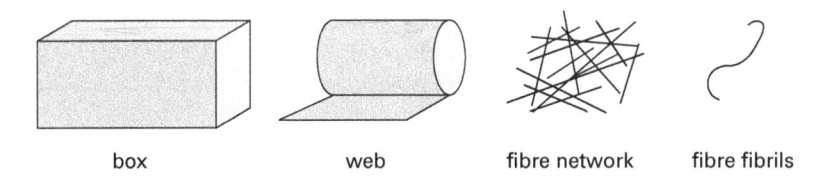

Figure 12.5. Mechanical properties of paper are present at different length scales in paper.

12.3.2 Continuum Models

Continuum modelling of paper is the most commonly used approach. Basically because in most applications it is the sheet properties of paper that are important, and not the properties of the individual fibres in the network of connected fibres. It is therefore sufficient to study how paper behaves as a continuum. In the continuum approach material properties of the paper are used as input in the models. Thus, experiments are conducted on paper specimens to determine, e.g. elastic and plastic properties.

Materials models for continuum mechanics modelling can in general be divided into four types of models, linear rate-independent models, linear rate-dependent models, non-linear rate-independent models and non-linear rate-dependent models.

In the rate dependent models stress and strain are functions of time, while they are not in rate independent models. In some applications time is an important aspect of the mechanical properties. While paper in room temperature is normally assumed to be rate independent, the rate-dependence might play an important role at increased moisture and temperature. If rate dependent models are used, the paper needs to be characterized with respect to time, i.e. creep (constant stress) or relaxation (constant strain) experiments need to be performed to gather the time dependence.

The concepts of linearity and non-linearity here refer to deformation as a function of load. In linear models the strain response is a linear function of stress, and an increase of the stress by a factor of two will result in a doubling of the strain, while this is not the case for non-linear models.

Example

Simulation of web tension in a printing press where different material models have been used, as shown in Figure 12.6. The material models used are from top to bottom: elastic, viscoelastic and viscoelastic with hygroexpansion models.

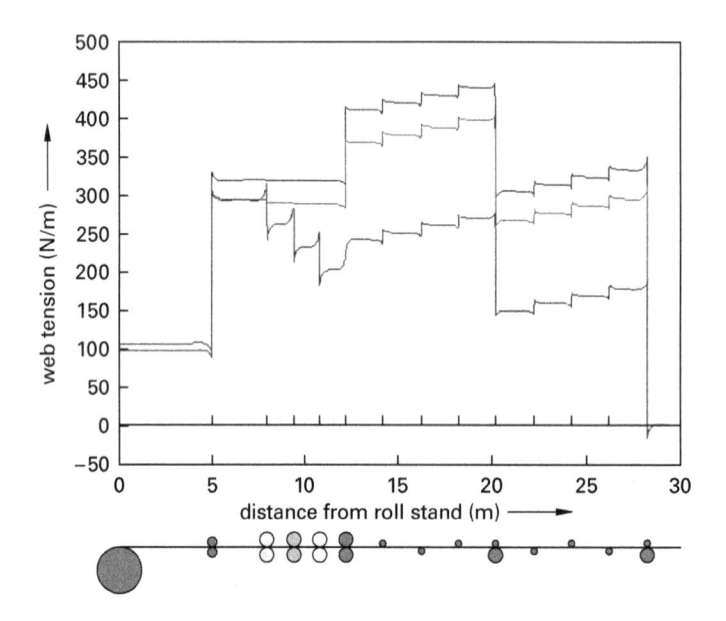

Figure 12.6. The web tension in a printing press simulated with three different material models. From top to bottom: elastic, viscoelastic and viscoelastic with hygroexpansion.

12.3.1.1 Elastic-plastic Behaviour

This section will highlight the elastic-plastic behaviour of paper, and will show the necessary equations to formulate a theory. In a continuum model constitutive equations must be formulated. In *Figure 12.4* above it can be observed that the initial part of the stress-strain curves can be approximated as linear. If the paper in this regime upon unloading returns to the initial configuration the material is said to be linear elastic. As seen in *Figure 12.4* the initial slope of the MD, CD and ZD curves differ considerably. This causes a problem when we later introduce the finite element method. It is generally difficult to solve numerical models that have large differences in the elastic constants in different principal directions. Therefore, problem can arise if constitutive equations developed for other material classes are used, such as metals.

Since paper behaves differently in the three principal directions it is said to be orthotropic. In the elastic regime the stresses and strain can be expressed with aid of 12 material constants. The components of the symmetric strain tensor are in a vectorial notation related to the components of the symmetric stress tensors as

$$\begin{bmatrix} \varepsilon_{MD} \\ \varepsilon_{CD} \\ \varepsilon_{ZD} \\ \gamma_{MD-CD} \\ \gamma_{CD-ZD} \\ \gamma_{MD-ZD} \end{bmatrix} = \begin{bmatrix} \dfrac{1}{E_{MD}} & -\dfrac{v_{CD-MD}}{E_{CD}} & -\dfrac{v_{ZD-MD}}{E_{ZD}} & 0 & 0 & 0 \\ -\dfrac{v_{MD-CD}}{E_{MD}} & \dfrac{1}{E_{CD}} & -\dfrac{v_{ZD-CD}}{E_{ZD}} & 0 & 0 & 0 \\ -\dfrac{v_{MD-ZD}}{E_{MD}} & -\dfrac{v_{CD-ZD}}{E_{CD}} & \dfrac{1}{E_{ZD}} & 0 & 0 & 0 \\ 0 & 0 & 0 & \dfrac{1}{G_{MD-CD}} & 0 & 0 \\ 0 & 0 & 0 & 0 & \dfrac{1}{G_{CD-ZD}} & 0 \\ 0 & 0 & 0 & 0 & 0 & \dfrac{1}{G_{ZD-MD}} \end{bmatrix} \begin{bmatrix} \sigma_{MD} \\ \sigma_{CD} \\ \sigma_{ZD} \\ \sigma_{MD-CD} \\ \sigma_{CD-ZD} \\ \sigma_{ZD-MD} \end{bmatrix} .(12.1)$$

The physical interpretation of components of the compliance matrix in Equation (12.1) follows in a straight-forward way by considering the strain response to each of the stress components. From any given stress state it is then possible to calculate the corresponding strains. Moreover, the equation above can be inverted if it is desired to calculate the stresses from a given strain state. Thermodynamic considerations require that the compliance matrix should be symmetric. Thus, Equation (12.1) can be complemented with three symmetry conditions

$$\frac{v_{MD-CD}}{E_{MD}} = \frac{v_{CD-MD}}{E_{CD}}, \quad \frac{v_{ZD-MD}}{E_{ZD}} = \frac{v_{MD-ZD}}{E_{MD}}, \quad \frac{v_{ZD-CD}}{E_{ZD}} = \frac{v_{CD-ZD}}{E_{CD}} \tag{12.2}$$

that reduces the number of elastic constants for an orthotropic linear elastic material to nine.

At some stress level the relations between stress and strain will no longer be linear, and the material starts to deform non-linearly. This deformation is usually described as elastic-plastic, and is not reversible. If a loaded specimen that has been deformed plastically is unloaded, a permanent deformation will remain upon unloading. For uniaxial loading plastic deformation (or yielding) is initiated at a stress level, σ_Y, known as the yield stress. For multiaxial loading the stress state that initiates plastic deformation will in general be a function of all components of the stress tensor. This function can be looked upon as a surface in stress space called the yield surface. If the current stress state is inside the surface the deformation is purely elastic. This yield surface needs to be defined. During the elastic-plastic deformation the stress components always fulfil this function, i.e. the stress state is on the yield surface. The yield surface is written on the form

$$f(\sigma_{MD}, \sigma_{CD}, \sigma_{ZD}, \tau_{MD-CD}, \tau_{MD-ZD}, \tau_{CD-ZD}) = 0, \tag{12.3}$$

where σ_i the normal are stress components in the normal direction and τ_i are the shear stress components. The yield surface is used to evaluate if yielding occurs, independently of loading conditions. A yield criterion that is commonly used for paper is the Tsai-Wu criteria. The criterion is anisotropic, and distinguishes tensile and compressive loads. The Tsai-Wu criterion is expresses as

$$f = D_1\sigma_{MD}^2 + D_2\sigma_{CD}^2 + D_3\sigma_{zD}^2 - D_4\sigma_{MD}\sigma_{CD} - D_5\sigma_{MD}\sigma_{ZD} - D_6\sigma_{CD}\sigma_{ZD} +$$
$$D_7\tau_{MD-CD} + D_8\tau_{MD-ZD} + D_9\tau_{CD-ZD} + D_{10}\sigma_{MD} + D_{11}\sigma_{CD} + D_{12}\sigma_{ZD} - 1 = 0, \quad (12.4)$$

where the constants D_i are functions of the yield stresses in the different loading directions.

After initial yielding the material deforms elastic-plastically, i.e. the deformation is build-up of both elastic reversible and inelastic irreversible deformation. The stress-strain response is then not as stiff as in the elastic regime, but will still show a hardening behaviour, i.e. stress increases as strain increase. Hardening functions that depend on the plastic strain, ε^p captures this behaviour. Thus each stress component is expresses as

$$\sigma_i = \sigma_i^s + f(\varepsilon^p), \quad (12.5)$$

where σ_i is either the normal stress components of the shear components, and σ_i^s is the yield stress in the corresponding direction.

12.3.3 Network Models

Network models can be used to study certain mechanisms in paper materials in more detail, an example of a network model by Heyden (2000) is shown in *Figure 12.7*. In this approach a fibre network is constructed by distribution of fibres, fines and other additives. Fibres that come in contact will bond to each other, which require a material model for the bonding. Moreover, material data on the fibre level also need to be submitted. There are two challenging tasks with network modelling. First, one can predict paper properties, i.e. continuum properties from properties at the fibre level, by assigning material data for the fibres and bonds. Second, mechanisms on the fibre level can be studied in detail; this can include mechanisms such as fibre bridging at delaminations, creep, fibre deformation etc. It is often the available computer resources that

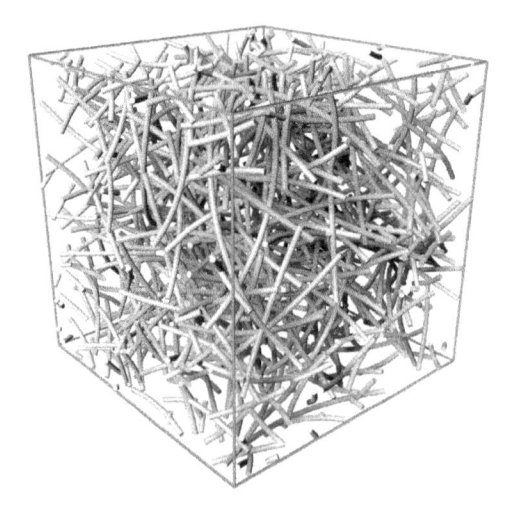

Figure 12.7. Example of three-dimensional network model, (Heyden, 2000).

limit the type of problem that is possible to study with network models. Since in general many fibre are incorporated in network models, in order to simulate a real paper, the models easily become computer demanding.

12.4 The Finite Element Method

The finite element method is a mathematical tool for solving ordinary and partial differential equations that frequently appear the modelling of continuum mechanics problems. There are several commercially available finite element programs. One of the problems with those is that there are normally no predefined models for paper material available in these programs. However, commercial programs in general have an option to define your own material models, although this is not an easy task and requires large experience. Several aspects should be considered when defining your own material, as discussed below.

The finite element method is based on the principle of virtual work. In the finite element method boundary conditions are used to constrain the model. These can be tractions, t^r, or displacements u on the outer surface, or body forces f or corresponding displacement fields u acting on the whole body. The energy given to the model by the boundary conditions need to be transformed to internal stresses, σ, and strains ε. An energy balance can then be formulated through the principle of virtual work

$$\int_{V_0} \sigma : \varepsilon \, dV_0 = \int_S t^T \cdot u \, dS + \int_V f^T \cdot u \, dV \;, \tag{12.6}$$

where the integrals indicate that the model as a whole need to fulfil the energy balance. Thus note that the formulation is weak since it does not require energy balance in each point separately, as long as the whole model fulfils energy balance.

In the finite element formulation a displacement field is proposed, from that information can the corresponding strain field be calculated, and thereafter the corresponding stress field is calculated. It can be shown that this will lead to an upper limit of energy needed, i.e. it is not an exact solution since the displacement field is assumed. But it will be a conservative solution. If the exact strain field is known, the solution will be exact. In general, an accurate strain field gives a better stress field.

12.4.1 Displacements

To be able to do an initial guess on the displacement field, space is divided into a mesh, i.e. discrete elements with nodes along their boundaries are created. The nodes are the points where neighbouring elements are connected to each other. In *Figure 12.8* are some common elements shown. Different types of elements are used depending on the dimensionality and geometry of the problem.

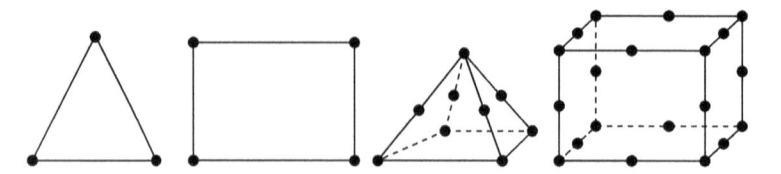

Figure 12.8. Common two and three-dimensional finite elements.

When the geometry is deformed the nodes in each element will move, as illustrated in *Figure 12.9*, where a two-dimensional continuum element with four nodes is shown.

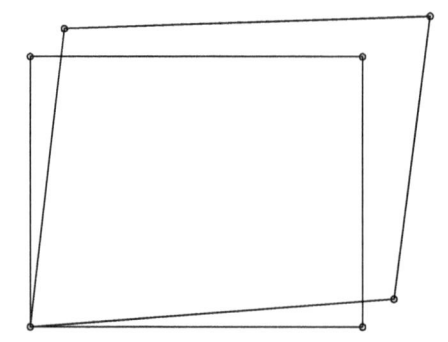

Figure 12.9. A four-noded continuum element. When deformation progress in a model, the nodes in the element move.

From the boundary conditions it is possible to assume how each node move. By assuming that the displacement between the nodes can be described by piecewise polynomial approximations of displacement fields it is possible to get the displacement of the whole element. In *Figure 12.9* it is shown how an initial square element deforms, since three of its nodes get new positions. In the figure is the deformation of the element edges assumed to be linear. It is then possible to calculate a displacement of an arbitrary point within the element with aid of the displacements of the corner nodes u^N. This is done by defining a tensor N_N, that contain interpolation functions that depend on some material coordinate system,

$$u = N_N u^N, \qquad (12.7)$$

where u is the incremental displacement at an arbitrary point in the element. In the same manner strain can be calculated from the displacements of the corner nodes. Hence,

$$\varepsilon = \beta_N u^N \qquad (12.8)$$

where β_N is a matrix that depends on position of the point being considered.

12.4.2 Stresses

Constitutive equations are needed to calculate stresses from strains, i.e.,

$$\sigma = F(\varepsilon). \quad (12.9)$$

If the constitutive equations are elastic Equation (12.9) reduces to Hooks generalized law, and the stresses are directly calculated by putting in the strains in the equation. However, if the constitutive equations are non-linear, e.g. elastic-plastic, the equation needs to be solved numerically in an iterative manner. This is because we need to check whether the calculated stress-state is elastic or elastic-plastic, and in the latter case also verify that the final stress-state fulfils the yield criteria as expressed generally by Equation (12.3).

12.4.3 Matrix Formulation

Above it is described how strain and stress can be calculated at arbitrary positions in an element. When practical models are constructed they consist of many elements. Then, strain is calculated in a certain number of points within each element. These are called integration points of Gauss points. As this is done for all elements in the model, a strain field is created. By aid of this information strain contours can be displayed in the model. Moreover, this creates a system of equations where displacements in the integration points, 1-M, can be calculated from the node displacements

$$\begin{bmatrix} u_1 \\ u_2 \\ \vdots \\ u_M \end{bmatrix} = \begin{bmatrix} N_{11} & \cdots & \cdots & N_{1M} \\ \vdots & \ddots & & \vdots \\ \vdots & & \ddots & \vdots \\ N_{N1} & \cdots & \cdots & N_{NM} \end{bmatrix} \begin{bmatrix} u_1^1 \\ u_2^1 \\ \vdots \\ u_M^1 \end{bmatrix} \quad (12.10)$$

In the same manner the strains can be calculated from the displacements. Hence,

$$\begin{bmatrix} \varepsilon_1 \\ \varepsilon_2 \\ \vdots \\ \varepsilon_M \end{bmatrix} = \begin{bmatrix} \beta_{11} & \cdots & \cdots & \beta_{1M} \\ \vdots & \ddots & & \vdots \\ \vdots & & \ddots & \vdots \\ \beta_{N1} & \cdots & \cdots & \beta_{NM} \end{bmatrix} \begin{bmatrix} u_1^1 \\ u_2^1 \\ \vdots \\ u_M^1 \end{bmatrix}. \quad (12.11)$$

Thus a matrix formulation comes in handy to solve the problems, and the two systems of equations above can be replaced with a matrix formulation,

$$u = Nu^N, \quad (12.12)$$

$$\varepsilon = \beta u^N. \quad (12.13)$$

If the position for each integration point is used as input to formulate the matrix β_N. In the same manner the stresses in the integration points, 1 – M, can be calculated

$$\begin{bmatrix} \sigma_1 \\ \vdots \\ \sigma_M \end{bmatrix} = \begin{bmatrix} F_1(\varepsilon_1,\ldots,\varepsilon_M) \\ \vdots \\ F_M(\varepsilon_1,\ldots,\varepsilon_M) \end{bmatrix}, \tag{12.14}$$

or in matrix notation

$$\sigma = F(\varepsilon) \tag{12.15}$$

Moreover the surface tractions and body forces can be expressed in vector notation. Hence,

$$t = \begin{bmatrix} t_1 \\ \vdots \\ t_M \end{bmatrix} f = \begin{bmatrix} f_1 \\ \vdots \\ f_M \end{bmatrix} \tag{12.16}$$

12.4.4 Principle of Virtual Work

To minimize the energy in the model is in general an iterative approach used. The aim is to find a solution to a given problem by minimizing the energy in the system. This can be written as

$$\int_{V_0} \beta_N : \sigma dV_0 = \int_S N_N^T \cdot t dS + \int_V N_N^T \cdot f dV \ . \tag{12.17}$$

This system of equations is the basis for the finite element method, and can be written on the form

$$F^N(u^M) = 0 \ . \tag{12.18}$$

The solution of this system of equations is straight forward if the equations are linear, i.e. if the material is linear elastic and small deformations are assumed. Then the problem has an analytical solution that can be found with e.g. Gauss elimination. If the system of equations instead is non-linear, e.g. the materials has been modelled as elastic-plastic, viscoelastic, viscoplastic or finite deformations are considered, the system of equations needs to be solve by a numerical solver that iterates until a solution has been found. This is done be repeating the steps described above. The following steps are then considered

- Displacements are proposed.
- Strains are calculated
- Stresses are calculated.

- If Equation (12.18) is fulfilled within given tolerances then the solution is accepted, otherwise are new displacements are proposed and the procedure is repeated.

12.5 Example: Creasing of Paperboard

12.5.1 Deformation Mechanisms

Before packages are folded the edged of what will be the box are creased. In *Figure 12.10* photographs of a creasing operation performed in a scanning electron microscope can be seen. The paperboard is placed above a female die, and a rule is pressed down into the paperboard. During this operation several mechanisms are activated in the paperboard. The most important mechanism will be:

1. Elastic deformation during loading in MD, CD and ZD.
2. Initiation of plastic deformation.
3. Plastic deformation in MD, CD and ZD.
4. Delamination between paper plies due to out-of-plane shear in MD and CD during loading.
5. Elastic deformation in MD, CD and ZD during unloading.
6. Delamination between paper plies due to normal stresses in ZD during unloading.

In the creasing operation it is a controlled amount of delamination of the paperboard that is desirable, since this creates edges and corners in the box that has better properties than paperboard that has not been creased. It is thus advantageous to cause unrecoverable damage in the material during the creasing operation. A numerical model of paperboard that is used to model creasing must therefore incorporate the delamination effect, since it is a feature that greatly influences the overall material and structural behaviour.

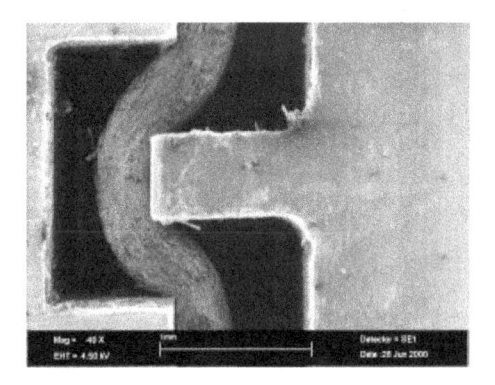

Figure 12.10. Scanning electron microscope pictures of the creasing process, (a) punch step and (b) after removal of punch. (Dunn, 2000).

12.5.2 Model Generation

In order to model the creasing operation a constitutive framework has been proposed by Xia (2000). The model consists of two parts, an in-plane model and a delamination model. The in-plane model accounts for elastic-plastic properties of each layer in a paperboard, while the delamination model accounts for delamination between or within layers. The delamination model accounts for the plastic deformation in the ZD. The theoretical framework will be outlined below; it will closely follow the work by Xia (2000). The constitutive equations account for large deformation. It is not possible to outline the whole large deformation formalism in this chapter; instead the interested reader is referred to for example Ottosen and Ristinmaa (1998).

12.5.3 In-plane Model

The in-plane model assumes that there is elastic-plastic deformation in the MD-CD plane, and elastic deformation in the ZD.

12.5.3.1 Stress-strain Relationship

In large displacement analysis the deformed state, x, is distinguished from the undeformed state, X. A line element, dx, in the deformed state depends on a line element in the undeformed state, dX, through the deformation gradient tensor, F. Hence,

$$dx = F\, dX \tag{12.19}$$

where

$$F = \frac{\partial x}{\partial X}. \tag{12.20}$$

It is assumed that the total deformation gradient F at a material point can be multiplicatively decomposed into elastic and plastic parts. Hence

$$F = F^e F^p \tag{12.21}$$

where F^p represents the accumulation of inelastic deformation. Evaluation of plastic strains is governed by

$$\dot{F}^p = L^p F^p, \tag{12.22}$$

where L^p is the plastic velocity gradient that is defined by the flow rule. In this case it assumed that

$$L^p = \dot{\gamma} K, \tag{12.23}$$

where γ is the equivalent plastic strain and K is the normalized flow directions.

The elastic strain is obtained by using the elastic Green strain measure,

$$E^e = \frac{1}{2}\left(F^{eT}F^e - I\right), \tag{12.24}$$

where I is the second order identity tensor. The second Piola-Kirchoff stress measure, T, is related to Green strain using the linear relationship

$$\bar{T} = C : E^e, \tag{12.25}$$

where C is the fourth-order elastic stiffness tensor, which is taken to be orthotropic.

To model the through-thickness nonlinear elastic stress-strain relationship under ZD compression, the through-thickness elastic modulus is taken to be exponential functions of the compression. Hence,

$$E_{ZD} = E_{ZD}^0 e^{-aE_{22}^e} \tag{12.26}$$

12.5.3.2 Yield Condition

In the model it is assumed that the yield surface can be constructed from six sub-surfaces, where N^I is the normal to the Ith sub-surface. The sub-surfaces are related to loading directions where it is possible to do experiments. In this case the sub-surfaces are: MD tension, CD tension, shear (positive), MD compression, CD compression and shear (negative). The criterion was proposed since no existing criteria successfully described the experimental paperboard data available in the literature. The yield criteria is expressed as

$$f(\bar{T}) = \sum_{I=1}^{6} \chi^I \left(\frac{\bar{T}:N^I}{S^I}\right)^{2k} - 1, \tag{12.27}$$

where k is a constant, normally $k = 2$, and S^I are the equivalent strengths corresponding to the sub-surfaces. In general S^I depends on the equivalent plastic strain to incorporate the hardening behaviour. Lastly, the switching parameter χ^I has the properties

$$\chi^I = \begin{cases} 1 & \text{if } \bar{T}:N^I > 0 \\ 0 & \text{otherwise} \end{cases}. \tag{12.28}$$

12.5.4 Delamination Model

The delamination model describes how two interfaces delaminate during loading. It is a cohesive model that has elastic and plastic parts. Delamination is assumed to initiate due to tensile loading in the out-of-plane direction of the paperboard, and due to out-of-plane shear loading.

12.5.4.1 Stress-strain Relationship

The relative displacement between two opposing surfaces can be divided into elastic and plastic parts. Thus each displacement component can be written as

$$\delta_i = \delta_i^e + \delta_i^p , \qquad (12.29)$$

where i represent the three loading direction: ZD tension, MD-ZD shear and CD-ZD shear in a local coordinate system. The total displacement increment during a time increment Δt thus becomes

$$\Delta \delta_i = \Delta \delta_i^e + \Delta \delta_i^p \qquad (12.30)$$

The change in the traction vector across the interface due to incremental relative displacements is governed by

$$\Delta T_\alpha = K_\alpha (\Delta \delta_\alpha - \Delta \delta_\alpha^p) , \qquad (12.31)$$

where K_α denotes the components of the instantaneous interface stiffness in the α-direction, which normally depend on the effective plastic displacement, $\bar{\delta}^p$. This dependence can be used to formulate a damage law that mimics the experimentally observed behaviour.

The evolution of plastic displacements is in the model formulated as

$$\Delta \delta_i^p = \chi M_i \Delta \bar{\delta}^p , \qquad (12.32)$$

where M_i are the normalised components of the unit flow, and χ behaves such that

$$\chi = \begin{cases} 1 & \text{if } f = 0 \text{ and } T_i \Delta \delta_i^p > 0 \\ 0 & \text{if } f < 0 \text{ or } f = 0 \text{ and } T_i \Delta \delta_i^p < 0 \end{cases}. \qquad (12.33)$$

12.5.4.2 Yield Criterion

Yielding occurs at the interface when

$$f(T, \bar{\delta}^p) = \sum \frac{S_1 T_\alpha^2}{S_\alpha^2} + T_1 - S_1 = 0 \qquad (12.34)$$

where S_α is the instantaneous interface strength which in general depends on the equivalent plastic displacement, $\bar{\delta}^p$, according to an expression that mimics the observed experimental behaviour.

12.5.5 Modelling

After the theoretical model has been implemented into the commercial finite element program ABAQUS (2004) it can be used to do simulations. Since the theoretical framework is general it can be used to study any combination of loading. Before a model like this is used it is important to verify it against experiments. It can be both macroscopic experiments, such as a tensile test, and microscopic experiments. From the macroscopic experiments the principal behaviour of the model can be verified. From the microscopic experiments it can be clarified that the same deformation mechanisms are present in the model as in the real loading case, e.g. it need to be verified that delamination occur in the same manner and at the same position in the model and the experimental set-up. When the model is verified against well defined experiments, the model is trust worthy, and it can be used to study new loading cases and configurations. In *Figure 12.11* shows two simulations of creasing of a five-ply paperboard. These simulate the experiments shown in *Figure 12.10*, both with respect to material properties and external loading.

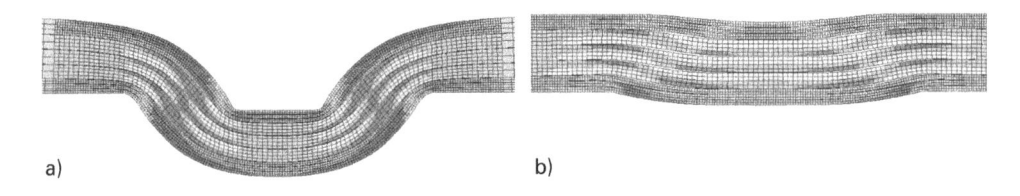

Figure 12.11. Finite element modelling of the creasing process, (a) punch step and (b) after removal of punch.

12.6 References

ABAQUS (2004) version 6.4, www.abaqus.com
Dunn, H. (2000) Micromechanisms of paperboard mechanics. *Master Thesis,* Department of Mechanical engineering, MIT, USA.
Heyden, S. (2000) Network modelling for the evaluation of mechanical properties of cellulose fibre fluff. *Doctoral Thesis*, Division of Structural Mechanics, Lund University, Sweden.
Lif, J. (2003) Analysis of the time and humidity-dependent mechanical behaviour of paper webs at offset printing press conditions. *Doctoral Thesis*, KTH Solid Mechanics, Sweden.
Persson, K. (2000) Micromechanical modelling of wood and fibre properties. *Doctoral Thesis*, Report, Division of Structural Mechanics, Lund University, Sweden.
Saabye Ottosen, N., and Ristinmaa, M. (1998) *The Mechanics of Constitutive Modelling Volume 2 Numerical and thermodynamical concepts*. Division of Solid Mechanics, Lund University, Sweden.

Index

A

absorption, Bristow method 212
absorption coefficient 151–152
absorptivity, surface 216
adhesion, paper strength 170–172
adipic acid, wet strength 192
adsorption, moisture, isotherms 117
air, moisture content 113–114
air-borne drying 106
air-leak methods, surface roughness 213–214
air permeance 66–67
alkaline-curing resins 192–196
amines, wet strength 193
anisotropy
– isotropic hand-sheets 96
– paper properties 92–93
azetidinium chloride, wet strength 194

B

bending curvature, humidity 136
bending resistance 56–57
bending stiffness
– laminate 303–305
– paper 2, 55–56
board, corrugated 259–261
bonded joint, drying 170
bonding mechanisms, fibres 73–74
book paper 246
Brecht-Knittweis method 62
brightness
– ISO 160, 167
Bristow absorption method 212
buckling, paper 45
bursting strength, paper 53–54

C

capillary forces, fibres 75–77
carton board
– coated, structure 20
– delamination 5
categories, paper 15

chemical forces, fibres 77–79
chromaticity coordinates 162–163
CIE whiteness 167
CIELAB-coordinates 165
climatology 114–116
coated carton board, structure 20
coated paper, light weight, structure 19
cockling 133–134
colour, separation 245
colour perception 161
– tristimulus values 161
colour printing 245
colour space, CIELAB 165
colour vision, human eye 156
compressibility, surface 214–215
compression index, moisture content 142
compressive collapse, paper boxes 3
condebelt drying 105
consolidation, bonded joint 170
contact angle, surface 217
contact area, surface 212
continuum models, mechanical modelling 321–324
converting
– mathematical modelling 315–333
– moisture influences 115–116
– packaging materials 258–270
converting operations, simulation 60–61
copy paper, structure 19
corrugated board 259–261
creasing
– deformation mechanisms 329
– delamination model 331–332
– in-plane model 330–331
– model generation 330
– modelling 333
– packaging materials 263–265
– paper 13
– paperboard 329–333
– stress-strain relationship 330–332
– yield condition 331

– yield criterion 332
creep, mechanosorptive 143
creep curves 50
curl
– laminate 308
– measurement 135, 139
– moisture influence 133–140
– polar diagram 309–310
cutting, packaging materials 263–265
cylinder drying 97–103

D
delamination 14
– carton board 5
delamination model, creasing 331–332
delamination resistance 58–60
density, paper 29
desorption, moisture 119
diethylenetriamine, wet strength 192
digital printing 238–242
– packaging 284
dimensional changes
– measurement 130–132
– moisture cycling 126–130
dimethylolurea, wet strength 189
displacement, finite element method 325–326
dry strength additives 172–182
– compression properties 178–181
– tensile properties 172–177
dry strength agents, mechanisms 169–183
dry strength mechanisms 169–183
drying 69–107
– bonded joint 170
– concepts 103–107
– cylinder 97–103
– hygroexpansion coefficient 123–127
– impingement 104
– multi-ply board 89–90
– observation 70
– paper sheet 92–97
– phases 97–102
– pressing 97
– stock preparation 90–92
drying stress 86

E
e-print, packaging 284
e-tags, packaging 284–285
enduse, mathematical modelling 315–333
epichlorohydrin, wet strength 193
excitation purity 163–164
eye, colour vision 156

F
fibre lifting, surface 224
fibre network, micro level 81–84
fibre textiles, properties compared to paper 22
fibres
– bonding mechanisms 73–74
– capillary forces 75–77
– chemical forces 77–79
– functional groups 203–204
– micro level 79–84
– nano level forces 73–79
– network forces 75
– tensile strength 34–36
fine paper, structure 17
finite element method 316, 325–329
– matrix formulation 327–328
– virtual work 328
flexible packaging 258
flexography 235–236
– inks 250
fluorescence 166–167
flute types 260
fold number 57–58
folded printing papers, leaning stacks 308–309
folding, packaging materials 265–266
folding endurance 57–58
folding failure 12
forces
– capillary, fibres 75–77
– chemical, fibres 77–79
– fibres, nano level 73–79
– laminate 299–301, 303
– network, fibres 75
formaldehyde, wet strength 189
fracture mechanics, paper 63–65
friction, surface 225

functional groups, fibres   203–204

G
g-PAM   *see* glyoxalated polyacrylamide
gloss
– measurement   229–230
– scale   230
– surface   225–230
– variations   230
gluing, packaging materials   265–266
glyoxalated polyacrylamide (g-PAM)   196–200
grammage, paper   26
graphic papers   247
gray scale, printing   242–244
greaseproof paper, structure   18
gurley-apparatus   66

H
hand-sheets, anisotropy   96
human eye, colour vision   156
humidity, bending curvature   136
hygroexpansion   120–127
– laminate, stress ans strain   314
– moisture content   129
– orientation   310
– polar diagram   310
– shear-free angle   311
– strain   130
hygroexpansion coefficient   120
– drying   123–127
– laminate   302
hysteresis, sorption isotherms   116–120

I
IGT tester   61
illuminants, standard   155
impingement drying   104
ink drying   248–251
– chemical   255
– offset printing   253–256
– physical   254
ink transfer   218–220
inkjet inks   251
inkjet paper   247–248

inks
– flexography   250
– printing   248–251
ISO-brightness   160, 167
ISO-hierarchy, optical measurements   158
isotropic hand-sheets, anisotropy   96
isotropic materials, laminate   295–296

J
joint strength, adhesion   170

K
Kelvin equation, sorption   118
Kubelka-Munk theory, light reflection   147–152
– applications   152
– limitations   151

L
laminate
– bending stiffness   303–305
– curl   308
– definition   289–292
– forces   299–301, 303
– hygroexpansion coefficient   302
– isotropic materials   295–296
– matrices   294–295
– moisture absorption, strain   296–299
– moments   299–301, 303
– orthotropic materials   295–296
– properties, moisture   307–308
– stiffness matrix   302
– strain transformation   301
– strain, mechanical influences   299
– strain, moisture absorption   296–299
– stress transformation   301
– tensile stiffness   303–305
– twist   308
– twist free angle   311–313
laminate layer   289–292
– angle   301
– strain   292–294
– stress   292–294
laminate theory, papermaker   287–314
length scales, mechanical modelling   320
level deviation, moisture ratio   141

lifetime, paper boxes 4
light absorption 146–147
light absorption coefficient 152
light reflection, Kubelka-Munk theory 147–152
light scattering 146–147
light scattering coefficient 152
light weight coated paper, structure 19
linerboard, structure 16
linting, surface 223–224
lithography, offset 234–235

M
machine level, wide web 97–107
macro level, paper sheet 84–97
magazine paper 246
materials, properties compared to paper 21–23
mathematical modelling
– converting 315–333
– enduse 315–333
matrices, laminate 294–295
matrix formulation, finite element method 327–328
mechanical modelling 317–320
– continuum models 321–324
– length scales 320
– network models 324
mechanical properties
– affecting factors 8–9
– moisture content 141–143
mechanical strain 86
mechanosorptive creep 143
melamine-formaldehyde, wet strength 190–192
modelling
– mechanical 317–320
– complex papers 5
– paper structural levels 5–7
moisture
– absorption, laminate 296–299
– absorption, paper 112
– curl 133–140
– laminate, properties 307–308
– mollier-diagram 113
– sorption isotherms 116–120
– twist 133–140
moisture content 111, 141–143
– air 113–114
– compression index 142
– hygroexpansion 129
– tensile stiffness 142
moisture cycling, dimensional changes 126–130
moisture ratio 111
– stiffness 141
mollier-diagram, moist air 113
monomethylol, wet strength 191
mottle, surface 223
Mullen test, bursting strength 53
multiply board, drying 89–90
multilayers, polyelectrolyte, paper strength 176

N
nano level forces, fibres 73–79
network, micro level 79–84
network forces, fibres 75
network models 324
newsprint 245
– structure 16

O
office paper 246
offset inks 248–249
offset lithography 234–235
offset printing, ink drying 253–256
optical measurements 159–160
– instrument calibration 158
– ISO-hierarchy 158
optical properties 145–167
– measurement conditions 156–158
– papermaking 146
– quality control 146
– standard observers 156
optical quantities, relationships 153–154
orthotropic materials, laminate 295–296

P
packages
– filling 266–268
– interaction 282–284

packaging 257–286
- compression requirements 278–282
- deformation 283
- design 276–284
- digital printing 284
- e-print 284
- e-tags 284–285
- environmental exposures 282
- future trends 284–286
- marketing performance 284
- performance, design 276–284
- performance, requirements 270–276
- performance, technical 277–284
- requirements 270–276
- smart packages 285
- supply chain 270–273
packaging materials 258–270
- barrier properties 269
- converting 258–270
- creasing 263–265
- cutting 263–265
- folding 265–266
- gluing 265–266
- new 286
PAE *see* poly(aminoamide)-epichlorohydrin
Page's theory, tensile strength 36
paper
- air permeance 66–67
- and printing 233–256
- bending resistance 56–57
- bending stiffness 55–56
- buckling 45
- bursting strength 53–54
- categories 15
- creasing 13
- creep curves 50
- delamination 14
- delamination resistance 58–60
- density 29
- directions 10
- drying 71–72
- elastic-plastic behaviour 322–324
- fold number 57–58
- folding endurance 57–58
- folding failure 12

- fracture mechanics 63–65
- grammage 26
- hygroexpansion 120–127
- interaction with water vapour 109–143
- mechanical aspects 320–325
- modelling of structure 1–24
- moisture adsorption 110, 112
- nature 9–10
- physical nature 8
- Poisson's ratio 30–32
- printing 245–248
- specific stress 30
- stiffness determination 40–41
- strain 30–32
- strain at break 37
- stress 30
- stress at break 46
- stress-strain curves 47
- structure 1–24
- structure of products 15–20
- tearing resistance 51
- tearing resistance, theory 52–53
- tensile energy absorption 37
- tensile stiffness 38–39
- tensile strength 32–34
- tensile test 31
- thickness 26–29
- thickness direction strength 58–60
- thickness measurement 27
- twist 137
paper machine, web break 4
paper physics 25–67
paper products, structure 15–20
paper properties
- affecting factors 8–9
- anisotropy 92–93
- compared to materials 21–23
- compression 45–49
- creep 43–45
- drying 87
- during drying 69–107
- elastic 37–38
- functional 1–5
- macro level 84–89
- modelling, complexity 5
- modelling, structural levels 5–7

– moisture content 141–143
– optical 145–167
– standardization 23
– strain 87–89
– stress-strain 42–43
– viscoelastic 49–51
paper sheet
– drying 92–97
– macro level 84–97
paper strength
– adhesion 170–172
– dry strength additives 172–182
– polyelectrolyte multilayers 176
paper surface *see* surface
paperboard, creasing 329–333
papermaker, laminate theory 287–314
papermaking
– moisture influences 115–116
– optical properties 146
– water content 71
permeance, surface 216
pick strength, surface 224
plastic behaviour, paper 322–324
Poisson's ratio, paper 30–32
polar diagram, mechanical properties 309–310
poly(aminoamide)-epichlorohydrin (PAE)
– crosslinked 195
– wet strength 192
polyacrylamide resin, glyoxalated 196–200
polyelectrolyte multilayers, paper strength 176
polyethylene film, properties compared to paper 22
porosity, surface 215
prepress 242–245
pressing, drying 97
print
– quality 152–253
– unevenness 253
print density 220, 252
– unevenness 253
print gloss 252
– unevenness 253
print through 221–222

printability, surface 218–223
printing
– colour 245
– conventional techniques 234–237
– digital 238–242
– gray scale 242–244
– resolution 242–244
printing inks 248–251
printing papers 233–256
– folded, leaning stacks 308–309
production control, methods 61–63
pulp, optical properties 145–167

Q
quality control, optical properties 146

R
radiance factor, fluorescence 167
reflectance, surface 228–229
reflectance factor, measurement 154–155
reflection, surface 226–228
reflectivity 150
reflectivity factor, measurement 154–155
resins
– alkaline-curing 192–196
– wet-strength 189–203
resolution, printing 242–244
rigid packaging 258
rotogravure 236–237
– inks 249
roughness
– measures 211–212
– surface 211–214

S
scattering coefficient 151–152
Scott bond test 62
set-off, surface 222–223
shear-free angle, hygroexpansion 311
short-span test 34–36
shrinkage, macro level 84–89
single fibre, network 79–81
smart packages, packaging 285
soak time, wet strength 187
sorption
– Kelvin equation 118

– moisture, isotherms 116–120
specific stress, paper 30
speckle photography 131
spectral power distribution, standard illuminants 155
spectral radiance factor, fluorescence 167
stacks, folded printing papers 308–309
standard illuminants 155
– spectral power distribution 155
standard observers, optical properties 156–158
standardization, paper properties 23
standardized measurement conditions, optical properties 156–158
starch, wet strength 200–203
STFI deformation 131
stiffness
– moisture ratio 141
– polar diagram 309–310
stiffness determination, ultrasonic techniques 40–41
stiffness matrix, laminate 302
stock preparation, drying 90–92
strain 30–32
– and paper properties 87–89
– laminate layer 292–294
– laminate, hygroexpansion 314
– laminate, mechanical influences 299
– laminate, moisture absorption 296–299
– mechanical 86
strain at break 37
strain transformation, laminate 301
stress 30
– drying 86
– finite element method 327
– laminate layer 292–294
– laminate, hygroexpansion 314
stress at break 46
stress-strain, and paper properties 42–43
stress-strain curves 47
stress-strain relationship, creasing 330–332
stress transformation, laminate 301
stretch, macro level 84–89
surface 209–232
– absorptivity 216

– air leak methods 213–214
– compressibility 214–215
– contact angle 217
– contact area 212
– creation 230–232
– friction 225
– gloss 225–230
– gloss measurement 229–230
– linting 223–224
– permeance 216
– porosity 215
– printability 218–223
– reflectance 228–229
– reflection 226–228
– roughness 211–214
– strength 223–224
– two-sidedness 210–211
– volume 212–213
– wettability 217–218
– wetting delay 218

T
tearing resistance 51
– theory 52–53
tensile energy absorption, paper 37
tensile stiffness 38–39
– laminate 303–305
– moisture content 142
– orientation 310
tensile strength 32–34
– fibres 34–36
– Page´s theory 36
tensile test 31
tension, creep properties 43–45
textiles, properties compared to paper 22
thickness, paper 26–29
– measurement 27
thickness direction strength 58–60
tissue, structure 17
tristimulus values, colour perception 161
twist 137
– laminate 308
– moisture influence 133–140
– polar diagram 309–310
twist free angle, laminate 311–313
two-sidedness, surface 210–211

U
ultrasonic techniques, stiffness determination 40–41
ultraviolet curable inks 250
unevenness
– print density 253
– print gloss 253
urea-formaldehyde, wet strength 189–190

V
vapour, interaction with paper 109–143
virtual work, finite element method 328

W
water content, papermaking 71
water vapour, interaction with paper 109–143
wavelength, dominant 163–164
web break, paper machine 4
wet strength
– chemicals 204–205
– mechanisms 185–188
– protection mechanism 187
– reinforcement mechanism. 188
– soak time 187
wet strength agents, mechanisms 185–208
wet-strength resins, chemistry 189–203
wettability, surface 217–218
wetting delay, surface 218
whiteness 165–166
– CIE 167
wide web, machine level 97–107

Z
zero-span test 34–36

HOLZFORSCHUNG

International Journal of the Biology, Chemistry, Physics, and Technology of Wood

Editor-in-Chief: Oskar Faix, Germany

Publication frequency: bi-monthly (6 issues per year).
Approx. 700 pages per volume. 21 x 29.7 cm
ISSN (Print) 0018-3830
ISSN (Online) 1437-434X
CODEN HOLZAZ
Language: Englisch

Holzforschung is an international scholarly journal that publishes cutting-edge research on the biology, chemistry, physics and technology of wood and wood components. High quality papers about biotechnology and tree genetics are also welcome. Rated year after year as the number one scientific journal in the category of Pulp and Paper (ISI Journal Citation Index), *Holzforschung* represents innovative, high quality basic and applied research. The German title reflects the journal's origins in a long scientific tradition, but all articles are published in English to stimulate and promote cooperation between experts all over the world. Ahead-of-print publishing ensures fastest possible knowledge transfer.

Indexed in: Academic OneFile (Gale/Cengage Learning) – Aerospace & High Technology Database – Aluminium Industry Abstracts – CAB Abstracts – Ceramic Abstracts/World Ceramic Abstracts – Chemical Abstracts and the CAS databases – Computer & Information Systems Abstracts – Copper Data Center Database – Corrosion Abstracts – CSA Illustrata – Natural Sciences – CSA / ASCE Civil Engineering Abstracts – Current Contents/Agriculture, Biology, and Environmental Sciences – Earthquake Engineering Abstracts – Electronics & Communications Abstracts – EMBiology – Engineered Materials Abstracts – Engineering Information: Compendex – Engineering Information: PaperChem – Journal Citation Reports/Science Edition – Materials Business File – Materials Science Citation Index – Mechanical & Transportation Engineering Abstracts – METADEX – Paperbase – Science Citation Index – Science Citation Index Expanded (SciSearch) – Scopus – Solid State & Superconductivity Abstracts.

All de Gruyter journals are now hosted on **Reference Global**, de Gruyter's new and integrated platform.
Please visit www.reference-global.com for more information and free TOC alerts.
Electronic sample copy at www.degruyter.com/journals/holz

de Gruyter
Berlin · New York

www.degruyter.com